CHEMICAL BONDING IN SOLIDS AND FLUIDS

MARK LADD DSc (Lond)
Formerly Reader in Chemical Crystallography,
University of Surrey

ELLIS HORWOOD
NEW YORK   LONDON   TORONTO   SYDNEY   TOKYO   SINGAPORE

First published 1994 by
Ellis Horwood Limited
Market Cross House, Cooper Street
Chichester
West Sussex, PO19 1EB
A division of
Simon & Schuster International Group

Printed and bound in Great Britain
by Hartnolls, Bodmin

British Library Cataloguing in Publication Data

A catalogue record for this book is available from the British Library

ISBN 0–13–474933–2   Hbk
ISBN 0–13–474925–1   Pbk

Library of Congress Cataloging-in-Publication Data

Available from the publisher

2 3 4 5   97 96 95 94

Table of contents

Preface xiii

Physical constants and other numerical data xv

Symbol notation xvii

1 Preamble 1
  1.1 Atomic nature of matter. 1
  1.2 Types of interatomic bonding 3
      1.2.1 Covalent bonding 3
      1.2.2 Ionic bonding 3
      1.2.3 Van der Waals' bonding 4
          1.2.3.1 Hydrogen-bonding. 4
      1.2.4 Metallic bonding. 5
  1.3 States of matter. 6
      1.3.1 Equipartition and energy states. 7
      1.3.2 Boltzmann distribution. 8
      1.3.3 Solid state 8
      1.3.4 Liquid state 9
      1.3.5 Gaseous state 10
          1.3.5.1 Molar arithmetic. 10
      1.3.6 Other forms of matter 12
          1.3.6.1 Glasses 12
          1.3.6.2 Liquid crystals. 14
          1.3.6.3 Solutions 16
          1.3.6.4 Colloids 16
    Problems for Chapter 1 17

2 Bonding between atoms. .18
  2.1 Introduction. 18
  2.2 Classical mechanics—success and failure 19
      2.2.1 Black-body radiation. 20
      2.2.2 Photoelectric effect. 24
      2.2.3 Compton effect 25
      2.2.4 Diffraction of electrons. 25

	2.2.5	Wave–particle duality 26
	2.2.6	Atomic spectra. 27
2.3	Wave equation 29
	2.3.1	Born interpretation of the wave equation 30
	2.3.2	Uncertainty principle. 31
	2.3.3	Normalization and quantization 32
2.4	Particle in a box—quantization of translational energy 33
	2.4.1	Tunnelling. 35
	2.4.2	Boxes of higher dimensions. 36
2.5	Vibrational and rotational motion 38
	2.5.1	Vibrational motion. 38
	2.5.2	Rotational motion in two dimensions. 40
	2.5.3	Rotational motion in three dimensions 40
	2.5.4	Space and spin quantization 41
	2.5.5	Quantum numbers. .42
2.6	Structure of the hydrogen atom 43
	2.6.1	Atomic orbitals 44
	2.6.2	Orbital terminology 48
	2.6.3	Selection rules for atoms 50
	2.6.4	Atoms with more than one electron. 51
	2.6.5	Effective atomic number—screening 51
		2.6.5.1 Slater's rules. 51
	2.6.6	Aufbau principle. .53
	2.6.7	Ionization energy .53
2.7	Simple molecules— theoretical principles 59
	2.7.1	Variation method .60
	2.7.2	Linear combination of atomic orbitals 62
		2.7.2.1 Overlap integrals 63
		2.7.2.2 Coulomb integrals. 64
	2.7.3	Orthogonality 64
2.8	Simple molecules – structures. 65
	2.8.1	Molecular orbital method 65
		2.8.1.1 Bonding and antibonding orbitals. 68
	2.8.2	Diatomic molecules 69
		2.8.2.1 Homonuclear diatomics 70
		2.8.2.2 Symmetry of orbitals. 71
		2.8.2.3 Heteronuclear diatomics 74
		2.8.2.4 Hybridization 74
2.9	Polyatomic molecules 75
	2.9.1	H_2O molecule—directed bonds. 77
	2.9.2	Methane. 79
	2.9.3	Delocalized systems 80
		2.9.3.1 Ethene and ethyne. 80
2.10	Hückel molecular orbital theory 82
	2.10.1	Delocalization energy 84
	2.10.2	Butadiene wave functions 84
	2.10.3	π Bond order 85
	2.10.4	Free valence 86

	2.10.5	Aromatic systems	88
	2.10.6	Charge distributions	88
		2.10.6.1 Aniline and pyridine	90
2.11	Valence-bond approximation.	91
	2.11.1	Valence-bond model	91
	2.11.2	Hydrogen molecule	91
		2.11.2.1 Spin factor	92
	2.11.3	Hydrogen fluoride	93
	2.11.4	Electronegativity.	94
2.12	MO and VB models compared	95
	2.12.1	VB and MO models with homonuclear diatomics	96
	2.12.2	Electron density	97
2.13	Ligand-field theory	97
2.14	Apparently abnormal valence	101
2.15	Valence-shell electron-pair repulsion theory.	101
2.16	Covalent states of matter.	103
	2.16.1	Covalent solids	104
	2.16.2	Structural and physical properties of the covalent bond . . .	104
2.17	Semi-empirical and *ab initio* molecular orbital theories	105
	Problems for Chapter 2	106

3	**Bonding between molecules**	109
3.1	Gases	109
	3.1.1	Kinetic theory of gases	109
		3.1.1.1 Pressure exerted by a gas	109
		3.1.1.2 Equipartition.	111
		3.1.1.3 Mixtures of ideal gases	111
		3.1.1.4 Maxwell–Boltzmann distribution of velocities	112
		3.1.1.5 Maxwell–Boltzmann distribution of speeds.	114
	3.1.2	Imperfect gases	116
		3.1.2.1. Van der Waals' equation of state	118
		3.1.2.2 Comparing gases	118
	3.1.3	Intermolecular attraction	120
		3.1.3.1 Molecule volume effect	120
		3.1.3.2 Collision frequency	121
		3.1.3.3 Mean free path	122
	3.1.4	Origin of intermolecular forces.	123
		3.1.4.1 Electric moments	123
		3.1.4.2 Polarizability.	123
		3.1.4.3 Ion–dipole attraction	124
		3.1.4.4 Dipole–dipole interaction	124
		3.1.4.5 Dipole–induced dipole interaction	128
		3.1.4.6 Induced dipole–induced dipole interaction	129
		3.1.4.7 Pairwise additivity	130
	3.1.5	Intermolecular potentials	130
		3.1.5.1 Lennard-Jones 12–6 potential function.	133
	3.1.6	Gas viscosity	134

3.2   Liquids. 137
    3.2.1   Liquid–gas equilibrium 137
    3.2.2   Some thermodynamics of equilibrium 139
    3.2.3   Radial distribution function 140
        3.2.3.1   Diffraction studies 140
        3.2.3.2   Solids and fluids 142
    3.2.4   Equation of state for a fluid 144
    3.2.5   Monte Carlo method 146
    3.2.6   Molecular dynamics. 147
    3.2.7   Computer simulation of liquid water. 149
    3.2.8   Use of models. 150
    3.2.9   Viscosity 152
        3.2.9.1   Liquid viscosity by molecular dynamics 152
    3.2.10   Quantum liquids 153
    3.2.11   Amorphous solids. 154
3.3   Molecular solids 156
    3.3.1   Packing of molecules 156
    3.3.2   Classification of molecular solids 157
        3.3.2.1   Noble gases. 157
        3.3.2.2   Elements 158
        3.3.2.3   Small inorganic molecules. 160
        3.3.2.4   Organic compounds 161
        3.3.2.5   Classification of organic compounds. 165
    3.3.3   Structural and physical characteristics of molecular compounds 165
        3.3.3.1   Solubility of molecular compounds 170
    3.3.4   Molecular mechanics 170
        3.3.4.1   Some types of procedure 171
        3.3.4.2   Importance of molecular mechanics 173
Problems for Chapter 3 174

4   Bonding between ions: I. 177
4.1   Introduction. 177
4.2   Attractive energy 177
    4.2.1   Madelung constant 178
4.3   Lattice energy—electrostatic model. 180
4.4   Lattice energy—thermodynamic model 183
    4.4.1   Precision of the thermodynamic lattice energy. 186
        4.4.1.1   Enthalpy of sublimation 186
        4.4.1.2   Ionization energy. 187
        4.4.1.3   Dissociation enthalpy. 188
        4.4.1.4   Electron affinity 188
        4.4.1.5   Enthalpy of formation 191
4.5   Polarization in ionic compounds—precision of the electrostatic
    model for lattice energy 192
    4.5.1   Approximate calculation of lattice energy 193

4.6   Uses of lattice energies. 194
        4.6.1   Electron affinities and thermodynamic parameters. 194
        4.6.2   Compound stability 194
        4.6.3   Charge distribution on polyatomic ions. 195
4.7   Aspects of crystal chemistry 198
        4.7.1   Ionic radii 200
        4.7.2   Radius ratio and AX structure types 203
        4.7.3   Radius ratio and AX_2 structure types. 206
        4.7.4   Polarization in AX and AX_2 structures. 207
4.8   Structural and physical characteristics of ionic compounds 209
4.9   Solubility of ionic compounds 211
        4.9.1   Reference states for solubility. 211
        4.9.2   Solubility relationships. 212
        4.9.3   Two example calculations 212
                4.9.3.1   Silver iodide 212
                4.9.3.2   Lithium fluoride 212
        4.9.4   Solubility and energy. 213
4.10   Vibrations and defects in ionic compounds 216
        4.10.1   Absorption spectra. 216
        4.10.2   Heat capacity 217
        4.10.3   Defects in crystals 218
                4.10.3.1   Schottky defect. 218
                4.10.3.2   Frenkel defect 220
                4.10.3.3   Doping and colour centres 221
        4.10.4   Defects and ion mobility 221
4.11   Molten salts. 222
Problems for Chapter 4 223

5   Bonding between ions: II 227
5.1   Introduction 227
5.2   Classical free-electron theory 227
        5.2.1   Electrical conductivity 228
        5.2.2   Heat capacity 229
                5.2.2.1   Classical particles 230
                5.2.2.2   Classical solids. 230
                5.2.2.3   Einstein and Debye solids 232
                5.2.2.4   Heat capacity paradox 233
5.3   Wave-mechanical free-electron theory 233
        5.3.1   Density of states 236
        5.3.2   Fermi–Dirac distribution 238
                5.3.2.1   Mean energy of electrons. 240
                5.3.2.2   Electronic heat capacity 240
5.4   Band theory 241
        5.4.1   Energy bands and Brillouin zones. 242
        5.4.2   Energy bands and molecular orbital theory 245
                5.4.2.1   Occupation of orbitals 246
                5.4.2.2   Semiconductors and insulators 247

5.5   Structures of metals. 249
    5.5.1   Metallic radii 251
    5.5.2   Interstitial sites. 252
5.6   Structural and physical characteristics of metallic compounds 252
5.7   Superconductivity. 255
    5.7.1   Simple theory of superconductivity 255
    5.7.2   Thermal properties below the critical temperature 257
    5.7.3   Superconduction in organic compounds. 257
5.8   Alloys 259
    5.8.1   Copper–gold system 259
    5.8.2   Silver–cadmium system 261
        5.8.2.1   Hume-Rothery rules 262
Problems for Chapter 5 264

Appendix 1   Problem solving and computers   266
A1.1   Introduction. 266
A1.2   Solving numerical problems 266
    A1.2.1   Approach to problems 267
    A1.2.2   Suggested procedure 267
    A1.2.3   Solving an example problem 268
    A1.2.4   Solution 268
A1.3   Computer methods. 270
    A1.3.1   Program design and availability. 270
    A1.3.2   Linear least squares. 270
    A1.3.3   Gaussian quadrature 270
    A1.3.4   Madelung constant 270
    A1.3.5   Radial wave functions. 271
    A1.3.6   Maxwell-Boltzmann distribution 271
    A1.3.7   Electron-in-a-box 271
    A1.3.8   Monte Carlo calculations 271
    A1.3.9   Angular wave functions 272
    A1.3.10   Eigenvalue problems 273
        A1.3.10.1   Eigenvalues and eigenvectors. 273
        A1.3.10.2   Diagonal matrices 274
        A1.3.10.3   Jacobi diagonalization 275
        A1.3.10.4   General matrix diagonalization. 277
    A1.3.11   Hückel molecular orbital calculations 277
    A1.3.12   Point-group recognition 278
    A1.3.13   Curve fitting and interpolation 281
Problems for Appendix 1. 281

Appendix 2   Stereoviewing   283

Appendix 3   The hypsometric formula—an example of the Boltzmann
              distribution   284

Appendix 4   **Gamma function** 286

Appendix 5   **Spherical polar coordinates** 287
A5.1   Coordinates 287
A5.2   Volume element, $d\tau$ 287
A5.3   Laplacian operator ∇^2 287

Appendix 6   **Reduced mass** 290

Appendix 7   **Solution of a second-order differential equation** 292

Appendix 8   **Separation of variables** 294

Appendix 9   **Least-squares line and propagation of errors** 295
A9.1   Propagation of errors 296

Appendix 10   **Overlap integrals** 299

Appendix 11   **Equation of state for a solid** 300

Appendix 12   **Numerical integration** 302
A12.1   Numerical and other methods 302
      A12.1.1   Simpson's rule 302
      A12.1.2   Direct weighing method 303

Appendix 13   **Calculation of the Madelung constant** 305

Appendix 14   **Some thermodynamics of solutions** 307
A14.1   Partial molar volume 307
A14.2   Partial molar entropy 308
A14.3   Measurement of the partial molar entropy of an ion 308
A14.4   Generalized description of partial molar quantities 309
A14.5   Partial molar enthalpy in solutions 310
A14.6   Standard state for solutions 311

Appendix 15   **Debye–Hückel limiting law** 314

Appendix 16   **Average classical thermal energies** 316
A16.1   Average kinetic energy 316
A16.2   Average vibrational energy 316

Appendix 17   **Fermi–Dirac statistics** 318

Solutions to problems 321

Index 340

To Valentia

A wife of noble character
is her husband's crown

(Proverbs 12:4)

Preface

This book has been developed from the author's earlier, successful work on *Structure and Bonding in Solid State Chemistry*, which has now been entirely rewritten, brought up to date, and enhanced by a chapter on interatomic forces and chemical bonding in fluids so as to widen the overall scope and appeal.

The aims of the present book are to explore the nature of interatomic and intermolecular forces and the structures and properties associated with the bonding to which they give rise in solid and fluid chemical species.

Its philosophy, style and level are consistent with those of the earlier book. Thus it is addressed to first- and second-year undergraduates in chemistry, chemical physics, materials science and other subjects where chemistry is a major ancillary subject.

The mathematical treatments used in the book lie within the scope of any chemistry undergraduate. Certain mathematical arguments are developed in appendices, because their inclusion in the body of the text could perturb the flow of the main subject matter. Results from chemistry, physics and mathematics, appropriate to first- and second-year studies, such as equilibrium thermodynamics, the variation of mass with speed according to special relativity, and results from the calculus are drawn upon as required. Copious references to specialized texts and journal papers have been provided, so that the reader can extend the scope of the material given herein.

Each chapter has been provided with a set of problems of varying difficulty, and the reader is encouraged to solve them, because it will assist in the process of gaining familiarity with the themes of the book, and in testing the reader's ability to apply them in new and novel situations. Detailed solutions are given at the end of the book.

In view of the importance of problem solving, Appendix 1 provides both a suggested scheme for this activity and, as an additional novel feature, a brief discussion of computer programs that may be used as an aid to the thorough study of the book, and which can be operated on IBM-compatible PCs. The reader is advised to consult Appendix 1 at an early stage.

The SI system of units has been used generally, and in several situations conversions between SI and c.g.s. units have been given. It is not yet, and probably never will be, possible to neglect the c.g.s. system of units, because of the amount of research published, past and present, in these units. Competency in more than one system of units should not be despised: indeed, such an ability will enhance a student's appreciation of both his subject and its literature. Tables of fundamental constants, conversion factors and notation are provided in the preliminary pages of the book.

The periodic table of the elements, Table 2.5 in the text, has been accorded the groupings given in the latest IUPAC recommendations. Thus, the groups 1 to 18 replace, in order, groups IA-VIIA, VIII (three sub-groups), IB-VIIB and 0. Hydrogen is placed in group 1, with the usual reservations.

In a study of the solid state, one needs an understanding of the three-dimensional nature of crystal and molecular structures. This aspect of solids is more unfamiliar than it is intrinsically difficult, despite the nature of our world. In order to assist in the appreciation of three-dimensional structures, many of the illustrations are stereoscopic pairs, and a stereoviewer may be obtained by completing the card at the front of the book.

The author has great pleasure in acknowledging the help of Professor N. H. March, Coulson professor of theoretical chemistry at the University of Oxford, and Professor T. J. Kemp, the series editor and professor of physical chemistry at the University of Warwick, for reading the book in manuscript and for making useful suggestions. It is of significance to note that Professor March is currently preparing a book entitled *Interatomic Forces in Solids and Liquids*, to be published in about one year's time, of third-year undergraduate and postgraduate level, that will form an excellent complement to the present work.

Thanks are recorded also to Professor S. F. A. Kettle of the University of East Anglia, for a critical reading of the page proofs; to Dr S. Motherwell, formerly of the University of Cambridge, for use of the program PLUTO in preparing the stereoviews; to the publishers and authors for permission to reproduce those figures that carry appropriate acknowledgements, to my son Dr A. J. C. Ladd of the Lawrence Livermore National Laboratory, for the illustrations of computer simulations of solids and liquids; to my wife, Valentia, for the time that was needed for the preparation, revision and checking of this book; and to Ellis Horwood Publishers whose interest and care has enabled this book to be brought to a state of completion.

University of Surrey, 1993

M. F. C. Ladd

Physical constants and other numerical data

These data have been selected, or derived, from the compilation of E. R. Cohen and B. N. Taylor (1988) *J Phys Chem Ref Data*, **17**, 1795–1803; the values are reported in SI units. The figures in parentheses after each value represent the standard deviation to be applied to its last two digits. Conversions to other units are straightforward, and some examples are given at the end of this tabulation. Although the data are presented here with their full precision, we shall rarely need to employ more than about the first four of five significant figures. It may be noted that c and ε_0 are now *defined* by the values listed.

Speed of light in a vacuum	c	2.99792458×10^8 m s^{-1}
Permittivity of a vacuum	ε_0	$8.854187817 \times 10^{-12}$ F m^{-1}
Planck constant	h	$6.6260755(40) \times 10^{-34}$ J Hz^{-1}
Elementary charge	e	$1.60217733(49) \times 10^{-19}$ C
Avogadro constant	L	$6.0221367(36) \times 10^{23}$ mol^{-1}
Atomic mass unit	u	$1.6605402(10) \times 10^{-27}$ kg
Bohr magneton	μ_B	$9.2740154(31) \times 10^{-24}$ J T^{-1}
Rydberg constant	R_∞	$1.0973731534(13) \times 10^7$ m^{-1}
Rydberg constant for hydrogen	R_H	$1.0967758772(13) \times 10^7$ m^{-1}
Bohr radius	a_0	$5.29177249(24) \times 10^{-11}$ m^{-1}
Boltzmann constant	k_B	$1.380658(12) \times 10^{-23}$ J K^{-1}
Molar gas constant	$\mathscr{R}$	$8.314510(70)$ J K^{-1} mol^{-1}
		$0.0820577(7)$ dm^3 atm K^{-1} mol^{-1}
Molar volume of ideal gas at 273.15 K and 101325 Pa[1]	V_m	$22.41410(19) \times 10^{-3}$ m^3 mol^{-1}
Compton wavelength (electron)	λ_C	$2.42631058(22) \times 10^{-12}$ m
Rest mass of electron	m_e	$9.1093897(54) \times 10^{-31}$ kg
Rest mass of proton	m_p	$1.6726231(10) \times 10^{-27}$ kg
Rest mass of neutron	m_n	$1.6749286(10) \times 10^{-27}$ kg
Reduced mass of proton and electron	μ	$9.1044313(54) \times 10^{-31}$ kg
Faraday	$\mathscr{F}$	$9.6485309(29) \times 10^4$ C mol^{-1}
Ice-point temperature	T_{ice}	$273.1500(01)$ K

[1]101325 Pa = 1 atmosphere.

In addition to these SI units, there are others in everyday use in the scientific literature that cannot be ignored. Some of the more important of them are listed hereunder.

Length
1 Å (angstrom) = 10^{-10} m = 10 nm

Energy
1 eV (electronvolt) = $1.60217733(49) \times 10^{-19}$ J
1 cal (calorie) = 4.184 J = 96.485309(29) kJ mol^{-1}
1 cm^{-1} = $1.9864475(4) \times 10^{-23}$ J = $1.196265(5) \times 10^{-2}$ kJ mol^{-1}†

Pressure
1 atm (atmosphere) = 101325 Pa (N m^{-2}) = 760 Torr = 760 mmHg

Dipole moment
1 D (Debye) = 3.33564×10^{-30} C m

Prefixed to units
The following prefixes to units are in common use.

femto	pico	nano	micro	milli	centi	deci	kilo	mega	giga
f	p	n	μ	m	c	d	k	M	G
10^{-15}	10^{-12}	10^{-9}	10^{-6}	10^{-3}	10^{-2}	10^{-1}	10^{3}	10^{6}	10^{9}

† The cm^{-1} is a pseudo-energy unit used in spectroscopy.

Symbol notation

It is not possible, within the normal scientific conventions, to avoid duplicate usage of symbols. The following list explains the usage in this book; where symbols are duplicated, the context will provide the necessary clarity.

In the text, the form $V(x)$, for example, means that V is a function of x; $V(\text{Al})$ means the parameter V specifically for aluminium; V_m means that m qualifies V, in this case per mole; and V_j means the value of V for the jth species.

Symbols

a	periodic repeat distance; unit-cell side (along x); constant of the van der Waals' equation of state; activity; acceleration
a_0	Bohr radius for hydrogen
A	area; (Helmholtz) free energy
$\mathscr{A}$	Madelung constant
aq	hydrated state of infinite dilution
b	unit-cell side (along y); constant of the van der Waals' equation of state
B	second virial coefficient; coefficient of repulsive potential
c	speed of light in a vacuum; unit-cell side (along z); coefficients of ψ in LCAO/HMO theory; molar concentration
c_V	constant volume heat capacity per atom
C	third virial coefficient; coefficient of dipole–dipole attractive potential
C_p	molar heat capacity at constant pressure
C_V	molar heat capacity at constant volume
d	distance; bond distance; interplanar spacing (in real space); molecular (collision) diameter; atomic diameter
d	(d) orbital (also d_{z^2}, d_{xz}, d_{yz}, d_{xy}, $d_{x^2-y^2}$); sub-shell notation (for $l = 2$)
d^*	distance in reciprocal space (**k**-space)
d	first differential operator
D	bulk density; constant of dipole-quadrupole attractive potential; diffusion coefficient
D	Debye unit; second differential operator
D_0	experimental dissociation energy/enthalpy
D_e	theoretical dissociation energy
D_π	delocalization energy
e	charge on the electron
e^-	electron
e, exp	exponential function

E	energy; emf; electron affinity
E_F	Fermi energy
E_H	Hartree
E_m	molar energy
E_π	total π-binding energy
$\mathscr{E}$	energy density; Rayleigh ratio; electrical field strength; configurational energy
f	vibration frequency (mechanical); Fermi–Dirac distribution function; general function; activity coefficient
$f_\pm$	mean ionic activity coefficient
f	(f) orbital; sub-shell notation (for $l = 3$)
F	force; force per unit area; crystallographic structure factor
$\mathscr{F}$	Faraday; free-valence parameter
g	density of states function; statistical energy state; gerade (*even*)
g	gram, gas
G	(Gibbs) free energy
G_m	molar (Gibbs) free energy
h	Planck constant; wave index; Miller index
H	enthalpy; Coulomb integral
H_m	molar enthalpy
$\mathscr{H}_1$	one-dimensional Hamiltonian operator
$\mathscr{H}$	three-dimensional Hamiltonian operator
i	complex operator
i	vector of unit magnitude
I	ionization energy; moment of inertia; radiation intensity; current; ionic strength (in solution)
j	current density; (jth) species
J	angular momentum; (momentum) flux
$\mathscr{k}$	force constant; restoring force per unit displacement;
k	magnitude of wave vector; Miller index
k_B	Boltzmann constant
k_F	radius of Fermi sphere
k	wave vector (in **k**-space)
K	shell notation (for $n = 1$); equilibrium coefficient
K	Kelvin
l	mean free path; total angular momentum quantum number; Miller index
l	liquid
L	Avogrado constant; shell notation (for $n = 2$)
m	mass
m^*	variable mass of electron
m_e	rest mass of electron
m_l	orbital angular momentum quantum number
m_n	rest mass of neutron
m_p	rest mass of proton
m_s	spin angular momentum quantum number
m	metre

M	excitance; shell notation (for $n = 3$); maximum bonding power
M_m	molar mass
M_r	relative molar mass
M_S	mass of a sample
n	principal quantum number; number of moles
n_e	number of electrons, *per se* and in bonding molecular orbitals
n_e^*	number of electrons in antibonding molecular orbitals
n_x, n_y, n_z	quantum numbers
$\mathcal{n}$	number of defect states
N	number of species; normalization constant; shell notation (for $n = 4$); density of states (number) function
$\mathcal{N}$	number of species per unit volume
p	pressure; momentum; mobile (π) bond order
p_c	critical pressure
p_r	reduced pressure
p	(p) orbital (also p_x, p_y, p_z); sub-shell notation (for $l = 1$)
P	total bond order; probability function
$\mathbb{P}$	configuration of momenta
q	numerical value of charge
Q	product qe
$r, \imath$	distance between species; radial coordinate; radius
r_e	equilibrium distance between species
R	radial wave function; resistance
R_H	Rydberg constant for hydrogen
R_∞	Rydberg constant
$\mathscr{R}$	molar gas constant
$\mathbb{R}$	configuration of positions; radius ratio
s	spin quantum number
s	second; (s) orbital; sub-shell notation (for $l = 0$); solid
S	entropy; overlap integral; magnitude of diffraction vector
t	time
T	temperature
T_b	boiling-point temperature
T_B	Boyle temperature
T_c	critical temperature
T_F	Fermi temperature
T_{ice}	ice-point temperature
T_m	melting-point temperature
T_r	reduced temperature
u	atomic mass unit; ungerade (*odd*)
U	energy (mechanical); internal energy (thermodynamic)
U_m	molar internal energy
v	speed; polar coordinate (velocity space); molecular volume
v	vibrational quantum number
v	velocity
V	volume; potential energy function; potential difference
V_c	critical volume

V_m	molar volume
V_r	reduced volume
w	work; probability function
W	joint probability function
x	(x) reference axis (perpendicular to y and z); distance along x–axis (fractional measure in crystallographic context)
X	distance along x-axis in absolute measure
$\bar{X}$	average value of a property; partial molar property
y	(y) reference axis (perpendicular to x and z)
Y	spherical harmonic function
z	(z) reference axis (perpendicular to x and y); collision frequency
Z	atomic number; number of collisions per unit space (area, volume); number of formula-entities per unit cell (crystallographic)
Z_{eff}	effective atomic number
$\mathscr{Z}$	compression factor
$\mathscr{Z}_c$	compression factor in critical state
α	spin state $\frac{1}{2}$; Coulomb integral (H_{ii}); polarizability; coefficient of thermal expansivity
α'	volume polarizability
β	spin state $-\frac{1}{2}$: resonance integral (H_{ij})
χ	electronegativity
δ	small quantity of a parameter; hard-sphere diameter
δ_+, δ_-	partial charges
δ_K	Kronecker delta
Δ	difference between two quantities or thermodynamic states; ligand-field-splitting energy parameter
∇^2	Laplacian operator
ε	energy per atom
ε_L	energy parameter in Lennard-Jones potential function
ε_r	relative permittivity
ε_0	permittivity of a vacuum
ϕ	probability function; polar coordinate; general angle
ϕ_M	work function (thermionic) of a species M
Φ	joint probability function; angular wave function; torsion angle
Γ	gamma function
η	viscosity coefficient
ψ	atomic wave function
ψ^*	atomic wave function conjugate to ψ
Ψ	molecular wave function; torsion angle
Ψ_π	π molecular wave function
κ	coefficient of compressibility; coefficient of thermal conductivity; net bonding parameter
λ	wavelength; fractional function-contribution descriptor
λ_C	Compton wavelength
μ	reduced mass of electron–proton system; dipole moment; chemical potential; mobility (of electron)

μ_B	Bohr magneton
ν	frequency (radiative)
$\bar{\nu}$	wave number $(1/\lambda)$
ν_+, ν_-	numbers of positive, negative ions (in solution)
π	ratio of the circumference of a circle to its diameter; bonding molecular orbital descriptor
π^*	antibonding molecular orbital descriptor
Π	product function
θ	polar coordinate; general angle
Θ	angular wave function; Debye temperature
ρ	electron density; electrical resistivity; exponent of radial wave function; exponent of repulsive potential
σ	screening constant; bonding molecular orbital descriptor; collision cross-section; electrical conductivity
σ^*	antibonding molecular orbital descriptor
$\sum$	summation sign
τ	relaxation time; $(d\tau)$ infinitesimal element of volume
τ_c	critical temperature (in metal)
ω	vibrational frequency $(2\pi\nu)$
Ω	ohm; polar coordinate

In addition to this use of symbols, several of them $(A, B, \ldots, a, b, \ldots \alpha, \beta, \ldots)$ are used from time to time as constants or other parameters, but are defined in the context.

Subscript notation

b	boiling-point
c	crystal
d	dissolution; diffusion; drift (velocity)
e	evaporation; electron; equilibrium
f	formation
h	hydration
m	molar; melting-point
s	sublimation
t	transition

1

Preamble

1.1 ATOMIC NATURE OF MATTER

Every material object around us is held together by interatomic bonding forces, and the aims of this book are to explore the nature of these forces and the structures and properties associated with the bonding to which they give rise in solid and fluid chemical species.

Matter consists of aggregates of atoms, of ions (atoms carrying an electric charge) or of molecules (discrete groups of bonded atoms). We accept the existence of atoms and molecules because we can, to some extent, see them. Fig. 1.1a is an electron density contour map of euphenyl iodoacetate ($C_{32}H_{53}IO_2$) that was obtained from an X-ray study on the crystalline material. The contour lines delineate regions of approximately spherical shapes, which we identify as atoms. The atoms are, in turn, linked to form a discrete group, the molecule, and the molecules are packed together in a precise and regular manner in the crystalline solid. Fig 1.1b, c, d show, respectively, the molecular skeleton, deduced from the electron density map, a chemical structural formula and the packing of the molecules in the crystal unit cell. Atoms may also be 'seen' by means of field-ion microscopy[1].

The atom, assumed spherical to a good approximation, has a nucleus at its centre, with a diameter of 10^{-15} m to 10^{-14} m. It contains protons, each with a unit positive charge, and neutrons, which carry no charge. These particles[2] have absolute masses of 1.6726×10^{-27} kg and 1.6749×10^{-27} kg, respectively. The nucleus is surrounded by electrons, each of which carries a unit negative charge and has a mass of 9.1094×10^{-31} kg. The number of electrons in an atom is equal to the number of its protons, so that the atom is, as a whole, electrically neutral. In relative terms, the proton and the neutron, approximately equal in mass, are each about 1837 times the mass of the electron, so that the mass of an atom resides mainly in its nucleus.

The laws of classical mechanics, which we know about from their applications to macroscopic objects, often fail with electrons and atoms. Quantum mechanics, of which classical mechanics is frequently seen to be a limiting case, must be used to describe microscopic situations and processes. For example, electrons may be considered to occupy certain regions of space in an atom called atomic orbitals, with

---

[1] See, for example, P.W. Atkins (1990) *Physical Chemistry*, OUP; K.M. Bowkett and D.A. Smith (1970) *Field-Ion Microscopy*, North-Holland.
[2] These and other fundamental units are listed with full precision on page xv.

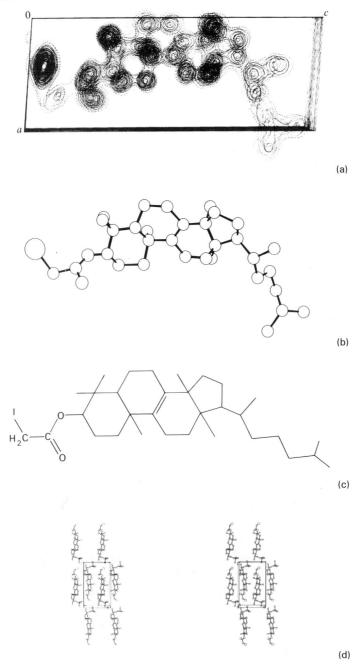

(a)

(b)

(c)

(d)

Fig. 1.1. Euphenyl iodoacetate, $C_{32}H_{53}IO_2$. (a) Electron density contour map; the closed contours join points of equal electron density in the unit cell. (b) Molecular model determined from the electron density map; for clarity, hydrogen atoms have been omitted. (c) Chemical structural formula; the eight-membered side chain is much less extended in this structure than in many steroids. (d) Packing of molecules in and around the unit cell: the viewing direction is along *a* towards the origin, approximately normal to that in (c) so as to avoid overlap of molecules in the viewing direction; only one unit cell length is shown in the direction of viewing and, again, hydrogen atoms have been omitted for clarity.

which are associated certain energy states for the electron. More precisely, we use the term atomic orbital to mean a wave function describing an electron, and the precise determination of the properties of the electron that would be demanded by a classical treatment gives way to a probability interpretation of electron position, momentum and density.

1.2   TYPES OF INTERATOMIC BONDING

Although we use terms such as covalent molecules, ionic crystals, molecular compounds and metallic solids, we recognize that these descriptions are imperfect. Nevertheless, they form a useful background against which the bonding of atoms in solids and fluids can be discussed, and we will review these four types briefly at this stage.

1.2.1   Covalent bonding

Covalent bonding forces involve an electron-sharing mechanism, which is exemplified most simply in molecular hydrogen. The electron of each hydrogen atom is shared between the two atoms to form a covalently bonded molecule of hydrogen, H_2. When the atoms involved are dissimilar, as in hydrogen chloride, for example, there will be a shift of electron charge in the direction of the more electronegative species, chlorine in this example, and the molecule is said to be polar, or to possess a dipole moment (see section 2.8.2.3). The best example of the covalent bond is probably in the diamond form of carbon, where covalent bonding forces link the carbon atoms to form a three-dimensional structure (Fig. 1.2); a note on stereoviewing is given in Appendix 2.

1.2.2   Ionic bonding

This form of bonding arises through Coulombic forces of attraction between ions, that is, between atoms that have lost or acquired one or more electrons. In sodium chloride, for example (Fig. 1.3), Na^+ and Cl^- ions exist, and they interact through forces that are proportional to $-1/r^2$, or to $-1/r$ in terms of potential energy, where r is the distance between the centres of the ions, and the negative sign indicates an attraction between them. In the solid, the attractive forces are balanced by repulsive forces between ions of like charge and, more strongly, especially under compressive stress, between other electrons in the ions. Sodium chloride is typical of an ionic structure.

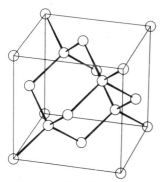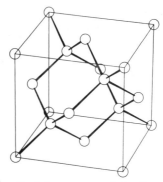

Fig. 1.2. Stereoview of the face-centred cubic unit cell of the structure of the diamond allotrope (polymorph) of carbon; covalent bonds link the atoms in three dimensions. Each carbon atom lies at the centre of a regular tetrahedron formed by four other atoms. In the structure, each corner atom is shared by four such tetrahedra.

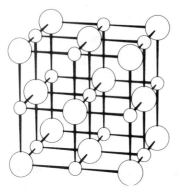 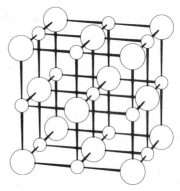

Fig. 1.3. Stereoview of the face-centred cubic unit cell of the sodium chloride structure; the circles in decreasing order of size represent Cl^- and Na^+. Each Na^+/Cl^- ion lies at the centre of an octahedron formed by $6Cl^-/6Na^+$ ions. What is the sharing principle here?

1.2.3   Van der Waals' bonding

At low temperatures, the noble gases can be crystallized. It is not difficult to envisage repulsive forces between the atoms in solid argon, arising from the proximity of filled electron shells on adjacent atoms, but what is the manner of attraction between these atoms?

The electrons in an atom may be said to be changing their positions with time. At any instant, the effective centres of the positive nucleus and the electrons may not coincide. Their separation constitutes a polarity within the atom; a similar polarity is then induced in a neighbouring atom, which leads to an electrostatic attraction between them, albeit of a relatively weak nature. Although such individual attractions are transient, over a period of time there are many of them, which serve to bind the argon atoms together in the solid (see also section 3.1.4.6).

These bonding forces exist between *all* species, but they are particularly important in the noble gases and other molecules that are not permanently polar, in bringing them together in the solid or liquid states.

1.2.3.1   Hydrogen-bonding

Hydrogen-bonding is a particular electrostatic type of interaction. Hydrogen is able to form a link between two atoms, most strongly with fluorine, oxygen and nitrogen, which enhances the total energy. We can find evidence for the strength of hydrogen-bonding by comparing, for example, the melting-point temperatures of the hydrides of group 15, 16 and 17 elements of the periodic table, as shown in Table 1.1; the

Table 1.1. Melting-point temperatures/K of some hydrides

H_3N	196	H_2O	273	HF	190
H_3P	140	H_2S	190	HCl	159
H_3As	157	H_2Se	207	HBr	186
H_3Sb	185	H_2Te	224	HI	222

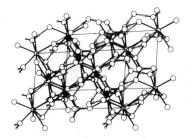 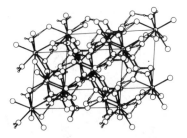

Fig. 1.4. Stereoview of the unit cell of the structure of Gypsum, $CaSO_4.2H_2O$; circles in decreasing order of size represent O, Ca, S and H. Note the importance of the hydrogen bonds (double lines – a hand-lens may help) to the cohesion of the structure; they link it across planes normal to the left–right direction in the diagram.

abnormally high-values for the first row compounds are clearly evident, indicating a stronger interaction between the species than would have been expected: normally, melting-point temperature increases with an increase in the relative molar mass.

Fig. 1.4 illustrates the structure of calcium sulphate dihydrate (Gypsum), $CaSO_4.2H_2O$. In this compound, covalent bonding is present in the SO_4^{2-} anions, ionic bonding exists between the Ca^{2+} cations and the SO_4^{2-} anions, with van der Waals' bonding acting between all species. In addition, hydrogen-bonding occurs between the oxygen atoms of the SO_4^{2-} anions and the water molecules; in fact, the hydrogen bonds are responsible for the coherence of the structure in one direction.

1.2.4 Metallic bonding

In metals, the free valence electrons of the atoms form a sort of electron 'sea' that serves to bind the positive ions together. The high electrical conductivity that is associated with metals arises from the mobility of the sea of electrons, whereas the malleability of metals, their flow under stress, depends upon the ease with which the electrons can re-bond to displaced cations. Gold is an excellent example of metallic bonding (Fig. 1.5). As we shall see, the existence of pure bond type is an ideal. Some substances conform to one type closely, but we must be prepared to find that more

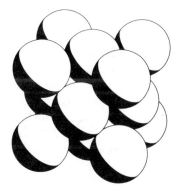 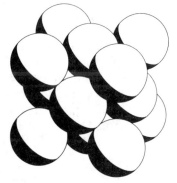

Fig. 1.5. Stereoview of a space-filling model of the close-packed cubic unit cell exhibited by many metals, including gold. How is this structure related to that of sodium chloride (Fig. 1.3)?

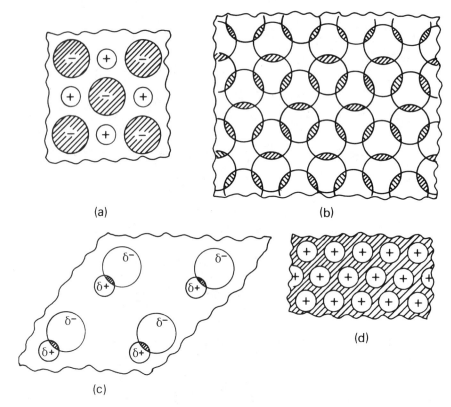

Fig. 1.6. Diagrammatic representations of types of interatomic bonding forces:
(a) ionic—Coulombic attraction between positive and negative ions;
(b) covalent—overlap of electron density; sharing of electrons;
(c) van der Waals'—dipolar (permanent or induced) attraction, with δ_+ and δ_- representing partial charges;
(d) metallic—positive ions in a 'sea' of electrons.

than one type is needed to describe the actual bond in a compound. The four main bond types outlined above are presented diagrammatically in Fig. 1.6.

1.3 STATES OF MATTER

Atoms may combine to form a substance that can exist, generally, in three different states. For example, sodium chloride is encountered usually as a colourless, crystalline solid. If pure, it is a non-conductor of electricity, but at 1074 K it melts to form a colourless liquid that conducts electricity, because the ions that are present in the solid are then free to move in the liquid state. At the higher temperature of 1690 K the liquid boils, and the vapour[1] consists of discrete molecules of NaCl and Na_2Cl_2, wherein the atoms are linked mainly by covalent bonding, different from that in the solid and the melt. The changes

$$solid \rightarrow liquid \rightarrow gas$$

[1] A gas below its critical temperature is called a vapour.

may occur for other substances at lower temperatures, and with much less change in the forces between the component atoms and molecules. Thus, in the equilibria:

$$\text{Ice} \underset{273\,\text{K}}{\overrightarrow{\xleftarrow{\hspace{1cm}}}} \text{Water} \underset{373\,\text{K}}{\overrightarrow{\xleftarrow{\hspace{1cm}}}} \text{Water vapour} \tag{1.1}$$

the covalently bonded water molecule is preserved in all three states, although the hydrogen-bonding between the molecules is progressively disrupted as the temperature is increased.

The state, solid, liquid or gas, of a substance that exists at a given temperature is determined by the result of a competition between the forces acting on its molecules (or atoms) and the thermal energy that the molecules possess.

Questions that may arise in the context of transformations between one state of a substance and another are 'Is it feasible?' and, if so, 'At what rate does it take place?' The first question is addressed by *thermodynamics* and, in general, transformations that occur are those that are accompanied by a decrease in the free energy of the system (ΔG negative). The second question lies in the realm of *rate processes*, which involve a reaction of one species with another, and where the energy needed to initiate the reaction is supplied either by collision or by the input of radiation. We shall not treat these topics specifically in this book, although we may draw upon them from time to time.

We shall, therefore, confine our study mainly to the forces between both atoms and molecules, in the different states of matter that are recognized.

1.3.1   Equipartition and energy states

The classical distribution (equipartition) of energy asserts that *the average energy associated with each degree of freedom in a system of a large number of particles in thermal equilibrium has the same value*, $\frac{1}{2} k_B T$, which depends only on the temperature T; k_B is the Boltzmann constant, 1.3807×10^{-23} J K^{-1}. A degree of freedom is a mode of motion of a particle, and generally depends upon the square of either a velocity or a coordinate. The average translational kinetic energy of a body of mass m and speed v, is $\frac{1}{2} m v^2$ and is equivalent to $\frac{1}{2} k_B T$. The same result of $\frac{1}{2} k_B T$ holds also for vibrational and rotational degrees of freedom (see Appendix 16).

The energy of a molecule (or atom) could involve some or all of the motions of translation, rotation and vibration, together with the energy of the electrons in the species. We may consider the vibrational energy of the hydrogen chloride molecule as an example. At low temperatures, almost all the molecules will be in the lowest available energy state (energy level). As the temperature is increased, some molecules will move to occupy higher energy states, corresponding to increased amplitudes of vibration in the molecules. Classical theory permitted all energies to exist—an energy continuum—but at the atomic level, quantum mechanical theory dictates that only discrete energy states are allowed.

As an example of a macroscopic system, to which classical mechanics applies, consider a light coil-spring carrying a heavy mass m, and let it be displaced slightly from its rest position and allowed to oscillate. It would then perform a simple harmonic motion, with a frequency of vibration f given by

$$f = \frac{1}{2\pi} \left(\frac{k}{m} \right)^{1/2} \tag{1.2}$$

where ℓ is the restoring force constant per unit displacement of the spring, that is, $\ell = F/x$, where F is the force exerted on the mass by the spring for a linear displacement x; $\pi = 3.14159 \ldots$ or $4 \tan^{-1}(1)$, the ratio of the circumference of a circle to its diameter. The energy U stored in the spring that has been extended by an amount x is given by

$$U = \tfrac{1}{2}\ell x^2 = 2\pi^2 f^2 mx^2. \tag{1.3}$$

If all energies are permitted, then all frequencies of oscillation of the spring are also possible, depending only upon ℓ and m.

Let the spring be replaced by a vibrating microscopic system of two atoms joined by a single bond, such as the hydrogen chloride molecule. Then, quantum mechanics shows that the permitted energy states are $\tfrac{1}{2}h\nu, \tfrac{3}{2}h\nu, \tfrac{5}{2}h\nu, \ldots$, where ν is the frequency of vibration along the bond and h is the Planck constant (6.6261×10^{-34} J Hz^{-1}).

With the spring, quantum effects are not noticed: if the frequency is, typically, 1 Hz, then the quantum energy step would be h J, equivalent to $\sim 4 \times 10^{-13}$ kJ mol^{-1}. This amount of energy is too small to be recognized experimentally, and energy is transferred in an effectively continuous manner. The hydrogen chloride molecule, however, has a vibration frequency of $\sim 10^{14}$ Hz; in this case, the quantum energy step $h\nu$ would amount to 39.9 kJ mol^{-1}, which is significant. Thus, with macroscopic bodies, quantum effects may be ignored, and classical mechanics gives a satisfactory account of their energies. At the atomic level, quantum mechanics must be invoked to give a true picture of the energy states.

1.3.2 Boltzmann distribution

When a substance at a very low temperature is heated, the molecular vibrations increase in amplitude and some molecules move into higher energy states. The higher the temperature, the greater the tendency for molecules to move in this way. The population, or average occupancy, N_j of an energy state ε_j at a temperature T is given by the Boltzmann distribution (see Appendix 3):

$$N_j = N_0 \exp(-\varepsilon_j/kT) \tag{1.4}$$

where N_0 is the population in the lowest energy state. If the energies are converted to molar terms $E_{m,j}$ ($E_{m,j} = L\varepsilon_j$, where L is the Avogadro constant, 6.0221×10^{23} mol^{-1}), then k_B is replaced by the gas constant $\mathscr{R}$ (8.3145 J K^{-1} mol^{-1}), since $Lk_B = \mathscr{R}$, and we write

$$N_j = N_0 \exp(-E_{m,j}/\mathscr{R}T). \tag{1.5}$$

For hydrogen chloride, with the first vibrational state frequency of $\sim 10^{14}$ Hz, the corresponding energy is 39.9 kJ mol^{-1}. Hence, N_1/N_0 becomes $\exp(-39.9/2.48)$, or approximately 10^{-7}, at 298 K. Evidently, hydrogen chloride molecules are effectively in the lowest energy state, or *ground state*, at this temperature.

1.3.3 Solid state

A solid is a substance of fixed volume and shape at a given temperature. The atoms or molecules of the solid are arranged in a regular manner on, or in a fixed relation to, the points of a lattice. Referring to the geometry of Fig. 1.3, we can say that the atoms vibrate about their mean positions, which are themselves invariant with time, and their vibrational energy makes the major contribution to the heat capacity of the solid, at normal temperatures.

In a solid, the atomic vibrations are anharmonic, such that an increase in temperature causes an increase in the distance between the mean positions of the atoms, and the material expands. Even if the vibrations were perfectly harmonic, the increase in free energy of a crystal with increase in temperature would lead to an increase in volume[1], but this effect is very small compared with that arising through the anharmonicity.

A qualification may be needed about the invariance with time of the mean atomic positions, with an increase in temperature. In some solids, certain groups of atoms may exhibit dynamic disorder (free rotation) or static disorder with respect to their mean positions. For example, sodium cyanide at room temperature has the ortho-rhombic structure shown in Fig. 1.7, but at or above 279 K its increased thermal energy causes it to transform to the structure type of sodium chloride (Fig. 1.3). Although each cyanide anion is linear, its random state orientation, averaged over space, simulates the envelope of a sphere, so that the higher-symmetry cubic structure obtains at the greater temperature.

1.3.4   Liquid state

At a given temperature, a liquid is fixed in volume but has no definite form. It takes up the shape of its containing vesssel, although not filling it completely because its boundary surface limits the movement of the bulk of the liquid. Evidently, the molecules in a liquid are close enough to one another for intermolecular forces to influence the relative spatial distribution of the molecules. However, the internal energy of a liquid is commensurate with that arising from the intermolecular forces, and order does not exist over distances of more than a few atomic dimensions, often called short-range order.

A liquid can be considered as being formed from the corresponding solid simply by heating it. As the temperature of the solid increases, the amplitudes of vibration of the

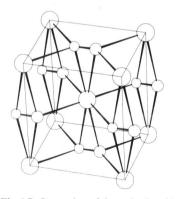 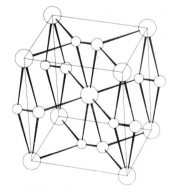

Fig. 1.7. Stereoview of the orthorhombic structure (below 279 K) of sodium cyanide, NaCN; the circles in decreasing order of size represent Na^+, N and C (together CN^-). Can you see how this structure is related to that of sodium chloride (Fig. 1.3)?

[1] See, for example, E.A. Guggenheim, (1963) *Boltzmann's Distribution Law*, North-Holland.

atoms about their mean positions increase. At the melting temperature they are free to move in the melt, and most of the structure is lost. Fig. 1.8 shows a computer simulation of the motion of argon atoms in the solid and in the corresponding melt; the loss of the lattice-based structure upon melting is indicated clearly.

At normal temperatures, the densities of liquids and solids are commensurate. Generally, the density of a liquid is less than that of the corresponding solid. Ice, silicon, germanium, grey tin and bismuth are some examples for which the density of the liquid is greater than that of the solid. It arises because these solids have a rather 'open' structure with low coordination number, often four, and we shall see that the average coordination number in liquids is greater. Liquids and solids have low compressibilities because, in most cases, the total space that they occupy is well filled. In this respect, gases are notably distinct.

An important property of a liquid is its ability to flow under an applied pressure, which may be simply a difference in hydrostatic level between one end of its container and the other; it depends upon the viscosity of the liquid. The greater the attractive forces between the molecules of the liquid, the higher will be its viscosity. At 298 K water has a viscosity approximately three times that of dimethyl ketone (acetone), because of strong hydrogen-bonding between the molecules of the water.

The viscosity of a liquid follows a Boltzmann distribution:

$$\eta = C \exp(E_{m,\eta}/\mathcal{R}T) \tag{1.6}$$

where η is the viscosity coefficient, $E_{m,\eta}$ is the molar energy required to overcome the attractive forces between the molecules of the liquid, and C is a constant; values for $E_{m,\eta}$ lie in the range 1 kJ mol^{-1} to 30 kJ mol^{-1}, for liquids of relatively low molecular mass.

1.3.5   Gaseous state

Gases are characterized by large volume changes consequent upon variations in their temperature and pressure, and by their ability to flow into the whole space available to them. Gases, therefore, are fixed in neither shape nor volume. They are miscible with one another in all proportions, and differ markedly from solids and liquids in that many of their properties are independent of their chemical nature, and may be determined by general gas laws. It has been shown experimentally that at low values of the pressure p or at high values of the temperature T (or both), gases follow the ideal gas equation:

$$pV = n\mathcal{R}T \tag{1.7}$$

where V is the volume containing n moles of gas, and T is the absolute temperature. For $n = 1$, V becomes the molar volume V_m.

1.3.5.1   Molar arithmetic

We may recall in passing that 1 mole of any substance contains as many entities of that substance as there are carbon atoms in 0.012 kg of ^{12}C, which number refers to the Avogadro constant L.

Confusion sometimes arises because of the use of both g mol^{-1} and kg mol^{-1} as units of molar mass. We set out comparisons in Table 1.2, using water (H_2O) as an example substance.

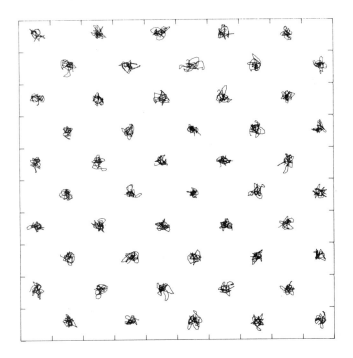

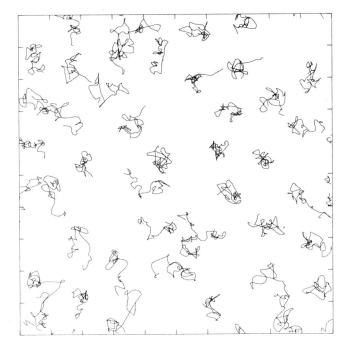

Fig. 1.8. Models of argon from computer simulation solutions of Newton's equations of motion: (a) solid state; (b) liquid state. (Reproduced with permission from 'Computer Simulations of Liquids and Liquid Interfaces', A. J. C. Ladd, PhD Thesis, Cambridge, 1977.)

Table 1.2. Comparisons of mass units

Mass of sample M_S		Mass of molecule m	
Molar mass M_m		Relative molar mass M_r	
Number of moles of substance n		Atomic mass unit u	

Quantity	g basis	kg basis
M_S	10 g	0.010 kg
M_r	18.02	18.02
n	0.555 mol	0.555 mol
M_m	18.02 g mol^{-1}	0.01802 kg mol^{-1}
m	$(18.02/L)$ g	$(0.01802/L)$ kg
u	$\left(\dfrac{18.02/L}{18.02}\right)$ g	$\left(\dfrac{0.01802/L}{10^3 \times 0.01802}\right)$ kg

Therefore, in general

$$n = M_S/M_m = M_S/(M_r \text{ g mol}^{-1}) = M_S/(10^{-3}M_r \text{ kg mol}^{-1}) \quad (1.8)$$

and

$$m = M_r u = M_m/L = M_r \text{ g mol}^{-1}/L = M_r \text{ kg mol}^{-1}/10^3 L. \quad (1.9)$$

The atomic mass unit (1.66054×10^{-24} g, or 1.66054×10^{-27} kg) might therefore be regarded as the mass of a hypothetical atom of unit relative molar mass.

From a structural point of view, gases are almost totally random assemblies of molecules in constant motion. The average distances between the molecules of a gas and of their travel between collisions are normally large, relative to the molecular diameter, so that the kinetic energy of translation of the molecules is significantly greater than the intermolecular attractions. At low pressure or high temperature then, the circumstances in which (1.7) is obeyed, interatomic forces are negligible, which is a postulate of the kinetic theory of gases. Fig. 1.9 shows diagrammatic simulations of argon gas at 1, 2 and 5 atmospheres pressure. It is evident that at the low pressure the atoms are widely separated, so that forces of attraction are not significant under these conditions.

1.3.6 Other forms of matter
None of the following forms of matter constitutes a distinct state of matter. Each is of importance and is given detailed consideration in dedicated treatises; we refer to them here briefly for the sake of completeness. The superfluid state is discussed briefly in Chapter 5.

1.3.6.1 *Glasses*
Glasses may be regarded as (supercooled) liquids of extremely high viscosity; the quantity E_η in (1.6) is very large at normal temperatures. The form of (1.6) shows that

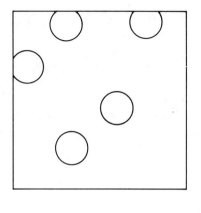

(a)

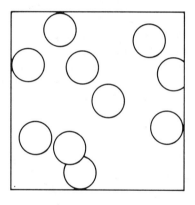

(b)

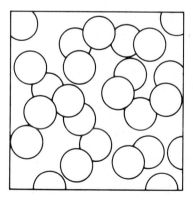

(c)

Fig. 1.9. Models of argon gas at 298 K: (a) at 1 atmosphere, (b) at 2 atmosphere, (c) at 5 atmosphere, as seen in a direction normal to the face of a cubic container. The smallness of the numbers of molecules is exaggerated, but the relative numbers in (a), (b) and (c) are in an approximately correct ratio.

the value of the viscosity decreases with increasing temperature, and most glasses begin to flow, or 'creep', at temperatures in excess of about 800 K.

A glass may also be regarded as an amorphous, or non-crystalline, solid. To the unaided eye, fragments of glass and quartz look similar to each other, yet only the quartz is crystalline. Glass has an atomic arrangement that displays short-range order, like that in liquids. Fig. 1.10 illustrates the structures of quartz and silica glass. Both of them are based on the same atomic group, the tetrahedral SiO_4 structural unit, but in quartz these units are arranged with periodic regularity in three dimensions.

1.3.6.2 *Liquid crystals*

Certain organic crystals, on heating carefully, pass into a form that is intermediate between the solid and liquid states. Such substances are known as liquid crystals.

Cholesteryl benzoate, $C_{34}H_{50}O_2$, melts sharply at 419 K to form an opaque liquid crystal which clears to a normal liquid at 452 K. Liquid crystals usually consist of large, elongated molecules that possess one or more polar groups, such as $=NH_2$ or $=CO$. In the crystalline state, these substances pack with their molecules aligned parallel to one another, and forces of attraction exist between the polar groups, in addition to the van der Waals' forces described above. On heating, the weaker van der Waals' forces are overcome first, by the thermal energy supplied to the crystal, and relative movement can occur. Increased heating breaks the links between the polar groups as well, and the substance then passes into the true liquid state.

$C_6H_5CO_2$

Several phases of liquid crystal are recognized (Fig. 1.11), but not all liquid crystals exhibit every phase. In the nematic phase, shown, for example, by ammonium oleate, $CH_3(CH_2)_7CH=CH(CH_2)_7CO_2NH_4$, the long molecules are arranged with their lengths parallel, rather like an army of descending parachutists. In the smectic phase, shown, for example, by 4-azoxyanisole, H_3CO—⬡—$N(O)=N$—⬡—OCH_3 the molecules are arranged on equally spaced planes, but without any lateral periodicity, rather like a crowd of shoppers in a departmental store.

The transitions between the phases are reversible, and occur at definite transition temperatures that vary with pressure according to the Clausius–Clapeyron equation [1]:

$$\frac{dp}{dT} = \frac{\Delta H_t}{T(V_2 - V_1)} \tag{1.10}$$

[1]This equation is discussed in any good treatment of thermodynamics.

Fig. 1.10. Arrangements of SiO$_4$ structural units (the darker spheres represent Si): (a) α-quartz; (b) silica glass. The long-range regularity of the quartz structure is clearly absent in the silica glass. (Crown copyright. Reproduced from NPL Mathematics Report Ma62 by R. J. Bell and P. Dean, with permission of the Director, National Physical Laboratory, Teddington, England.)

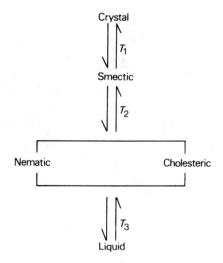

Fig. 1.11. Possible phases of liquid crystals, although a given liquid crystal would probably not show all of them; T_1, T_2 and T_3 represent transition temperatures.

where V_1 and V_2 are volume quantities in two phases and ΔH_t is the enthalpy of transition at the temperature T.

The cholesteric liquid crystals are optically anisotropic, that is, their optical properties vary with direction in the crystal, and exhibit colours that depend upon temperature. This property is very marked in nematic liquid crystals, and forms the basis of digital display functions in instruments and watches.

1.3.6.3 Solutions
Solutions consist of a solid dissolved in a solvent. The process of dissolution involves the breaking of the bonds that hold the entities together in the solid, simultaneously bringing about an interaction (solvation) with the solvent. As well as the breaking of linkages in the solid, there will be changes (breaking and making) of the structure of the solvent and solution; both enthalpic and entropic effects are involved in the solution process. We shall have a little more to say about solubility in the chapter dealing with ionic compounds.

1.3.6.4 Colloids
Colloids consist of small particles, less than about 500 nm in diameter, called the disperse phase, distributed in a fluid, known as the dispersion medium. The particles are too small to be seen with an ordinary microscope, and they pass through fine filters; they can be detected by light-scattering experiments.

Colloids are classified as sols, for dispersions of solids in liquids, such as precipitated sulphur and inorganic sulphides; as aerosols, for dispersions of liquids in gases, such as sprays, or of solids in gases, such as smokes; or as emulsions, for dispersions of liquids in liquids, such as milk. Where the particles of the disperse phase are attracted strongly to those of the dispersion medium, the colloid is called lyophilic, otherwise it is termed lyophobic.

The structure of a colloid depends upon the electrical characters of both the disperse phase and the dispersion medium, as well as on the actual size of the particles.

PROBLEMS FOR CHAPTER 1

1.1 From the data in Table 1.1, predict the melting-point temperature for hydrogen astitide, HAt.

1.2 A coil-spring carries a mass of 250 g, and it is set into gentle oscillation. If the spring constant k is 0.5 kg s^{-2}, calculate the frequency of oscillation of the spring, and the energy stored in the spring when it is extended by 0.01 m.

1.3 The hydrogen chloride molecule has a ground state vibration frequency of 8.96×10^{13} Hz at 298 K. If the effective (reduced) mass of the molecule is 1.63×10^{-27} kg, calculate the bond force constant (k of equation (1.2)).

1.4 The first vibration state above the ground state in the carbon monoxide molecule has a frequency of vibration of 6.3×10^{13} Hz. What is the ratio of the number of molecules in this state to that in the ground state, at 298 K?

1.5 How many sodium chloride entities are there in its face-centred cubic unit cell (Fig.1.3)? If the side of the cube is 0.564 nm, calculate the density of crystalline sodium chloride.

1.6 Calculate the relative molar mass and the absolute mass of euphenyl iodoacetate (Fig. 1.1).

2

Bonding between atoms

2.1 INTRODUCTION

It is not difficult to accept the idea of attraction between oppositely charged chemical species, like Na^+ and Cl^-. It is not so obvious, however, how two neutral hydrogen atoms unite to form the hydrogen molecule, H_2. Yet we expect that the forces involved in bonding will be electrical, because such is the nature of matter.

Most students of chemistry will, at some time, have used the dot diagrams of Lewis (1916). He was aware of the particular stability of the noble gases, and postulated that, in compound formation, atoms could attain an effective noble-gas configuration by an electron-sharing process, each atom providing one electron to a bond pair. Thus, methane was written as

$$
\begin{array}{ccc}
& H & \\
& \cdot\cdot & \\
H & : C : H & \\
& \cdot\cdot & \\
& H &
\end{array}
\qquad \text{or} \qquad
\begin{array}{ccc}
& H & \\
& | & \\
H & \!\!-\!\!C\!\!-\!\! & H \\
& | & \\
& H &
\end{array}
$$

In this compound, carbon and hydrogen obtain, formally, the configurations of the noble gases neon and helium, respectively, and each linkage is a single bond. Lewis explained multiple bonds by the sharing of two or more pairs of electrons, and ethene was pictured as

$$
\begin{array}{cc}
H & H \\
\cdot\cdot & \cdot\cdot \\
C & :: C \\
\cdot\cdot & \cdot\cdot \\
H & H
\end{array}
\qquad \text{or} \qquad
\begin{array}{cc}
H & H \\
| & | \\
C & =\!\!=C \\
| & | \\
H & H
\end{array}
$$

and both single and double bonds are present in this molecule. The particular geometries shown here for these molecules have been drawn for convenience, and do not necessarily imply geometrical reality.

It is a straightforward matter to write a Lewis structure for a given molecule, ethanal (CH_3CHO), for example. The total number of bonding or valence electrons are four for each carbon, one for each hydrogen and two for each oxygen atom, or fourteen electrons in all. We begin by drawing one shared pair between each pair of bonded atoms:

$$
\begin{array}{ccc}
& H & & O \\
& \cdot\cdot & & \cdot\cdot \\
H & : C : C & & \\
& \cdot\cdot & & \cdot \\
& H & H &
\end{array}
\qquad \text{or} \qquad
\begin{array}{c}
H \qquad\qquad O \\
| \qquad\quad / \\
H\!-\!C\!-\!C \\
| \qquad\quad \backslash \\
H \qquad\qquad H
\end{array}
$$

Twelve electrons have been used, and the other two are then distributed so as to form a double bond with oxygen:

$$
\begin{array}{ccc}
& H & & O \\
& \cdot\cdot & & \cdot\cdot \\
H & : C : C & & \\
& \cdot\cdot & & \cdot \\
& H & H &
\end{array}
\qquad \text{or} \qquad
\begin{array}{c}
H \qquad\qquad O \\
| \qquad\quad /\!/ \\
H\!-\!C\!-\!C \\
| \qquad\quad \backslash \\
H \qquad\qquad H
\end{array}
$$

Again, each atom has achieved a noble-gas configuration, or octet of electrons; there are also two 'lone pairs' on the oxygen atom.

Certain atoms behave as though they have an expanded octet of ten or twelve electrons, as in PCl_5 or SF_6, respectively. An explanation of the expanded octet was not possible with Lewis's hypothesis; nor could it deal with compounds such as B_2H_6, which is apparently deficient of two electrons.

The basic idea of the shared pair of electrons can still be a useful, formal concept, but it is qualitative in character. A satisfactory theory of bonding needs to show, for example, how it is that methane forms four equivalent C—H bonds, each of length 0.109 nm, and with H—C—H angles of 109.47°, whereas in ethene there are, apparently, only three bonds from carbon, and with H—C—H (and H—C—C) angles of 120°. Lewis's description was unable to supply these details.

We shall be concerned intimately with electrons in all types of bonding between atoms, so it is useful to consider first some of the fundamental properties of the electron, and how classical mechanics failed to explain certain important experimental results involving electrons, and radiation.

2.2 CLASSICAL MECHANICS — SUCCESS AND FAILURE

Until the beginning of the twentieth century, it was assumed that atoms, like larger particles, behaved physically according to Newtonian mechanics. However, no satisfactory quantitative atomic theory could be derived on a classical basis, and experimental results had been amassed which indicated that a more refined theory was needed. It was not until 1926 that a new system, wave mechanics, was devised and applied successfully to atomic systems, and in this chapter we shall consider the development of this theory.

The behaviour of a classical particle can be described by two basic equations. In the first of them, for one dimension, along x, the total energy E of a particle of mass m and speed v, at a position x and time t with respect to a given origin, is equal to the sum of its kinetic energy $\frac{1}{2} mv^2$ and potential energy $V(x)$:

$$E = \tfrac{1}{2} mv^2 + V(x) \tag{2.1}$$

where v and x are both dependent upon t. Since the momentum p of the particle is given by

$$p = mv \tag{2.2}$$

(2.1) may be written as

$$E = \frac{p^2}{2m} + V(x). \tag{2.3}$$

The second basic equation depends on Newton's second law (force = mass × acceleration):

$$F(x) = m \frac{\mathrm{d}v}{\mathrm{d}t} \tag{2.4}$$

or, using (2.2)

$$F(x) = \frac{\mathrm{d}p}{\mathrm{d}t}. \tag{2.5}$$

These equations permit both a continuous range of energies for the particle, and the simultaneous determination of its position and momentum, or trajectory. However, we indicated in the first chapter that the nature of the energies of atomic systems differed from those of macroscopic bodies; this feature emerged from experiments, some of which will be considered next.

2.2.1   Black-body radiation
When objects are heated they emit radiation. As the temperature of the radiator is increased, the frequency of the emitted radiation increases from the infrared end of the spectrum, through the visible range and into the ultraviolet region. A good approximation to a black-body radiator is a heated container with a pin-hole in one wall through which radiation is emitted, and may be sampled. An indication of the temperature of the radiator is given by its colour; the radiation emitted from the hole is in thermal equilibrium with the container.

Fig. 2.1 shows the energy distribution $E(v)$ as a function of the frequency v. The value v_{max} at which $E(v)$ is a maximum moves to higher frequencies as the temperature region of the spectrum. Similar curves are obtained in terms of wavelength λ, but with λ_{max} moving to lower wavelengths as the temperature is increased.

Experiments by Wien (1894) showed that

$$T\lambda_{max} = \text{constant} \tag{2.6}$$

where the constant had the experimental value of 2.9×10^{-3} m K.

In another set of experiments, Stefan (1879) had shown that the energy density $\mathscr{E}$, the total energy density per unit volume emitted over all wavelengths, followed the equation

$$\mathscr{E} = aT^4 \tag{2.7}$$

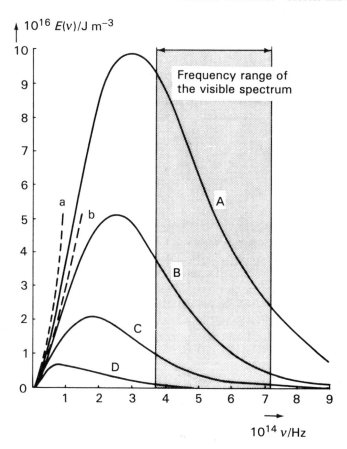

Fig. 2.1. Black-body radiation curves.
    Rayleigh–Jeans equation: *a* 5000 K, *B* 4000 K; the curves continue upwards without limit.
Planck equation: *A* 5000 K, *B* 4000 K, *C* 3000 K, *D* 2000 K. Similar curves, with maxima,
are obtained if $E(\lambda)$ is plotted as a function of wavelength λ. The movement of E_{max} is readily
seen, and the 'ultraviolet catastrophe' would arise if, in reality, the energy distribution
                        followed the Rayleigh–Jeans equation.

where the constant *a* was independent of the nature of the material of the body. This
relationship was also derived from thermodynamics by Boltzmann (1884), in the form

$$M = \sigma T^4 \tag{2.8}$$

where *M* is known as the *excitance* of the body, or its total power emitted per unit
area, and the Stefan–Boltzmann constant σ had the experimental value of 5.67×10^{-8}
W m^{-2} K^{-4}; (2.8) forms the basis of optical radiation pyrometry.

   Rayleigh considered that radiation was emitted by classical molecular oscillators,
one for each frequency v, the intensity of the radiation being proportional to the
amplitude of oscillation. He determined the number $N(v)dv$ of oscillators with
frequency lying between v and $v + dv$ inside a cubical enclosure of side c/λ, where *c* is
the speed of light in a vacuum $(2.9979 \times 10^8 \text{ m s}^{-1})$. The result, as amended by Jeans, is

$$N(v)\,dv = 8\pi v^2/c^3\,dv. \tag{2.9}$$

We show later (Appendix 16), or by equipartition (section 1.3.1), that the mean energy of a classical oscillator at a temperature T is $k_B T$ (two degrees of freedom, from the kinetic and potential energies), where k_B is the Boltzmann constant. Hence, the required energy density, as given by the Rayleigh–Jeans equation, is

$$E(v)\, dv = N(v) k_B T\, dv = 8\pi v^2 k_B T/c^3\, dv. \tag{2.10}$$

In terms of wavelength λ, remembering that $dv = (-c/\lambda^2)\, d\lambda$, and neglecting the negative sign because we are interested in the absolute value of the energy density, we obtain

$$E(\lambda)\, d\lambda = 8\pi k_B T/\lambda^4\, d\lambda. \tag{2.11}$$

However, (2.10), or (2.11), conflicts with Wien's observation (2.6) and, as can be seen from Fig. 2.1, would lead to an accumulation of radiation in the high-energy region, the so-called ultraviolet catastrophe, since $E(v)$ would increase without limit as the frequency increased. Since it would be true at any temperature, it follows that objects could glow in the dark, except that there would then be no darkness at all!

Planck (1900) made a major and fundamental revision to the Rayleigh–Jeans equation. He was able to account for the experimental results on radiation by postulating that an oscillator of frequency v could radiate energy only in integral multiples of hv, or quanta, and that the energy E of a single quantum is given by

$$E = hv. \tag{2.12}$$

Radiation may be considered to consist of particles called *photons*, each of energy hv. Consider, for example, a 100 W tungsten lamp; it emits light at a wavelength of approximately 556 nm. The frequency of the light is, therefore, 5.39×10^{14} Hz. Hence, the rate of generation of photons in the operation of the lamp is 100 W/$(6.6261 \times 10^{-34}$ J Hz^{-1} $\times 5.39 \times 10^{14}$ Hz), or 2.8×10^{20} s^{-1}.

Classical theory permits electromagnetic oscillators of all frequencies, even very high values, to be activated in the cavity of the black body by absorbing radiation from its walls. Quantum theory, however, allows oscillators to be activated only if they can acquire energy nhv, $(n = 1, 2, 3, \ldots)$. The effect is to damp the high-frequency oscillators, because they cannot acquire sufficient quanta of energy from the black body.

The Planck equation reconciled the equations of Rayleigh–Jeans, Wien and Stefan–Boltzmann, which now emerge as special cases:

$$E(v)\, dv = (8\pi h v^3/c^3)\{\exp(hv/k_B T) - 1\}^{-1}\, dv. \tag{2.13}$$

In terms of wavelength,

$$E(\lambda)\, d\lambda = (8\pi h c/\lambda^5)\{\exp(hc/\lambda k_B T) - 1\}^{-1}\, d\lambda \tag{2.14}$$

and the exponential term in (2.13) or (2.14) leads to the damping of the high-energy oscillations.

At this stage we might enquire what amount of energy density of light in the wavelength range 590 nm to 600 nm exists in a cavity of volume 10 cm^3 at a temperature of 500° C, according to (a) the Planck equation, (b) the Rayleigh–Jeans equation?

The mean wavelength may be taken as 595 nm, and the wavelength range as 10 nm. Hence, from (2.14)

$$E(\lambda)\,d\lambda = \{6.6947 \times 10^7 \text{ J m}^{-4}/[\exp(31.275) - 1]\}\,(10 \times 10^{-9} \text{ m})$$
$$= 1.75 \times 10^{-14} \text{ J m}^{-3}$$

whereas from (2.11)

$$E(\lambda)\,d\lambda = \{2.1402 \times 10^6 \text{ J m}^{-4}\}\,(10 \times 10^{-9} \text{ m})$$
$$= 2.14 \times 10^{-2} \text{ J m}^{-3}$$

whereupon the damping effect of the exponential term is very evident.

Classical conditions correspond to no quantization, that is, to the limit of $E(v)$ as h tends to zero. From (2.13), we have

$$E(v) = (8\pi h v^3/c^3)\{(hv/k_B T) + (hv/k_B T)^2/2! + (hv/k_B T)^3/3! + \ldots\}^{-1}$$
$$= (8\pi h v^3/c^3)(k_B T/hv)\{1 + (hv/k_B T)/2! + (hv/k_B T)^2/3! + \ldots\}^{-1} \quad (2.15)$$

Hence, dividing the numerator and denominator by h,

$$\underset{h\to 0}{\text{Limit }} E(v) = 8\pi v^2 k_B T/c^3 \qquad (2.16)$$

which is the Rayleigh–Jeans equation (2.10) for $E(v)$. Since h is really a constant, it is preferable to consider that (2.16) is reached as the dimensionless quantity $(hv/k_B T)$ tends to a very small value. Thus, the classical condition is attained for $(hv/k_B T) \ll 1$, that is, at low frequencies and high temperatures; typically, $v = 10^{11}$ Hz and $T = 3000$ K, whence $(hv/k_B T) = 0.0016$. At high frequencies and low temperatures $(hv/k_B T) \gg 1$; typically, $v = 5 \times 10^{14}$ Hz and $T = 500$ K, whence $(hv/k_B T) = 48$. Then, (2.13) may be approximated to

$$E(v)\,dv = (8\pi h v^3/c^3)\exp(-hv/k_B T)\,dv \qquad (2.17)$$

In terms of wavelength, to provide a link with (2.6), differentiation of (2.14) with respect to λ, and setting the derivative to zero, gives

$$hc/(5\lambda_{max}k_B T) = 1 - \exp(hc/\lambda_{max}k_B T) \qquad (2.18)$$

and the solution, by successive approximations, is

$$\lambda_{max}T = 0.20140\,hc/k_B = 2.898 \times 10^{-3} \text{ m K}, \qquad (2.19)$$

in very good agreement with experiment. So Wien had observed quantum behaviour, whereas Rayleigh–Jeans addressed the black body under classical conditions.

It may be shown[1] that the power radiated per unit area from a black body at a wavelength λ is given by

$$E(\lambda)\,d\lambda\,(c/4) = 2\pi hc^2\lambda^{-5}\{\exp(hc/\lambda k_B T) - 1\}^{-1}\,d\lambda. \qquad (2.20)$$

---

[1]See, for example, F. H. Crawford (1963) *Heat, Thermodynamics and Statistical Physics*, Harcourt, Brace and World Inc.

The total excitance M is then given by

$$M = c/4 \int_0^\infty E(\lambda) \, \mathrm{d}\lambda \tag{2.21}$$

which, on integration through (2.20), gives

$$M = 2\pi^5 k_B^4 T^4 / 15 h^3 c^2 = \sigma T^4 \tag{2.22}$$

and σ evaluates to 5.670×10^{-8} W m^{-2} K^{-4}, in excellent agreement with experiment.

2.2.2 Photoelectric effect

Fig. 2.2 is a schematic diagram of an apparatus for obtaining the photoelectric effect. A monochromatic light source incident upon a metal cathode in vacuo results in an electron flow from the cathode, provided that the external emf E is suitably adjusted. Experimentally, several conditions were found to exist:

    (a) the magnitude of the current flowing was proportional to the intensity of the light source;

    (b) the kinetic energy of the electrons emitted was independent of the light intensity;

    (c) the number of electrons emitted was proportional to the light intensity;

    (d) the mean kinetic energy of the electrons emitted was proportional to the frequency of the incident light;

    (e) no electrons were emitted unless the frequency of the incident light exceeded a certain minimum value, whereupon even a low light intensity was sufficient to cause emission; there was no delay in emission, provided that the incident energy exceeded the threshold value.

Conditions (d) and (e) are particularly important, and they indicate that the incident radiation, if of sufficient energy, will expel an electron from the metal. If the light source comprises packets $h\nu$ of energy then, by conservation of energy, the

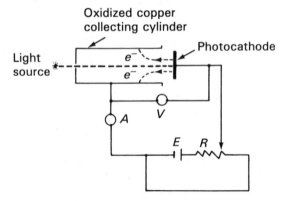

Fig. 2.2. Schematic diagram of apparatus for demonstrating the photoelectric effect. A contact potential difference exists between the dissimilar cathode and cylinder metals: it is assumed here that this potential difference makes the cathode positive with respect to the cylinder; A ammeter, V voltmeter, R variable resistor, E external battery.

kinetic energy of the electrons emitted should follow the law

$$\tfrac{1}{2}mv^2 = h\nu - \phi_M \tag{2.23}$$

as was pointed out by Einstein (1905). When the incident energy $h\nu$ transferred to an electron exceeds ϕ_M, an electron is ejected with kinetic energy equal to the excess of $h\nu$ over ϕ_M. The term ϕ_M is the energy required to dislodge an electron from the metal of the cathode, and is called the (thermionic) work function of the metal. The values of ϕ_M for a few metals are listed hereunder in eV (1 eV $=$ 1.6022 $\times$ 10^{-19} J).

	Li	Na	K	Mg	Cu	Ag
ϕ_M/eV	2.42	2.3	2.25	3.7	4.8	4.3

The light beam appears to be particulate, and the photons each have momentum p given by

$$p = mc \tag{2.24}$$

where m is the mass associated with a light photon. Using the Einstein equation that relates mass and energy

$$E = mc^2 \tag{2.25}$$

together with (2.12), for a single quantum, we obtain

$$p = E/c = h\nu/c. \tag{2.26}$$

2.2.3   Compton effect

When light interacts with electrons in a material, it is scattered with an increase in wavelength. The shift in wavelength is independent of the wavelength of the incident light, but depends on the angle of scatter (Compton, 1923). If we assume that a photon is a particle of momentum $h\nu/c$, then collision (scattering) between a photon and an electron of mass m_e, with conservation of both energy and momentum, results in an increase in wavelength given by

$$\delta\lambda = (h/m_e c)(1 - \cos\theta_C) \tag{2.27}$$

where θ_C is the angle of Compton scattering and $h/m_e c$ is the Compton wavelength λ_C. In Fig. 2.3, the incident photon has momentum h/λ_i and the scattered photon h/λ_s. For $\theta_C = 60°$ (illustrated), $\delta\lambda$ is $0.5\lambda_C$. The magnitude of λ_C is 2.426 pm, so that the maximum wavelength shift, at $\theta_C = 180°$, is 4.85 pm.

2.2.4   Diffraction of electrons

The experiments of Davisson and Germer in 1925 led to the view that electrons, believed to be particulate in nature, were also capable of behaving like waves. These workers found that electrons could be diffracted from single crystals of metallic nickel, similar to the way in which light is diffracted from a ruled grating. The experiment was repeated successfully with many other substances, and led to the view that electrons, in this type of experiment, behaved similarly to a wave.

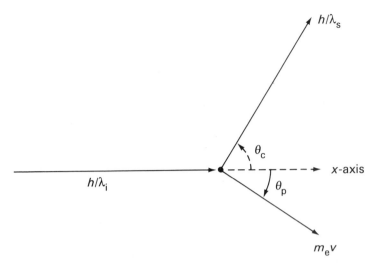

Fig. 2.3. Momenta and directions in the Compton effect: λ_i and λ_s are the wavelengths of, respectively, the incident and scattered light; for θ_c (the angle of scatter) = 60° (shown), the wavelength shift $\delta\lambda$ is equal to $\frac{1}{2}\lambda_c$ where λ_c is the Compton wavelength for the electron. Conservation of energy requires:

$$hc/\lambda_s + p^2/2m = hc/\lambda_i$$

and by conservation of momentum, we have:

$$h/\lambda_s \cos \theta_c + m_e v \cos \theta_p = h/\lambda_i$$
$$h/\lambda_s \sin \theta_c - m_e v \sin \theta_p = 0$$

where θ_p is the angle of scatter of the photon. From the diagram, the x-component of momentum p_x is $h/\lambda_i - h/\lambda_s \cos \theta_c \approx h/\lambda(1 - \cos \theta_c)$, where $\lambda = \lambda_i \approx \lambda_s$. If we could use a microscope to measure the position and momentum of an electron, then θ_c must lie within the range of the microscope objective, that is, between the limits $90° \pm \varepsilon$, where ε is the angle between the axis of the objective and the direction of the scattered photon, so that $\varepsilon = 90° - \theta_c$. Hence, p_x lies within $h/\lambda . (1 \pm \sin \varepsilon)$, or the uncertainty $\Delta p_x = h/\lambda \sin \varepsilon$. The resolving power of a microscope leads to an uncertainty in position $\Delta x = \lambda/\sin \varepsilon$. Hence,
$$\Delta p_x \Delta x \approx h,$$ which may be compared with (2.44).

2.2.5   Wave–particle duality

The photoelectric effect and the Compton effect are best explained by attributing a particulate nature to light, whereas diffraction experiments indicate its wave-like nature. Similarly, Thomson's well-known experiments on cathode rays (electrons) demonstrated their particulate nature, whereas the diffraction of electrons showed their wave-like properties.

One is led to the conclusion that, on the atomic scale, particles can exhibit the features of waves, and that waves can show the characteristics of particles, according to the types of experiments performed. In Table 2.1, some wave–particle properties are compared.

In 1924, de Broglie had suggested that any particle travelling with a linear momentum p had associated with it a wave character, where

$$p = h/\lambda. \tag{2.28}$$

This relationship is implicit in (2.26), since $v'c = 1/\lambda$, and it can be used as a starting point in formulating the wave mechanics of an electron.

Table 2.1. Wave-particle duality in light and matter

Light	Matter
|	|
Light, a manifestation of energy, emitted by a heated filament appears to be continuous.	Matter, a manifestation of mass, appears to be continuous.
|	|
In the photoelectric effect, light appears to be particulate in nature: $p = h/\lambda$	According to the atomic theory, matter has a particulate nature: $p = mv$
|	|
In diffraction experiments, light behaves like a wave motion: $\lambda = c/v$	In water waves, matter appears to execute a wave motion: $\lambda = h/p$

From the de Broglie equation, we can easily calculate the wavelength of electrons that have been accelerated by a given potential difference V. The energy acquired by an electron is then $e \times V$, for which reason the electron-volt is often used as a convenient energy unit. At the end of the acceleration, the energy is wholly kinetic. Hence, from (2.1) and (2.2)

$$p = m_e v = (2m_e E)^{1/2} = (2m_e \tfrac{1}{2} m_e v^2)^{1/2} = (2m_e eV)^{1/2}.$$

As an example, suppose that the electrons under consideration are in an accelerator operating at 1 GeV (10^9 eV). From the above equation, p evaluates to 1.709×10^{-20} N s^{-1}. Then, from (2.28), λ is 3.88×10^{-14} m, or 0.0388 pm.

2.2.6 Atomic spectra

When hydrogen atoms are excited by an electric discharge, their emission spectrum includes light at four frequencies in the visible region, as shown in Fig. 2.4. Balmer (1895) showed that this spectrum fitted an equation of the type

$$\bar{v} = R_H \left(\frac{1}{m^2} - \frac{1}{n^2} \right) \tag{2.29}$$

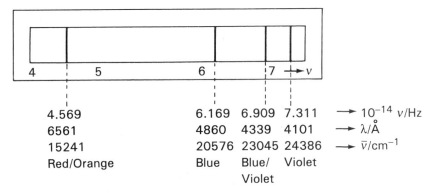

Fig. 2.4. Balmer series in the spectra of atomic hydrogen; other series exist in the infrared and ultraviolet regions, and are fitted by (2.29) with different values of the integer m.

where $\bar{v}$ is the wavenumber $(1/\lambda)$ of the spectral line, R_H is the Rydberg constant for hydrogen (experimentally, 109677.6 cm^{-1}), m is equal to 2 and n is another integer, greater than 2. It was a feature of Bohr's atomic theory that it explained the spectra of atomic hydrogen in terms of transitions between two energy levels, with the assumption that only when moving from one energy level E_2 to a level of lower energy E_1 was an electron able to emit radiation (Fig. 2.5), with a frequency given by

$$v = (E_2 - E_1)/h. \qquad (2.30)$$

This equation represented a fundamental break with the classical (Rutherford) theory, which required a charge (electron) moving along a circular path (orbit), thus under acceleration, to emit energy continuously. The theory predicted the energy levels correctly, and gave a value for the Rydberg constant in excellent agreement with experiment. The energy levels E_n in atomic hydrogen are given by the equation

$$E_n = \mu e^4/(8n^2h^2\varepsilon_0^2) \qquad (n = 1, 2, 3, \ldots,) \qquad (2.31)$$

where μ is the reduced mass of the system of one proton and one electron (see Appendix 6) and ε_0 is the permittivity of a vacuum. Evaluating $\{\mu e^4/(8h^2\varepsilon_0^2)\}/(hc)$ gives 109677.59 cm^{-1} for R_H.

Bohr's theory of the atom was based on classical mechanics, but with quantum conditions imposed in order both to fit the experimental results and to explain how electrons did not follow a spiral path into the nucleus, as implied in Rutherford's earlier theory. It was, however, a patchwork theory and, more seriously, it could not explain normal spectral results other than those of atomic hydrogen.

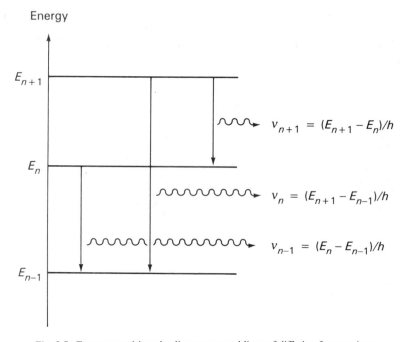

Fig. 2.5. Energy transitions leading to spectral lines of differing frequencies v.

The Bohr theory supposed that electrons moved around the nucleus in orbits, rather like the planets in our solar system. The introduction of elliptical orbits by Sommerfeld did little to improve the theory. The main objections to applying planetary theory to electrons in an atom is that it defines precisely both the position and momentum of an electron, and follows its orbit. The uncertainty principle, which arises from wave mechanics (see section 2.3.2), shows that these parameters cannot all be determined with precision simultaneously: that they can be for planets, with satisfactory precision, depends on the differing sizes of planets from electrons in relation to the methods used for their observation.

If, then, we are not to obtain the required success by following descriptions based on the particulate nature of the electron, it seems not unreasonable to consider next a theory based upon its wave properties, and we turn our attention to wave mechanics.

2.3 WAVE EQUATION

The Schrödinger equation (sometimes called the 'new' quantum theory[1], to distinguish it from the 'old' quantum theory of Planck/Bohr), proposed in 1926, leads, in principle, to a wave function for any system of electrons in an atom or molecule. For an electron that is free to move in one dimension, Schrödinger's equation may be written as

$$-\frac{h^2}{8\pi^2 m_e}\frac{d^2\psi(x)}{dx^2} + V(x)\psi(x) = E\psi(x) \tag{2.32}$$

where $\psi(x)$ is the one-dimensional wave function, $V(x)$ is the potential energy and E the total energy of the electron. Equation (2.32) may be conveniently put in the form

$$\mathscr{H}_1\psi = E\psi \tag{2.33}$$

with the one-dimensional Hamiltonian operator $\mathscr{H}_1$ given by

$$\mathscr{H}_1 = -\frac{h^2}{8\pi^2 m_e}\frac{d^2}{dx^2} + V; \tag{2.34}$$

V and ψ are understood to mean $V(x)$ and $\psi(x)$, respectively.

Although we cannot derive (2.32), it is possible to justify its form. If we assume that a particle of mass m is moving freely in a region of constant potential energy $V(x)$, or just V, then we write

$$-\frac{h^2}{8\pi^2 m}\frac{d^2\psi}{dx^2} = (E - V)\psi. \tag{2.35}$$

A simple solution of (2.35) is

$$\psi = \exp(ikx) = \cos kx + i \sin kx \tag{2.36}$$

where

$$k = \{8\pi^2 m(E - V)/h^2\}^{1/2}. \tag{2.37}$$

Those who wish to revise the solution of this type of differential equation may consult Appendix 7.

---

[1] The terms 'wave mechanics' and 'quantum mechanics' are often used synonymously.

If the potential energy is zero,

$$k = \{8\pi^2 mE/h^2\}^{1/2} \tag{2.38}$$

and from (2.3) it follows then that

$$p = \frac{kh}{2\pi}. \tag{2.39}$$

Since $\cos kx$ (or $\sin kx$) represents a wave of wavelength λ equal to $2\pi/k$, we have

$$\lambda = h/p \tag{2.40}$$

which is equivalent to de Broglie's equation (2.28). If V is non-zero, then at a fixed total energy we have, from (2.37)

$$\lambda = h/\{2m(E - V)\}^{1/2} \tag{2.41}$$

and as the kinetic energy decreases the wavelength increases, becoming infinite when the particle is at rest.

If $V(x)$ changes linearly with x, a particle would be subject to a force that is proportional to $-dV(x)/dx$. Since $V(x)$ is decreasing with increase in x, the kinetic energy increases, λ decreases and momentum increases. The scenario is reminiscent of a particle under acceleration from an imposed constant force, the motion of which is determined by Newton's second law. Newtonian mechanics, as we have already remarked, is a special case of wave mechanics.

We may note here that the wave equation (2.32) may be obtained from the classical equation (2.3) if we make the substitution

$$p = i\frac{h}{2\pi}\frac{d}{dx} \tag{2.42}$$

where the term on the right-hand side of (2.42) is the linear momentum operator.

Many solutions to (2.32) exist. For a free particle, $\psi = \alpha \exp(ikx)$, where α is a constant, is a solution, as double differentiation can confirm. Hence, k (and E) can take on any value. However, in applying (2.32) to the electron, we need to interpret ψ in such a way that certain values of E are eliminated, as is required by quantization.

2.3.1   Born interpretation of the wave equation

In the corpuscular theory of light, the intensity of the light at any instant is determined by the number of photons present, whereas in wave theory it is governed by the square of the amplitude of the wave function. In Born's interpretation of the electronic wave function, $\psi\psi^*$ dx, where ψ^* is the complex conjugate of ψ, represents the probability of finding the electron between the limits x and $x + dx$: in three dimensions, it may be given as $\psi\psi^*$ dτ, where dτ (dxdydz) is an infinitesimal element of volume in Cartesian space. If the wave function is real, $\psi = \psi^*$

In order to investigate the meaning of this interpretation, consider a wave function for atomic hydrogen in the form $\psi(r) = (1/\pi a_0^3)^{1/2} \exp(-r/a_0)$, where r is the distance of the electron from the nucleus, and a_0 is the Bohr radius for hydrogen, approximately 52.9 pm. We will determine the probability of locating the electron within a 1 pm volume centred (a) at the nucleus, (b) 25 pm from the nucleus, and (c) 100 pm from the nucleus.

(a)$|\psi|^2 \, d\tau = 1/(\pi \, 52.9^3 \text{ pm}^3) \exp(-2 \times 0/52.9) \times 1 \text{ pm}^3 = 2.1 \times 10^{-6}$
(b) In a similar manner: 8.3×10^{-7}
(c) 4.9×10^{-8}

As the distance from the nucleus increases, the probability of finding the electron falls off rapidly (see also section 2.6.1).

2.3.2   Uncertainty principle

Suppose that an electron occupies a precise position; then its wave function must have a large amplitude at that position and zero value elsewhere. Such a wave function can be built up by a superposition of waves having different amplitudes and wavelengths, but of correct relative phases. The situation is similar to the summation for electron density $\rho(r)$ in X-ray crystallography, considered here in one dimension:

$$\rho(x) = \frac{1}{a} \sum_h F(h)\cos 2\pi \frac{hX}{a} \tag{2.43}$$

where a is a repeat distance along the x-axis, and $F(h)$ is the structure factor for the wavelength represented by h. Fig. 2.6 shows the result of evaluating (2.43) for three and eight waves; it shows an atom located at about $X/a = 0.24$. The determination of the position of the atom has militated against the precise determination of its momentum: each individual wave in (2.43) has a definite momentum, given by the de Broglie equation (2.28) but, in superposition, the sum of the waves presents an indefinite wavelength and, hence, an indefinite momentum. The more waves that are included in (2.43) the better the precision of the position and the more the momentum tends to indeterminacy. Equation (2.43) is one form of a Fourier series, and the theoretical,

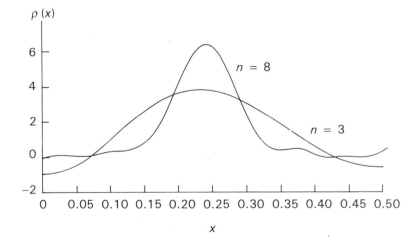

Fig. 2.6. Fourier summation of electron density (one-dimensional) in X-ray crystallography: an atom at x $(X/a) \approx 0.24$ is located more precisely by the superposition of the greater number of waves; the momentum becomes more imprecise at the same time.

ideal localization of a position requires an infinite number of waves, whereupon the momentum of the resultant is totally indeterminate. The result of the superposition of waves is the formation of a *wave packet*, and the size of the packet may be taken as the width of the base of the peak on the x-axis; this length is, in fact, the uncertainty in position, δx.

Position and momentum are *complementary*, that is, only one of them may be specified with precision at a given time. This result is one statement of Heisenberg's uncertainty principle, which may be summarized as

$$\delta x.\delta p_x \geq h/4\pi \tag{2.44}$$

where δx and δp_x are the uncertainties in position and momentum respectively, along the x-direction. This result is true generally, but its effect is negligible in dealing with macroscopic objects (see also the legend to Fig. 2.3).

2.3.3 Normalization and quantization

The Born interpretation of the wave function implies that

$$N^2 \int_{-\infty}^{\infty} \psi\psi^* \, d\tau = 1 \tag{2.45}$$

since the probability of finding the electron under consideration somewhere in space must be unity. Wave functions that satisfy this criterion are called normalized; N is the normalization constant for the wave function ψ. The normalized functions also satisfy the conditions of being finite, continuous and single valued. It is implicit in (2.45) that not all wave functions that are solutions of the Schrödinger equation are physically meaningful. The normalization condition imposes limits (boundary conditions) that lead directly to quantization of the electron energies, as we shall see in the next section.

The important solutions of (2.32) are called *stationary states*: E is invariant with time (a *conservative* system) and, hence, $\psi\psi^*$ is also time independent. The stationary states are our main interest; their wave functions are called *eigenfunctions*, and the corresponding energies are *eigenvalues*. We shall see that discrete values for energies follow directly from the wave equation, whereas Bohr had to introduce them empirically.

As an example of the application of (2.45), consider a radial wave function that may be written as $6ar \exp(-ar/2)$, where a is a constant. Since this wave function is real, we shall use $|\psi|^2$ in (2.45) and, with reference to Appendix 5 we have

$$36N_0^2a^2 \int_0^{\infty} r^4 \exp(-ar) \, dr = 1.$$

Following Appendix 4, we may write this expression in the form

$$\frac{1}{N^2} = \frac{36}{a^3} \int_0^{\infty} t^4 \exp(-t) \, dt$$

$$= \frac{36}{a^3} \Gamma(5) = \frac{36}{a^3} 4!$$

where $t = ar$. Hence, $N = a^{3/2}/(12\sqrt{6})$, which is the normalizing constant for the given wave function. Do not be put off by the use of the Γ-function. The underlying mathematics may be complex, but we are using the Γ-function as a calculating device — like a computer.

2.4 PARTICLE IN A BOX — QUANTIZATION OF TRANSLATIONAL ENERGY

Let a particle of mass m be confined to motion in a one-dimensional box (something like a single bead on a bead-frame) of length a, and let the box be terminated by a potential barrier of infinite height, such that the potential energy is zero for $0 \le x \le a$, but infinite for $0 > x > a$ (see Fig. 2.7a). The wave equation for this system is given by (2.32) with $V = 0$, and a more general solution than (2.36) is

$$\psi = A \exp(ikx) + B \exp(-ikx) \tag{2.46}$$

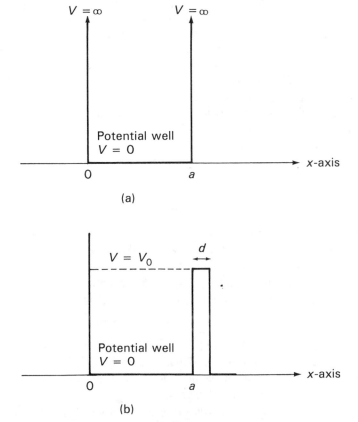

(a)

(b)

Fig. 2.7. Particle in a box: (a) an infinite potential barrier—the particle is confined to the box at all times; (b) a finite potential barrier of thickness d—there is a finite probability that the particle will tunnel through the barrier.

where A and B are constants, and k is given by (2.38); this solution, too, can be confirmed by its double differentiation. Using de Moivre's theorem

$$\exp(i\theta) = \cos\theta + i\sin\theta \tag{2.47}$$

(2.46) can be expanded to

$$\psi = C\cos kx + D\sin kx \tag{2.48}$$

where C and D are constants, equal to $(A + B)$ and $i(A - B)$, respectively. Since the particle is confined to the box, ψ must be zero at $x = 0$, which means that $C = 0$, and at $x = a$, which means that $ka = n\pi$, where n is an integer. Hence, generally

$$\psi_n = D\sin n\pi x/a. \tag{2.49}$$

Using (2.38), it follows that

$$E_n = \frac{n^2 h^2}{8ma^2}. \tag{2.50}$$

Thus, the energy is quantized, by the presence of the boundary walls, and determined by the single quantum number n. It follows that the lighter the particle or the closer together the walls become, the greater the separation of successive energy levels.

The lowest state ψ_1 has an energy E_1, the zero-point energy[1] of magnitude $h^2/8ma^2$. It is kinetic energy, since $V = 0$; even in the lowest energy state, the particle is in motion. This property is entirely wave-mechanical, in complete accord with the uncertainty principle: for if the minimum energy was zero, h would also be zero and the energy would not be quantized. It follows that the momentum would be exactly known (zero) and that $\delta x\,\delta p_x$ would be zero, too.

The probability of finding the particle lying in the interval x to $x + dx$, somewhere between 0 and a, is unity; hence, from (2.45)

$$\int_0^a D^2\sin^2 n\pi x/a\,\mathrm{d}x = 1 \tag{2.51}$$

from which it follows that

$$D = (2/a)^{1/2}. \tag{2.52}$$

Fig. 2.8 illustrates the solutions (2.49) for $n = 1$ to 5; they are similar to the fundamental (ψ_1) and first four overtone vibrations $(\psi_2$ to $\psi_5)$ of a stretched string. As the number of half-waves in the well increases, the curvature (second derivative) of the wave function increases, which implies an increase in kinetic energy as n increases.

The separation of any two neighbouring energy levels is given by

$$\Delta E = E_{n+1} - E_n = (2n + 1)h^2/8ma^2. \tag{2.53}$$

As a increases ΔE becomes smaller, and in the limit as $a \to \infty$ so $\Delta E \to 0$. It follows then that a completely free, or unbound, particle has unquantized energy. It is because of this situation that atoms and molecules encountered in laboratory experiments behave with unquantized translational energy.

The probability of finding the particle in the interval dx at the position x is the integrand of (2.51). Hence, it varies with x, particularly at low values of n. At high n,

---

[1] The translation of the original German word 'nullpunktsenergie' is, strictly, 'zero energy'.

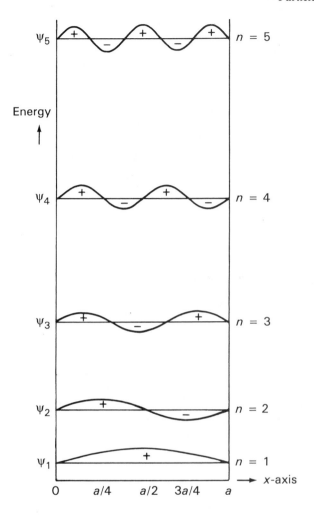

Fig. 2.8. Particle in a box: the wave functions of the first five solutions of the wave equation; the number of nodes at the nth level is $n - 1$, and the energy level above zero is proportional to n^2. The wave functions are standing waves, so the momentum of the particle is not well defined. However, each wave function is a superposition of the eigenfunctions $\exp(ikx)$ and $\exp(-ikx)$, and the measure of the particle momentum would be $\pm nh/2a$, which corresponds to the 'bead on the wire' moving to the right and left of a central zero position.

the probability is approximately equal at all values of x. In the limit, as $n \rightarrow \infty$, the probability is the same for all values of x. In other words, the particle exhibits classical behaviour at very high quantum numbers, which is a statement of the correspondence principle. The reader might like to plot $|\psi|^2$ from (2.49) for several increasing values n (larger than those given in Fig. 2.8), so as to verify this result.

2.4.1 Tunnelling
Let us take a tennis ball and lock it in a well-constructed safe. Newtonian mechanics tells us that the probability of the tennis ball being found outside the safe (unless

someone takes it out) is zero. In the case of the electron in a box, wave mechanics reveals a different situation.

Let the electron be confined to a box by a wall of height V_0 and thickness d (Fig. 2.7b). Within the box, the kinetic energy term, corresponding to $(E - V)$ in (2.35), is positive, and the solution to this equation may be written, from (2.49) and (2.37), in the form $D \sin\{(2\pi/h)[2m_e(E - V_0)]^{1/2}x\}$ or, more generally as $D \exp\{(i2\pi/h)[2m_e(E - V_0)]^{1/2}x\}$, since C in (2.48) is zero. In the region within the potential barrier V_0 is greater than E, and the general solution of the wave equation for this region is $D \exp\{-(i2\pi/h)[2m_e(V_0 - E)]^{1/2}x\}$. Thus, the probability ($\propto |\psi|^2$) of finding the electron in the region of negative kinetic energy is not zero, but falls off exponentially with the distance x of penetration within the barrier. As long as the barrier is neither infinitely high nor infinitely wide, there is a finite probability that the electron will tunnel through the barrier, which is an entirely quantum-mechanical situation.

2.4.2   Boxes of higher dimensions

We can extend the discussion of section 2.4 to a two-dimensional box (Fig. 2.9), and the appropriate wave equation (for $V = 0$) may be written as

$$-\frac{h^2}{8\pi^2 m}\left\{\frac{\partial^2\psi}{\partial x^2} + \frac{\partial^2\psi}{\partial y^2}\right\} = E\psi. \tag{2.54}$$

This differential equation is separable, that is, the eigenfunctions can be written as products in x and y, which means that, by analogy with (2.49), and including boundary conditions as before, we may write

$$\psi_{n_x, n_y} = D' \sin(n_x \pi x/a)\sin(n_y \pi y/b) \tag{2.55}$$

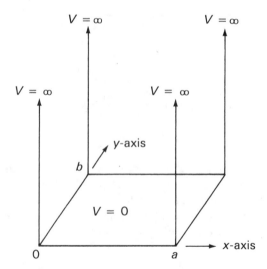

Fig. 2.9. Two-dimensional potential well (box); in the general case, a and b have different values. The vibrations may be likened to those of a square drum.

where n_x and n_y are quantum numbers, and the total energy E_{n_x,n_y} is given by

$$E_{n_x,n_y} = E_{n_x} + E_{n_y}. \tag{2.56}$$

Some of the mathematics of the separation of variables is given in Appendix 8.
From (2.50), and following (2.51)

$$E_{n_x,n_y} = (h^2/8m)(n_x^2/a^2 + n_y^2/b^2) \tag{2.57}$$

and

$$D' = 2/(ab)^{1/2}. \tag{2.58}$$

The lowest energy state ($n_x = n_y = 1$) is given by

$$E_{1,1} = (h^2/8m)(1/a^2 + 1/b^2) \tag{2.59}$$

and the corresponding wave function is

$$\psi_{1,1} = D' \sin(\pi x/a) \sin(\pi y/b). \tag{2.60}$$

The next highest states are $\psi_{1,2}$ and $\psi_{2,1}$ (Fig. 2.10), with energies

$$E_{1,2} = (h^2/8m)(1/a^2 + 4/b^2) \tag{2.61}$$

and

$$E_{2,1} = (h^2/8m)(4/a^2 + 1/b^2) \tag{2.62}$$

For a square two-dimensional box, $a = b$ and $E_{1,2} = E_{2,1}$; the energies are then said
to be degenerate (Fig. 2.11).

In a three-dimensional box, the energy levels are characterized by three integer
quantum numbers; it is not difficult to show that, for a rectangular box,

$$E_{n_x,n_y,n_z} = (h^2/8m)(n_x^2/a^2 + n_y^2/b^2 + n_z^2/c^2). \tag{2.63}$$

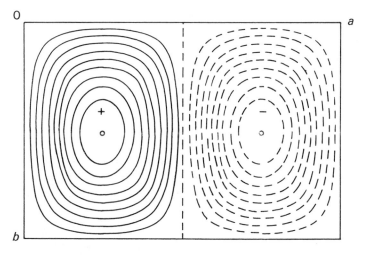

Fig. 2.10. Two-dimensional particle in a box: the wave function $\psi_{2,1}$ with $a/b = 1.5$. The
contours rise to a maximum (full lines) and descend to a minimum (dashed lines).

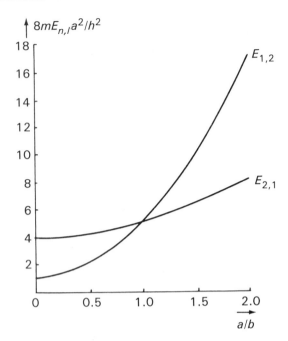

Fig. 2.11. Two-dimensional particle in a box: the energies $E_{2,1}$ and $E_{1,2}$ (scaled by $8ma^2/h^2$) are shown as functions of a/b; the curves cross at $a/b = 1$, when the energies become degenerate.

For a cube, $a = b = c$, and multiply degenerate energy states now exist. For example, for quantum numbers n_x, n_y and n_z equal to 1, 2 and 3 respectively, six degenerate states exist with energies given by

$$E_{n_x, n_y, n_z} = (h^2/8ma^2)(n_x^2 + n_y^2 + n_y^2) = 7h^2/4ma^2 \qquad (2.64)$$

2.5 VIBRATIONAL AND ROTATIONAL MOTION

In this section, we shall introduce some aspects of the quantum mechanics of vibration and rotation that are important in our subsequent study of molecules.

2.5.1 Vibrational motion
The energies associated with both vibrational and rotational motions are quantized. In section 1.3.1, we considered simple harmonic motion with a stretched spring. In a similar way, a particle undergoing simple harmonic motion about the position $x = 0$ experiences a restoring force $- \mathscr{k}x$, where $\mathscr{k}$ is a force constant, and $\mathscr{k}x^2$ is the potential energy of the particle. The Schrödinger equation for this situation takes the form

$$\frac{-h^2}{8\pi^2 m}\frac{\mathrm{d}^2\psi}{\mathrm{d}x^2} + \tfrac{1}{2}\mathscr{k}x^2\,\psi = E\,\psi. \qquad (2.65)$$

As with the particle in the box, the energies are quantized because a boundary condition requires that ψ tends to zero as x becomes large, since the potential energy rises sharply as x increases. The acceptable solutions of (2.65) give for the energies of the harmonic oscillator

$$E_v = (v + \tfrac{1}{2}) \, h\omega/2\pi \qquad\qquad (2.66)$$

where the vibrational quantum number v can take the values $0, 1, 2, \ldots$, and $\omega = (k/m)^{1/2}$, or $2\pi v$. The separation of all successive vibrational energy levels is the same, whatever the value of v (Fig. 2.12):

$$\Delta E = E_{v+1} - E_v = h\omega/2\pi. \qquad\qquad (2.67)$$

The zero-point energy of vibration corresponds to $h\omega/4\pi$: classical mechanics allows the stretched spring to be completely at rest, but wave mechanics requires that it has a residual vibrational energy, albeit negligible on an experimental scale. To return to our examples in section 1.3.1, the values of $h\omega/4\pi$ are approximately 3.3×10^{-34} J and 3.3×10^{-20} J for the spring and the HCl molecule, respectively. Converting these quantities to molar terms gives 2×10^{-13} kJ mol^{-1} and 20 kJ mol^{-1}, respectively; it is evident that the former quantity is experimentally negligible, but that the second of them is highly significant.

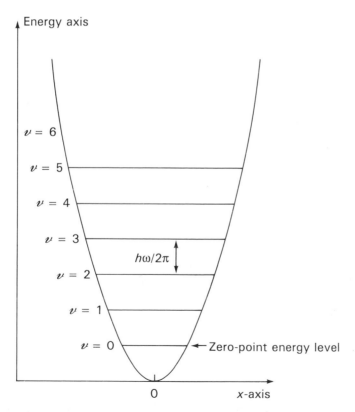

Fig. 2.12. Diagrammatic representation of the first few energy levels ($v = 0$ to 6) of the harmonic oscillator: the curve is parabolic about $x = 0$, and the spacing between any pair of adjacent levels is $h\omega/2\pi$; the level at $v = 0$ corresponds to the zero-point energy of vibration.

2.5.2   Rotational motion in two dimensions

A particle of mass m moving along a circular path of radius r has an angular momentum J equal to pr, where p is the linear momentum in a direction tangential to the circle at any point. Since the kinetic energy is given by $p^2/2m$, this value can be replaced by $J^2/2I$, where I is the moment of inertia mr^2 of the system; the (kinetic) energy now becomes

$$E = J^2/2I. \tag{2.68}$$

Since only certain values for the wavelength λ of the motion are permitted, E will be quantized; the number of wavelengths must fit the circumference $2\pi r$ of the circular path (cyclic boundary condition). Thus, λ must take the values $2\pi r/n$, which implies that $p = h/\lambda = nh/2\pi r$. Hence, $J = nh/2\pi$, and the permitted, quantized energies are given by

$$E = n^2 h^2/8\pi^2 I. \tag{2.69}$$

However, the momentum p may be directed in either one of two ways, corresponding to clockwise or anticlockwise rotation of the particle. Conventionally, angular momentum is indicated by a vector along the z-axis, perpendicular to the plane of rotation, and we may conclude that angular momentum is quantized according to

$$J_z = m_l h/2\pi \qquad (m_l = 0, \ \pm 1, \ \pm 2, \ldots). \tag{2.70}$$

Positive values of m_l correspond to anticlockwise rotation as viewed in the direction of $-z$, towards the plane of rotation. The energies of the particle are given by

$$E_{m_l} = m_l^2 h^2/8\pi^2 I \tag{2.71}$$

from which it is evident that the rotational energy is necessarily independent of the direction of rotation. These results may be confirmed by solving the Schrödinger equation (2.54), sensibly in polar coordinates (Appendix 5):

$$\frac{-h^2}{8\pi^2 m} \frac{1}{r^2} \frac{\partial^2 \psi}{\partial \phi^2} = E\,\psi \tag{2.72}$$

for which the solutions are

$$\psi_{m_l} = N \exp(im_l \phi) \tag{2.73}$$

where $m_l = 0, \ \pm 1, \ \pm 2, \ldots$ satisfy the equation. It is left as an exercise to the reader to derive (remember that ψ_{m_l} is complex) the normalization constant N in (2.73) and to show that the quantized energies derived from (2.72) and (2.73) are as given in (2.71). Again, ψ must fit the circumference of the circle in order to satisfy the boundary condition, which means that ψ must be the same at ϕ and $\phi + 2\pi$. This situation is achieved with the integral values for m_l.

Some of the aspects of this discussion of rotational motion are illustrated in Fig. 2.13.

2.5.3   Rotational motion in three dimensions

Rotation of a particle in three dimensions requires that the wave function must match up for any permitted path across the poles, as well as along the equator. The kinetic energy and angular momentum are determined by rotation about the three Cartesian

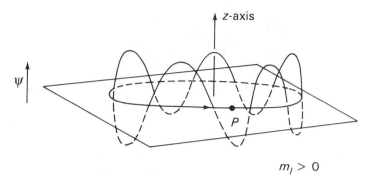

Fig. 2.13. Particle on a ring: the vertical axis corresponds to ψ in the case of the wave, and the z-axis, the direction of the J_z angular momentum vector, in the case of the particle P; for the direction of motion indicated, m_l is, by convention, positive.

axes x, y and z, and the new boundary condition requires a further quantum number governing the total angular momentum.

Solution of the Schrödinger equation for this situation leads to the following results:

Quantization of rotational energy

$$E_l = l(l + 1)h^2/8\pi^2 I. \tag{2.74}$$

Quantization of total angular momentum

$$J = [l(l + 1)]^{1/2}h/2\pi \qquad (l = 1, 2, 3, \ldots). \tag{2.75}$$

Quantization of angular momentum in the z-direction

$$J_z = m_l h/2\pi \qquad (m_l = 0, \pm 1, \pm 2, \ldots). \tag{2.76}$$

Separable wave functions

ϕ-dependent part: $\exp(im_l\phi)$, as in the particle on a ring;
θ-dependent part: $Y_{l, m_l} (\theta, \phi)$, the spherical harmonics (Table 2.2).

2.5.4   Space and spin quantization
Since J_z is quantized, the vector representing angular momentum, shown in Fig. 2.13, is restricted in its spatial position, by the integral nature of m_l. The momentum vector is normal to the plane of the path of the rotating particle, so it follows that this plane can have only certain orientations in space.

This property was confirmed by the Stern–Gerlach (1921, 1922) experiments, in which a beam of silver atoms was directed through an inhomogeneous magnetic field. The field set up across the beam interacts with the field set up by the rotating charged particles in the silver atoms and, because of the imhomogeneity of the field, the direction followed by an atom depends upon orientation. Classical theory predicts a continuous band of atoms emerging from the confines of the applied magnetic field, whereas two discrete sets of narrow bands were observed, in line with the wave-mechanical prediction of quantized angular momentum orientation.

Table 2.2 Some $Y_{l,m_l}(\theta, \phi)$ functions — spherical harmonics

l	m_l	$Y_{l,m_l}(\theta, \phi)$ (normalized)
0	0	$(1/4\pi)^{1/2}$
1	0	$(3/4\pi)^{1/2} \cos\theta$
1	+1	$(3/8\pi)^{1/2} \sin\theta \quad \exp(i\phi)$
1	−1	$(3/8\pi)^{1/2} \sin\theta \quad \exp(-i\phi)$
2	0	$(5/16\pi)^{1/2} (3\cos^2\theta - 1)$
2	+1	$(15/8\pi)^{1/2} \cos\theta \sin\theta \exp(i\phi)$
2	−1	$(15/8\pi)^{1/2} \cos\theta \sin\theta \exp(-i\phi)$
2	+2	$(15/32\pi)^{1/2} \sin^2\theta \quad \exp(i2\phi)$
2	−2	$(15/32\pi)^{1/2} \sin^2\theta \quad \exp(-i2\phi)$

However, with $2l + 1 = 2$, two bands could arise only if the quantum number l was equal to $\frac{1}{2}$. This dilemma was resolved by the suggestion that the Stern–Gerlach experiments revealed not an orbital angular momentum, but a spin angular momentum of the electron spinning on an axis through its centre of mass.

For an electron, the spin angular momentum quantum number s is $\frac{1}{2}$, corresponding to a spin angular momentum magnitude of $[s(s+1)]^{1/2}h/2\pi$. The component of the spin angular momentum along the z-axis is quantized in units of $m_s h/2\pi$, where $m_s = s, s-1, \ldots, -s$. An electron with $m_s = +\frac{1}{2}$ is often called an α electron, and that with $m_s = -\frac{1}{2}$, a β electron. Two electrons with their spins antiparallel (one + and one −) are called paired, and their total spin angular momentum is zero. Different fundamental particles have differing spins; for example, a photon has a spin of unity.

Spin angular momentum is a fundamental and fixed property of the electron. Later, the experiments of Uhlenbeck and Goudschmidt (1925) associated the electron spin with its magnetic moment. Each silver atom has one unpaired electron and, hence, a spin angular momentum that can assume only two spatial orientations, those actually observed earlier by Stern and Gerlach.

2.5.5 Quantum numbers

In our discussion of wave mechanics so far, we have introduced quantum numbers to specify a particle or electron, and we summarize them here for convenience.

Principal quantum number n: to specify the energy of hydrogen-like particles; n can take the positive values 1, 2, 3, ...

Orbital angular momentum (total) quantum number l: to specify motion around a central point; l can take the n positive values 0, 1, 2, 3, . . . , $n-1$.

Magnetic quantum number m: to specify the component of angular momentum resolved along the z-axis: orbital—m_l can take the $(2l-1)$ values: $l, l-1, l-2, \ldots, -l$; spin—m_s can take the $(2s-1)$ values: $s, s-1, s-2, \ldots, -s$.

Spin quantum number s: to specify the spin angular momentum of an electron ($s = \frac{1}{2}$); the orientation is given by m_s (the component resolved along the z-axis) $+ \frac{1}{2}$ ('spin up' or α) or $- \frac{1}{2}$ ('spin down', or β). The property of spin arises from a relativistic treatment of the Schrödinger equation, which requires time as a fourth dimension.

2.6 STRUCTURE OF THE HYDROGEN ATOM

Where we have developed ideas in terms of a particle, we could have used an electron instead. Hence, in a more chemical vein, we look next at the structure of the hydrogen atom.

The wave equation for the hydrogen atom may be written as

$$\mathscr{H} \, \psi = E \, \psi \tag{2.77}$$

and the three-dimensional Hamiltonian operator $\mathscr{H}$ for this case is given by

$$- (h^2/8\pi^2\mu)\nabla^2 - V(r) \tag{2.78}$$

where μ is the reduced mass of an electron-proton pair of separation r, $V(r)$ is the potential energy of the electron in the field of the proton, and ∇^2 is the Laplacian operator:

$$\nabla^2 = \frac{\partial^2}{\partial x^2} + \frac{\partial^2}{\partial y^2} + \frac{\partial^2}{\partial z^2} \tag{2.79}$$

The potential energy of the electron of charge $-e$ in the field of a proton of charge $+e$ is the Coulomb energy $-e^2/4\pi\varepsilon_0 r$. It is desirable to transform (2.79) to polar coordinates (Appendix 5), because of the spherical symmetry of the hydrogen atom, and we obtain for the desired wave equation:

$$- \frac{(h^2/8\pi^2\mu)}{r^2} \left\{ \frac{\partial}{\partial r} \left[r^2 \frac{\partial \psi}{\partial r} \right] + \frac{1}{\sin^2\theta} \frac{\partial^2 \psi}{\partial \phi^2} + \frac{1}{\sin \theta} \frac{\partial}{\partial \theta} \left[\sin \theta \frac{\partial \psi}{\partial \theta} \right] \right\} - \frac{e^2}{4\pi\varepsilon_0 r} \psi = E \, \psi. \tag{2.80}$$

This equation is separable into three terms, depending respectively on r, θ and ϕ. The wave function $\psi(r, \theta, \phi)$ then takes the form $R(r) \, \Theta(\theta) \, \Phi(\phi)$. The $\Theta\Phi$ products are the spherical harmonics $Y(\theta, \phi)$ in Table 2.2, whereas the $R(r)$ functions refer to the radial distribution in the wave function.

The exact solution of (2.80) is complex and will not be addressed here. In fact, the hydrogen atom is the only normal system for which the wave equation can be solved exactly, and the reader is referred to an appropriate quantum-mechanical text[1] for the necessary detail. We note here that the solution of (2.80), within the Born interpretation (2.45), leads to expressions for the wave function involving the three quantum numbers, n, l and m_l summarized in section 2.5.5, so that the wave function can be denoted ψ_{n,l,m_l}. Normally, energies depend upon all the quantum numbers, but in the particular case of the hydrogen atom only n is involved, and we obtain for the energies of the hydrogen atom

$$E_n = \frac{- \mu e^4}{8n^2 h^2 \varepsilon_0^2} \tag{2.81}$$

[1]See, for example, L. Pauling and E. B. Wilson (1935) *Introduction to Quantum Mechanics*, McGraw-Hill.

Table 2.3 Normalized hydrogenic radial functions
$R_{n,l}(r)$, to $n = 3$ ($\rho = 2Zr/na_0$)

n	l	$R_{n,l}(r)$	
1	0	$(Z/a_0)^{3/2}$ 2	$\exp(-\rho/2)$
2	0	$(Z/a_0)^{3/2}$ $1/2\sqrt{2}\,(2-\rho)$	$\exp(-\rho/2)$
2	1	$(Z/a_0)^{3/2}$ $1/2\sqrt{6}\,\rho$	$\exp(-\rho/2)$
3	0	$(Z/a_0)^{3/2}$ $1/9\sqrt{3}\,(6-6\rho+\rho^2)$	$\exp(-\rho/2)$
3	1	$(Z/a_0)^{3/2}$ $1/9\sqrt{6}\,(4\rho-\rho^2)$	$\exp(-\rho/2)$
3	2	$(Z/a_0)^{3/2}$ $1/9\sqrt{30}\,\rho^2$	$\exp(-\rho/2)$

in complete agreement with the values given by (2.31). The negative sign here indicates that we are dealing with bound states of the electron, for which energy is convention-ally a negative quantity, falling to zero at infinite value of r. One should not be tempted to regard (2.81) as evidence supporting the Bohr theory; the hydrogen atom is a special case, because of the simple form of the potential energy function, and the integer parameter n does not play the same role in the two theories.

The normalized radial wave functions for hydrogen-like (one-electron) species are given in Table 2.3, where $\rho = 2Zr/na_0$. The Bohr radius a_0 is given by $h^2\varepsilon_0/\pi\mu e^2$, which is approximately 52.92 pm, and Z is the atomic number; for hydrogen itself $Z = 1$.

In wave mechanics, energies are often expressed in terms of the Hartree (energy) E_H. Then, we have from (2.81)

$$E_n = -1/2n^2 \; E_H \tag{2.82}$$

where

$$E_H = e^2/4\pi\varepsilon_0 a_0. \tag{2.83}$$

The lowest permitted energy for the hydrogen atom ($n = 1$) is then $-\frac{1}{2} E_H$, or -2.1799×10^{-18} J.

The product of the $R(r)$ and $Y(\theta, \phi)$ functions gives the total wave functions, or atomic orbitals, for the hydrogen atom. For example, for $n = 2$, $l = 1$ and $m_l = +1$, we have

$$\psi_{2,1,\pm1} = 1/2\sqrt{6}\,(Z/a_0)^{3/2}\,\rho\exp(-\rho/2)\,(3/8\pi)^{1/2}\sin\theta\exp(\pm i\phi)$$
$$= 1/8\sqrt{\pi}\,(Z/a_0)^{3/2}\,\rho\exp(-\rho/2)\sin\theta\exp(\pm i\phi).$$

However, before proceeding along these lines, we will look a little further at atomic orbitals.

2.6.1 Atomic orbitals

An atomic orbital is the wave function of an electron in a bound state in an atom. The solutions of the wave equation for the hydrogen atom are important because they form a basis for discussing the structures of atoms and molecules generally. The form of the Born interpretation makes it desirable to visualize the hydrogen atom as an electron cloud around the central nucleus. This cloud has different shapes for different energy states of the electron, and we can discover the shape of any given cloud by plotting the wave function in a suitable form. In order to represent ψ_{n,l,m_l}, it would be necessary to plot in four dimensions; more conveniently, we can separate $R_{n,l}$ for constant θ and ϕ, and Y_{l,m_l} for constant r.

The atomic orbital $\psi_{1,0,0}$ is spherically symmetrical: it has no angular dependency, and it decreases in amplitude exponentially with the distance r of the electron from the nucleus. For hydrogen, the probability of finding an electron between distances r and $r + dr$ from the nucleus is given by (2.45), with ψ from Table 2.3, that is, $a_0^{-3/2}\,2\,\exp(-r/a_0)$, and with $d\tau$ equal to $4\pi r^2\,dr$, the volume enclosed by a spherical shell of radii r and $r + dr$. Fig. 2.14 shows R, the radial wave function itself, R^2, the probability density, and $4\pi r^2 R^2$, which may be called the radial distribution function, each plotted as a function of r for $R_{1,0}$ (designated 1s), $R_{2,0}$ (designated 2s) and $R_{2,1}$ (designated 2p).

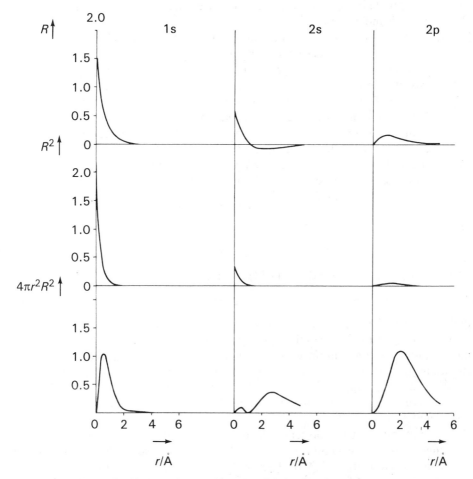

Fig. 2.14. Normalized hydrogenic (one-electron) radial eigenfunctions $R_{n,l}$ ($Z = 1$), plotted as a function of atomic number Z; 1 Å $= 10^{-10}$ m, so that $a_0 = 0.5292$ Å. All diagrams are on the same scale. The maximum in the curve of $4\pi r^2 R^2$ for the 1s orbital occurs at $r = a_0$, or 1.06 Å. The reader may care to determine the corresponding maxima for the 2s and 2p curves, using the wave functions of Table 2.3.

The exponential form of $R_{1,0}$ indicates that the most probable place for the electron should be at nucleus ($r = 0$). That it is not actually there (excluding electron capture) depends upon the balance of the kinetic energy of the radial motion of the electron and its potential energy with respect to the nucleus. The actual position of the maximum in the 1s radial distribution function may be found by differentiating $4\pi r^2 R_{1,0}^2$ with respect to r and setting the derivative to zero, which shows that the maximum in the function occurs at r equal to a_0, the Bohr radius.

The next function $R_{2,0}$ behaves in a similar manner (Fig. 2.14), except that the 'size' of the function is increased, and the radial function has more than one maximum: the numbers of nodes, or zeros of R, is equal to $n - l - 1$.

When we consider $R_{2,1}$, we have to take account of an accompanying angular function, the spherical harmonic. In energy states having finite angular momenta ($l > 0$), the z-component is given a precise orientation (section 2.5.2). From the uncertainty principle, the angular position of the electron around the z-axis, that is, in the xy-plane, is indefinite.

The state with $m_l = 0$ has zero angular momentum along the z-axis. The corresponding wave function is designated $2p_z$; a nodal surface through the nucleus exists in the xy-plane (Fig. 2.15a); the sign of the function is indicated by the $\pm$ signs in the lobes. This situation does not arise immediately for the other degenerate states, with $l = 1$, $m_l = \pm 1$. The corresponding wave functions are complex, but it is possible to form other real solutions that are linear combinations of those first given.

If two or more functions are eigenfunctions of a linear operator[1], then linear combinations of the functions, on one centre, are also eigenfunctions of the operator, provided that they are degenerate; the appropriate normalizing constants must be applied. Addressing the degenerate functions $Y_{1,1}$ and $Y_{1,-1}$ in Table 2.2, we can write

$$Y_{1,1} = \Phi^+ = K(\cos \phi + i \sin \phi) \qquad (2.84)$$

and

$$Y_{1,-1} = \Phi^- = K(\cos \phi - i \sin \phi) \qquad (2.85)$$

where K is $\sqrt{(3/8\pi)} \sin \theta$. Adding (2.84) and (2.85) gives

$$(\Phi^+ + \Phi^-) = 2K \cos \phi \qquad (2.86)$$

and this new function with its *additional* normalizing constant of $1/\sqrt{2}$ (the normalizing equation involves the square $\Phi\Phi^*$ of the wave function), can be built into the complete hydrogen atom wave functions, and is designated $2p_x$, with a nodal surface in the yz-plane (Fig. 2.15b). Subtraction of (2.85) from (2.84) gives similarly a function in $\sin \phi$ ($2p_y$, with an xz nodal plane). The complete hydrogenic[2] wave functions (up to $n = 3$) are listed) in Table 2.4; the p- and d-functions are strongly directional.

A simplification of the presentation of atomic orbitals is generally achieved by drawings, such as the example in Fig. 2.15, the three-dimensional surfaces of which are assumed to enclose about 95% (or 99%) of the electron cloud; in other words, there is a 95% probability of finding the electron within the surface described.

---

[1] If $\alpha(f + g) = \alpha f + \alpha g$, where f and g are any two functions, then the operator α is linear; d/dx is an example of a linear operator.

[2] The term 'hydrogenic' (hydrogen-like) refers to species such as He^+ and Li^{2+}, as well as to hydrogen itself.

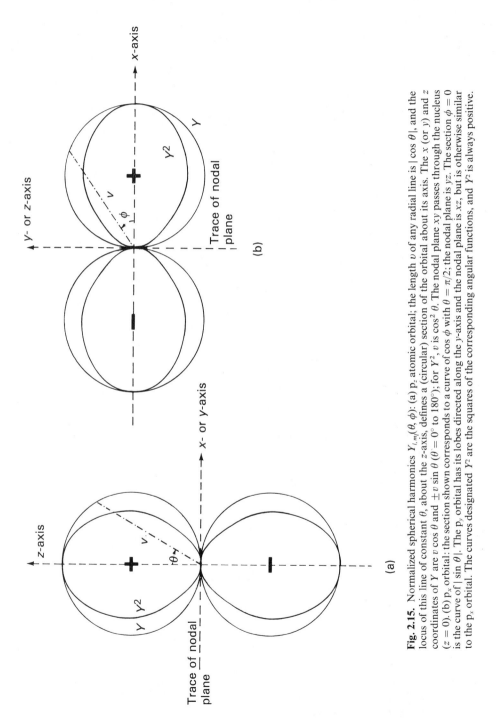

Fig. 2.15. Normalized spherical harmonics $Y_{l,m_l}(\theta, \phi)$: (a) p_z atomic orbital; the length v of any radial line is $|\cos \theta|$, and the locus of this line of constant θ, about the z-axis, defines a (circular) section of the orbital about its axis. The x (or y) and z coordinates of Y are $v \cos \theta$ and $\pm v \sin \theta$ ($\theta = 0°$ to $180°$); for Y^2, v is $\cos^2 \theta$. The nodal plane xy passes through the nucleus ($z = 0$). (b) p_x orbital: the section shown corresponds to a curve of $\cos \phi$ with $\theta = \pi/2$; the nodal plane is yz. The section $\phi = 0$ is the curve of $|\sin \theta|$. The p_y orbital has its lobes directed along the y-axis and the nodal plane is xz, but is otherwise similar to the p_x orbital. The curves designated Y^2 are the squares of the corresponding angular functions, and Y^2 is always positive.

Table 2.4. Normalized hydrogenic (one-electron) functions ψ_{n,l,m_l}

Function[a]	$R_{n,l}(r) \cdot Y_{l,m_l}(\theta, \phi)$		Orbital type
$\psi_{1,0,0}$	$1/\sqrt{\pi}\,(Z/a_0)^{3/2}$	$\exp(-\rho/2)$	1s
$\psi_{2,0,0}$	$1/4\sqrt{2\pi}\,(Z/a_0)^{3/2}$	$(2 - \rho)\exp(-\rho/2)$	2s
$\psi_{2,1,0}$	$1/4\sqrt{2\pi}\,(Z/a_0)^{3/2}$	$\rho\,\exp(-\rho/2)\cos\theta$	$2p_z$
$\psi_{2,1,\pm1}$	$1/4\sqrt{2\pi}\,(Z/a_0)^{3/2}$	$\rho\,\exp(-\rho/2)\sin\theta\cos\phi$	$2p_x$
$\psi_{2,1,\pm1}$	$1/4\sqrt{2\pi}\,(Z/a_0)^{3/2}$	$\rho\,\exp(-\rho/2)\sin\theta\sin\phi$	$2p_y$
$\psi_{3,0,0}$	$1/18\sqrt{3\pi}\,(Z/a_0)^{3/2}$	$(6 - 6\rho + \rho^2)\exp(-\rho/2)$	3s
$\psi_{3,1,0}$	$1/18\sqrt{2\pi}\,(Z/a_0)^{3/2}$	$(4\rho - \rho^2)\exp(-\rho/2)\cos\theta$	$3p_z$
$\psi_{3,1,\pm1}$	$1/18\sqrt{2\pi}\,(Z/a_0)^{3/2}$	$(4\rho - \rho^2)\exp(-\rho/2)\sin\theta\cos\phi$	$3p_x$
$\psi_{3,1,\pm1}$	$1/18\sqrt{2\pi}\,(Z/a_0)^{3/2}$	$(4\rho - \rho^2)\exp(-\rho/2)\sin\theta\sin\phi$	$3p_y$
$\psi_{3,2,0}$	$1/36\sqrt{6\pi}\,(Z/a_0)^{3/2}$	$\rho^2\exp(-\rho/2)(3\cos^2\theta - 1)$	$3d_{z^2}$
$\psi_{3,2,\pm1}$	$1/18\sqrt{6\pi}\,(Z/a_0)^{3/2}$	$\rho^2\exp(-\rho/2)\sin\theta\cos\theta\cos\phi$	$3d_{xz}$
$\psi_{3,2,\pm1}$	$1/18\sqrt{2\pi}\,(Z/a_0)^{3/2}$	$\rho^2\exp(-\rho/2)\sin\theta\cos\theta\sin\phi$	$3d_{yz}$
$\psi_{3,2,\pm2}$	$1/18\sqrt{2\pi}\,(Z/a_0)^{3/2}$	$\rho^2\exp(-\rho/2)\sin^2\theta\cos2\phi$	$3d_{x^2-y^2}$
$\psi_{3,2,\pm21}$	$1/36\sqrt{2\pi}\,(Z/a_0)^{3/2}$	$\rho^2\exp(-\rho/2)\sin^2\theta\sin2\phi$	$3d_{xy}$

[a] m_l values may not be assigned to the real functions of x, y, xz, yz, xy and $x^2 - y^2$ because, as quoted, they are linear combinations of $+m_l$ and $-m_l$, formulated as in (2.84) to (2.86), for example.

An even more descriptive illustration of an orbital is the density contour diagram, Fig. 2.16, shown for a $2p_z$ orbital. We shall not normally employ this type of representation, but we may note that it is similar to the electron density maps obtained in X-ray crystallographic studies of crystal and molecular structure (Figs. 1.1a and 2.17). X-ray methods do not resolve electron density for the given species in X-ray map for an atom represents an averaged electron density for the given species in the structure. For this reason, the X-ray electron density contour maximum of an atom does not usually coincide with the position of its nucleus, a significant factor in a discussion of the meaning of bond lengths between atoms.

2.6.2   Orbital terminology

The atomic orbitals for a given value of the principal quantum number n comprise a *shell*, whereupon the orbitals of varying l within a shell form a *sub-shell*. Thus, we obtain the common terminology, shown here for n up to 4:

n	1	2	3	4
Shell	K	L	M	N
l	0	1	2	3
Sub-shell	s	p	d	f

For $n = 2$, the L shell, the atomic orbitals in sub-shells are 2s (one orbital) and 2p (three degenerate orbitals). In general, a shell of number n contains a total of n^2

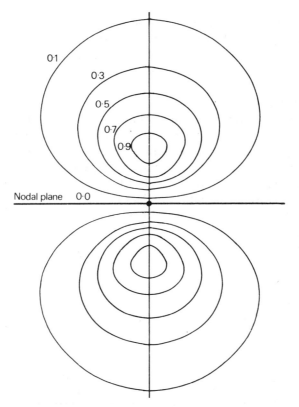

Fig. 2.16. Electron density contours for a $2p_z$ atomic orbital of carbon, as fractions of ψ_{max}^2: the 0.1 contour surface encloses about 66% of the $2p_z$ electron density; 90% would be enclosed by the 0.03 contour (not shown). The nodal plane of zero density is the xy-plane.

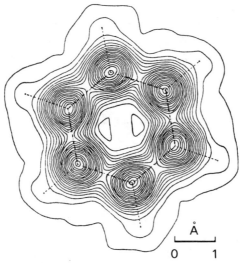

Fig. 2.17. Electron density map of a molecule of benzene, obtained from X-ray diffraction studies on the solid. The contour intervals are 0.25 e Å^{-3}, and there is a tendency for the hydrogen atoms to be just resolved (after Cox, Cruickshank and Smith, and reproduced by permission of the Royal Society, London).

orbitals. The alphabetic notation derives from spectroscopic usage. In particular, s, p, d and f were used to describe spectral transitions (*sharp, principal, fundamental* and *diffuse*) involving the sub-shells with $l = 0$ to 3, and are not just a random selection from the alphabet.

2.6.3   Selection rules for atoms

We saw in section 2.2.6 that the spectra of atomic hydrogen could be explained by transitions of electrons between energy levels. However, not all possible transitions are allowed. If a photon is expelled from an atom, conservation of angular momentum requires that the electron angular momentum must change by unity, because a photon has an intrinsic spin angular momentum of unity. Thus, a p electron ($l = 1$) can fall to an s orbital ($l = 0$), with emission of radiation; changes in n are not so restricted because n governs energy rather than angular momentum. Thus, we obtain the selection rules

$$\Delta n = 1, 2, 3, \ldots \tag{2.87}$$

$$\Delta l = \pm 1 \tag{2.88}$$

$$\Delta m_l = 0, \pm 1 \tag{2.89}$$

Possible transitions that conform to these selection rules may be depicted on a Grotrian diagram, of which Fig. 2.18 is a simple example.

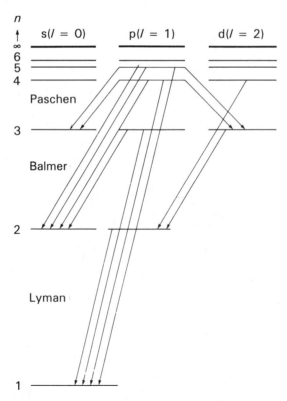

Fig. 2.18. Grotrian diagram showing some of the spectral transitions of atomic hydrogen; changes in n are unrestricted, but those in l (and m_l) must be ± 1. The transitions shown would not all be of the same intensity.

2.6.4   Atoms with more than one electron

The next atom after hydrogen in the periodic table of the elements, helium, has two electrons, and the Schrödinger equation cannot be solved exactly for this (or other) species. It becomes necessary to make an approximation to the solution.

Each electron may be represented by its hydrogenic wave function; then, if the wave function for helium is $\psi(1, 2)$, the approximation becomes

$$\psi(1, 2) = \psi(1)\,\psi(2). \tag{2.90}$$

The individual atomic orbitals $\psi(1)$ and $\psi(2)$ are hydrogenic, but the nuclear charge in each is modified to take account of both electrons. If the ground state of hydrogen is written as (1s), or $(1s)^1$, that for helium becomes $(1s)^2$.

In lithium we anticipate a start with $(1s)^2$: the K shell is now full, or *closed*, and the third electron occupies the next highest (energy) orbital $(2s)^1$. This arrangement accords with the Pauli exclusion principle, arising from wave mechanics, which states that *an atomic orbital can accommodate a maximum of two electrons, with opposed, or paired spins.*

Each electron is fully determined by the quantum numbers n, l, m_l and m_s, the latter being $\pm\frac{1}{2}$, as we have seen. Thus, the 1s wave function of hydrogen may be written either as

$$\psi_{1,0,0,+1/2} = 1/\sqrt{\pi}(Z/a_0)^{3/2}\exp(-\rho/2)\alpha \tag{2.91}$$

or

$$\psi_{1,0,0,-1/2} = 1/\sqrt{\pi}(Z/a_0)^{3/2}\exp(-\rho/2)\beta \tag{2.92}$$

so that the configuration $(1s)^2$ in helium implies a full 1s orbital, with paired (α, β) spins.

2.6.5   Effective atomic number—screening

The order of energies of atomic orbitals for low atomic numbers is

$$1s < 2s < 2p < 3s < 3p < 4s < 3d\ldots. \tag{2.93}$$

However, electron-electron repulsion can modify this order in situations where the energy levels of orbitals are close, as in the 3d/4s levels of the first transition series of elements, for example. Fig. 2.19 shows how the relative energies E of atomic orbitals vary with atomic number Z.

Electron–electron repulsion reduces Z to an *effective atomic number* Z_{eff}, given by

$$Z_{eff} = Z - \sigma \tag{2.94}$$

where σ is a quantum mechanical *screening* (shielding) constant. Screening constants may be calculated according to Slater's rules.

2.6.5.1   *Slater's rules*

In wave mechanical calculations it is often sufficiently accurate, with principal quantum numbers up to 4, to replace Z by Z_{eff}, given by (2.94), where the values for σ are found by the following empirical rules:

First, the atomic orbitals which are occupied are divided into the groups

1s | 2s, 2p | 3s, 3p | 3d | 4s, 4p | 4d | 4f | ,

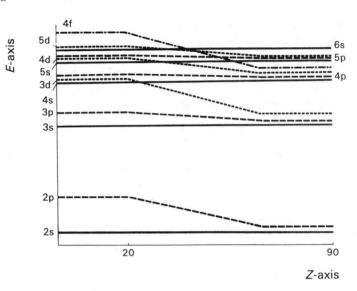

Fig. 2.19. Variation of the energies E with atomic number Z for atomic orbitals up to 4f. The energies of all orbitals except s vary with Z. In particular, the nd orbitals lie between the $(n + 1)$s and $(n + 1)$p for $n \geq 3$ ($Z < 20$), and actually cross the $(n + 1)$s to lower energies as Z becomes larger.

and so on. Then σ is formed by summing the following contributions:

(a) From any orbital group of energy higher than that of the group containing the electron under consideration, *zero*.

(b) From each *other* electron in the group containing the electron under consideration, 0.35 *per electron*, or 0.30 *per electron* if the group being considered is 1s.

(c) From the electron group of next lowest energy to that containing the electron under consideration, 0.85 *per electron*, if the electron under consideration is s or p, and 1.00 *per electron* for all lower energy groups. If the electron under consideration is d or f, 1.00 *per electron* for all lower energy electrons.

Some examples follow to show the working of these rules.

Atom	Z	Electron considered	σ	Z_{eff}
He	2	1s	1×0.30	1.70
Be	4	2s	$1 \times 0.35 + 2 \times 0.85$	1.95
C	6	1s	1×0.30	5.70
C	6	2s, 2p	$3 \times 0.35 + 2 \times 0.85$	3.25
Na	11	3s	$8 \times 0.85 + 2 \times 1.00$	2.20
Na$^+$	11	2s, 2p	$7 \times 0.35 + 2 \times 0.85$	6.85
Ni	28	3d	$7 \times 0.35 + 8 \times 1.00 + 8 \times 1.00 + 2 \times 1.00$	7.55

2.6.6   Aufbau principle

The description of atomic orbitals is linked closely with the periodic table of the elements (Table 2.5). Continuing the process that we have carried out with hydrogen helium and lithium, we find that beryllium and boron present no difficulties, being written as $(1s)^2(2s)^2$ and $(1s)^2(2s)^2(2p)$, respectively. With carbon, however , the two p-electrons could occupy either a single orbital with paired spins, or two orbitals with unpaired spins. The dilemma is resolved by Hund's multiplicity rule which states that *in the ground state, degenerate or near-degenerate orbitals tend to be occupied singly*, which means that the configuration with the greatest number of unpaired spins is preferred. Hund's rule depends upon the wave mechanical property of *spin correlation*: that electrons with unpaired spins tend to stay further apart than those with paired spins and, hence, repel each other less strongly. Thus, spin correlation diminishes the effect of electron-electron repulsion. Hence, carbon may be written $(1s)^2(2s)^2(2p_x)^1(2p_y)^1$, or just $(1s)^2(2s^2)(2p)^2$, with the above understanding.

Continued application of the aufbau[1] principle, taking account of the order of energies given in (2.93) leads to a description of the electronic configurations of the atoms. Fig. 2.20 illustrates the result of applying this procedure to the elements hydrogen to argon. Sometimes a shorthand notation is incorporated into the written configurations. For example, sodium may be given either as $(1s)^2(2s)^2(2p)^6(3s)^1$ or as $(Ne)(3s)^1$, where (Ne) implies an inner configuration equivalent to that of neon.

The periodic table is not completed exactly in the manner indicated by Fig. 2.20. Between calcium, $(Ar)(4s)^2$, and zinc, $(Ar)(3d)^{10}(4s)^2$, for example, there exists the first transition series of the elements. Here, the 3d orbitals are filled progressively from scandium to zinc, since the energy of 3d lies below that of 4s (Fig. 2.19). Even this process is not as simple as it seems; Cu is written as $(Ar)(3d)^{10}(4s)^1$, rather than $(Ar)(3d)^9(4s)^2$, as one might at first imagine. In fact, detailed calculations, including the effects of spin corrrelation, indicate that the magnetic properties of metallic copper may be best explained by the configuration $(Ar)(3d)^{9.5}(4s)^{1.5}$.

We may recall that the stability of the noble-gas configuration was the pillar of Lewis's electron-pair bond hypothesis. From Fig. 2.20, we see that this configuration is realized by a system of fully occupied orbitals up to and including those governed by the principal quantum number n, that is, with the outermost configurations $(ns)^2(np)^6$. Table 2.5 shows a completed periodic table, giving atomic numbers, relative atomic masses and electronic configurations.

2.6.7   Ionization energy

The ionization energy, sometimes called the ionization potential (energy), is the energy required to just remove an electron from an atom in its ground state, at $T = 0$. The first ionization energy I_1 refers to the process

$$M(g) \xrightarrow{\quad I_1 \quad} M^+(g) + e^- \qquad (2.95)$$

where M is a chemical species and e^- is an electron; subsequent ionization energies are defined in a similar manner.

We may note, in passing, that at any other temperature, (2.95) is characterized by an ionization enthalpy, since the gaseous products will contain an additional energy.

---

[1] *Aufbau*: German for 'building up'.

Table 2.5   Periodic table of elements

This periodic table numbering follows the most recent[a] recommendations of IUPAC. The elements in groups 1 and 2 form the s block, those in groups 13–18 the p block, the lanthanides and actinides the f block and the remaining, transition elements the d block. Groups 1–7 were formerly IA–VII A, 8–10 were group VIII, 11–17 were I B–VIII B and 18 was group 0. Each box contains the chemical symbol of the element, its atomic number, relative atomic mass

1	2	3	4	5	6	7	8	9
1 1.0079 **H** $(1s)^1$								
3 6.941(2) **Li** $(2s)^1$	4 9.0122 **Be** $(2s)^2$							
11 22.990 **Na** $(3s)^1$	12 24.305 **Mg** $(3s)^2$							
19 39.098 **K** $(4s)^1$	20 40.078(4) **Ca** $(4s)^2$	21 44.956 **Sc** $(3d)^1(4s)^2$	22 47.88(3) **Ti** $(3d)^2(4s)^2$	23 50.942 **V** $(3d)^3(4s)^2$	24 51.996 **Cr** $(3d)^5(4s)^1$	25 54.938 **Mn** $(3d)^5(4s)^2$	26 55.847(3) **Fe** $(3d)^6(4s)^2$	27 58.933 **Co** $(3d)^7(4s)^2$
37 85.468 **Rb** $(5s)^1$	38 87.62 **Sr** $(5s)^2$	39 88.906 **Y** $(4d)^1(5s)^2$	40 91.224 **Zr** $(4d)^2(5s)^2$	41 92.906 **Nb** $(4d)^4(5s)^1$	42 95.94 **Mo** $(4d)^5(5s)^1$	43 98.906 **⁹⁹Tc** $(4d)^5(5s)^2$	44 101.07(2) **Ru** $(4d)^7(5s)^1$	45 102.91 **Rh** $(4d)^8(5s)^1$
55 132.91 **Cs** $(6s)^1$	56 137.33 **Ba** $(6s)^2$	71 174.97 **Lu** $(4f)^{14}(5d)^1(6s)^2$	72 178.49(2) **Hf** $(5d)^2(6s)^2$	73 180.95 **Ta** $(5d)^3(6s)^2$	74 183.85(3) **W** $(5d)^4(6s)^2$	75 186.21 **Re** $(5d)^4(6s)^2$	76 190.2 **Os** $(5d)^6(6s)^2$	77 192.22(3) **Ir** $(5d)^7(6s)^2$
87 223.02 **²²³Fr** $(7s)^1$	88 226.03 **²²⁶Ra** $(7s)^2$	103 262.11 **²⁶²Lr** $(5f)^{14}(6d)^1(7s)^2$	104 (260) **Ku** $(5f)^{14}(6d)^2(7s)^2$	105 (261) **Ha** $(5f)^{14}(6d)^3(7s)^2$				

57 138.91 **La** $(4f)^0(5d)^1(6s)^2$	58 140.12 **Ce** $(4f)^1(5d)^1(6s)^2$	59 140.91 **Pr** $(4f)^3(5d)^0(6s)^2$	60 144.24(3) **Nd** $(4f)^4(5d)^0(6s)^2$	61 146.92 **¹⁴⁷Pm** $(4f)^5(5d)^0(6s)^2$	62 150.36(3) **Sm** $(4f)^6(5d)^0(6s)^2$	63 151.96 **Eu** $(4f)^7(5d)^0(6s)^2$
89 227.03 **²²⁷Ac** $(5f)^0(6d)^1(7s)^2$	90 232.04 **Th** $(5f)^0(6d)^2(7s)^2$	91 231.04 **Pa** $(5f)^2(6d)^1(7s)^2$	92 238.029 **U** $(5f)^3(6d)^1(7s)^2$	93 237.05 **²³⁷Np** $(5f)^4(6d)^1(7s)^2$	94 239.05 **²³⁹Pu** $(5f)^6(6d)^0(7s)^2$	95 241.06 **²⁴¹Am** $(5f)^7(6d)^0(7s)^2$

[a]See, for example, *Nomenclature of Inorganic Chemistry* (1989), Butterworth. [b] *Pure & Applied Chemistry* Vol 63, pp. 987–988 (1991).

(atomic weight) and outermost electronic configuration. The elements are arranged by group number and period (principal quantum number of outermost electron/s). The atomic masses are those recommended by IUPAC 1989[h]; they are relative values, being scaled to $^{12}C = 12$. The precision is ± 1 in the last digit quoted, unless otherwise indicated.

10	11	12	13	14	15	16	17	18
								2 4.0026 **He** $(1s)^2$
			5 10.811(5) **B** $(2s)^2(2p)^1$	6 12.011 **C** $(2s)^2(2p)^2$	7 14.007 **N** $(2s)^2(2p)^3$	8 15.999 **O** $(2s)^2(2p)^4$	9 18.998 **F** $(2s)^2(2p)^5$	10 20.180 **Ne** $(2s)^2(2p)^6$
			13 26.982 **Al** $(3s)^2(3p)^1$	14 28.086 **Si** $(3s)^2(3p)^2$	15 30.974 **P** $(3s)^2(3p)^3$	16 32.066(6) **S** $(3s)^2(3p)^4$	17 35.453 **Cl** $(3s)^2(3p)^5$	18 39.948 **Ar** $(3s)^2(3p)^6$
28 58.963 **Ni** $(3d)^8(4s)^2$	29 63.546(3) **Cu** $(3d)^{10}(4s)^1$	30 65.39(2) **Zn** $(3d)^{10}(4s)^2$	31 69.723 **Ga** $(4s)^2(4p)^1$	32 72.61(2) **Ge** $(4s)^2(4p)^2$	33 74.922 **As** $(4s)^2(4p)^3$	34 78.96(3) **Se** $(4s)^2(4p)^4$	35 79.904 **Br** $(4s)^2(4p)^5$	36 83.80 **Kr** $(4s)^2(4p)^6$
46 106.42 **Pd** $(4d)^{10}(5s)^0$	47 107.87 **Ag** $(4d)^{10}(5s)^1$	48 112.41 **Cd** $(4d)^{10}(5s)^2$	49 114.82 **In** $(5s)^2(5p)^1$	50 118.71 **Sn** $(5s)^2(5p)^2$	51 121.76 **Sb** $(5s)^2(5p)^3$	52 127.60(3) **Te** $(5s)^2(5p)^4$	53 126.90 **I** $(5s)^2(5p)^5$	54 131.29(2) **Xe** $(5s)^2(5p)^6$
78 195.08(3) **Pt** $(5d)^9(6s)^1$	79 196.97 **Au** $(5d)^{10}(6s)^1$	80 200.59(2) **Hg** $(5d)^{10}(6s)^2$	81 204.38 **Tl** $(6s)^2(6p)^1$	82 207.2 **Pb** $(6s)^2(6p)^2$	83 208.98 **Bi** $(6s)^2(6p)^3$	84 209.98 ^{210}**Po** $(6s)^2(6p)^4$	85 209.99 ^{210}**At** $(6s)^2(6p)^5$	86 222.02 ^{222}**Rn** $(6s)^2(6p)^6$

64 157.25(3) **Gd** $(4f)^7(5d)^1(6s)^2$	65 158.93 **Tb** $(4f)^9(5d)^0(6s)^2$	66 162.50(3) **Dy** $(4f)^{10}(5d)^0(6s)^2$	67 164.93 **Ho** $(4f)^{11}(5d)^0(6s)^2$	68 167.26(3) **Er** $(4f)^{12}(5d)^0(6s)^2$	69 168.93 **Tm** $(4f)^{13}(5d)^0(6s)^2$	70 173.04(3) **Yb** $(4f)^{14}(5d)^0(6s)^2$	**Lanthanides**
96 244.06 ^{244}**Cm** $(5f)^7(6d)^1(7s)^2$	97 249.08 ^{249}**Bk** $(5f)^9(6d)^0(7s)^2$	98 252.08 ^{252}**Cf** $(5f)^{10}(6d)^0(7s)^2$	99 252.08 ^{252}**Es** $(5f)^{11}(6d)^0(7s)^2$	100 257.10 ^{257}**Fm** $(5f)^{12}(6d)^0(7s)^2$	101 259.10 ^{258}**Md** $(5f)^{13}(6d)^0(7s)^2$	102 259.10 ^{259}**No** $(5f)^{14}(6d)^0(7s)^2$	**Actinides**

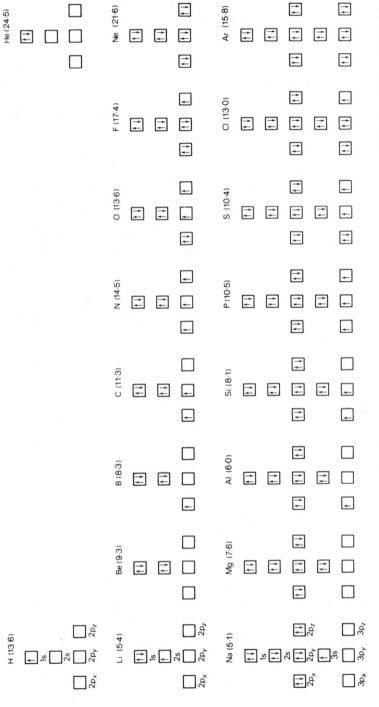

Fig. 2.20. The elements of hydrogen to argon, with atomic orbitals filled according to the aufbau principle, subject to the Pauli exclusion principle. The sign ↑↓ indicates paired electrons in a fully occupied orbital, whereas ↓ indicates occupancy by a single, unpaired electron. α (or β). The numbers in parentheses after the element symbols are first ionization energies I, in eV (1 eV = 1.6022×10^{-19} J). The significance of the p_x, p_y, p_z notation has been discussed in section 2.6.1.

In thermodynamic notation, I_1 could be replaced by $\Delta U_i^{\ominus}$, and we have for the process of (2.95) at 298 K

$$\Delta H_i^{\ominus} = \Delta U_i^{\ominus} + \Delta n \mathcal{R} T \qquad (2.96)$$

where $\ominus$ refers to the thermodynamic standard state, and Δn is the number of moles of products *minus* the number of moles of reactants.

Ionization energies may be obtained from spectral measurements, and we shall consider how this may be done for hydrogen. If we take $m = 1$ in (2.29), we refer to the Lyman series of spectra for atomic hydrogen. Some of the transitions in this series (see Fig. 2.18) are as follow:

n	2	3	4	5	6
$\bar{v}/\text{cm}^{-1}$	82258	97491	102822	105290	106631

From (2.29) it is clear that a graph of $\bar{v}$ against $1/n^2$ should be a straight line of slope $-R_H$ and intercept $\bar{v}_{\infty}$, the *series limit*. Fig. 2.21 is the appropriate plot, from which, by least squares, we obtain $R_H = 109676.9 \text{ cm}^{-1}$ and $\bar{v}_{\infty} = 109677.2 \text{ cm}^{-1}$. The latter quantity corresponds to the ionization energy, which we obtain by multiplication by hc, giving 2.1787×10^{-18} J, or 1312.0 kJ mol^{-1} [cf. the calculated E_H from (2.83)]. It may be noted that, in spectroscopy, it is quite common to refer to measurements on spectral lines in wavenumbers (cm^{-1}) as their 'frequencies'.

The energy of an atomic orbital is associated with the ionization energy of the atom; values of first ionization energies are given in parentheses after the element symbol in Fig. 2.20. They show an upward trend from lithium to neon, from sodium to argon, and so on. We can calculate the screening constant for the outer electron in each element across a period, using Slater's rules:

	Li	Be	B	C	N	O	F	Ne
Z	3	4	5	6	7	8	9	10
σ	1.70	2.05	2.40	2.75	3.10	3.45	3.80	4.15
Z_{eff}	1.3	1.95	2.60	3.25	3.90	4.55	5.20	5.85

The ionization energies I_1 for lithium to neon are plotted in Fig. 2.22 as a function of atomic number Z. The decrease at boron occurs because a higher energy 2p orbital becomes occupied; at oxygen the drop arises because a 2p atomic orbital is doubly occupied, thus increasing the electron–electron repulsion. Note that the higher (*more positive*) the energy, the less the work (*smaller value of I_1*) to produce ionization.

Table 2.6 completes the data in Fig. 2.20 on first ionization energies for the elements other than those of the transition series. The reader is invited to construct plots similar

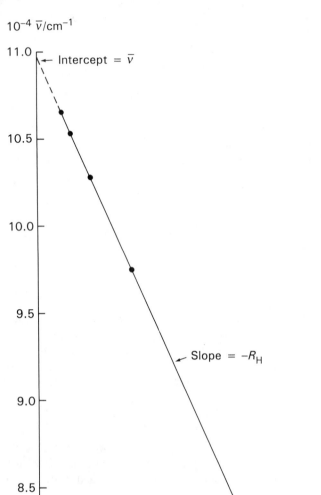

Fig. 2.21. Graph of wavenumber/cm^{-1} as a function of $1/n^2$ for the Lyman spectral series of
the hydrogen atom; the line has been fitted by the method of least squares (see Appendix 9).
The slope is $-R_H$: extrapolation to $1/n^2 = 0$ corresponds to $\bar{\nu} = \infty$, that is, at the series limit,
when ionization is achieved and the electron has just left the atom; the spectra changes from
a line spectrum to a continuum.

Table 2.6. First ionization energies/eV for lithium to radon, excluding the transition
series

K	Ca	Ga	Ge	As	Se	Br	Kr
4.34	6.11	6.00	7.90	9.82	9.75	11.82	14.00
Rb	Sr	In	Sn	Sb	Te	I	Xe
4.18	5.70	5.79	7.35	8.64	9.01	10.45	12.13
Cs	Ba	Tl	Pb	Bi	Po	At	Rn
3.89	5.21	6.11	7.42	7.29	8.4	9.6	10.7

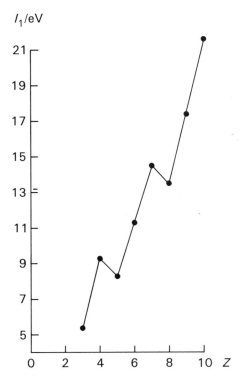

Fig. 2.22. Graph of the first ionization energy as a function of atomic number Z for the first short row (Li–Ne) of the periodic table. The fall at boron arises because of occupancy of a higher orbital (2p); that at oxygen because two p electrons have to be paired.

to that of Fig. 2.22, so as to observe the periodicity in I_1, and to consider any other variations in the light of the electron configurations of the elements.

The general upward trend in I_1 from lithium to neon reflects the increased attraction of the nucleus for an outermost electron; Z_{eff} increases from 0.43 Z in lithium to 0.59 Z in neon. We discuss ionization energies further in section 4.4.12.

2.7 SIMPLE MOLECULES—THEORETICAL PRINCIPLES

In a molecule, each electron comes under the influence of the potential fields of both the nucleus and the other electrons. In the simplest example, the hydrogen molecule, the electrostatic interactions that arise are indicated in Fig. 2.23.

In applying (2.77), we note that $\mathscr{H}$ is now given by

$$\mathscr{H} = -\frac{h^2}{8\pi^2 M}\{\nabla_A^2 + \nabla_B^2\} - \frac{h^2}{8\pi^2 \mu}\{\nabla_1^2 + \nabla_2^2\}$$

$$-\frac{e^2}{4\pi\varepsilon_0}\left[\frac{1}{r_{A1}} + \frac{1}{r_{A2}} + \frac{1}{r_{B1}} + \frac{1}{r_{B2}} - \frac{1}{r_{12}} - \frac{1}{r_e}\right] \tag{2.97}$$

where M is the mass of the nucleus. The first term on the right-hand side of (2.97) describes the kinetic energy of the nuclei A and B, the second term contains the kinetic

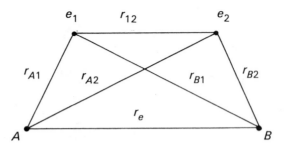

Fig. 2.23. Schematic arrangement of the two nuclei A and B, the two electrons e_1 and e_2, and the six interparticle distances in the hydrogen molecule.

energy of the electrons 1 and 2, and the third term lists the interparticle Coulombic attractions and repulsions. Now M/μ is approximately 1837, and we may neglect the term involving $1/M$ in comparison with the other two terms. This treatment is known as the *Born–Oppenheimer approximation*, and will be assumed to apply hereinafter. Effectively, it separates the kinetic energy of the nuclei from the electronic energy, and enables one to calculate the electronic energy as a function of r, the internuclear distance, for any given nuclear configuration. The electrons move so quickly that, effectively, they see the nuclei as standing still.

For a diatomic molecule, the curve of potential energy as a function of internuclear distance is of the type shown in Figure 2.24.

2.7.1   Variation method
The wave equation (2.77) cannot be solved exactly for species with two or more electrons. Approximate methods must be employed in order to obtain the configura-

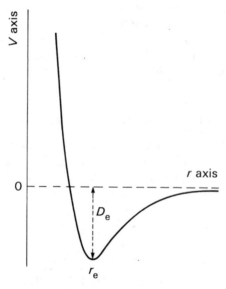

Fig. 2.24. Variation in potential energy V with internuclear separation r for a diatomic molecule. At the equilibrium distance r_e, the Coulombic forces of attraction balance the forces of repulsion. The quantity D_e is the theoretical dissociation energy of the molecule.

tion of minimum energy for a system. The variation method, an adaptation of Rayleigh's (1880) principle, is one way in which we can proceed.

First, we multiply both sides of (2.77) by ψ, or by ψ^*, if ψ is complex (we shall use ψ), and integrate over all space:

$$E = \frac{\int_{\text{all space}} \psi \mathcal{H} \psi \, d\psi}{\int_{\text{all space}} \psi^2 \, d\tau}. \tag{2.98}$$

We note in passing that $\mathcal{H}$ is an operator acting on ψ, and we cannot interchange the order $\psi \mathcal{H} \psi$ to $\mathcal{H} \psi \psi$.

A trial function ψ_t is adopted for ψ, and from (2.98), we write the Rayleigh ratio $\mathscr{E}$ as

$$\mathscr{E} = \frac{\int \psi_t \mathcal{H} \psi_t \, d\tau}{\int \psi_t^2 \, d\tau}. \tag{2.99}$$

where the integration limits are as before. In general, we expect that $\mathscr{E}$ will be greater (more positive) than the true value E, and ψ_t is subjected to modifications in order to seek the best estimate of E.

Let us practice the variation procedure on the ground state of the hydrogen atom, the result for which we know already. We shall postulate a trial function

$$\psi_t = \exp(-\zeta r) \tag{2.100}$$

where ζ is a constant. Since this function is independent of θ and ϕ, we can extract the radial part of (2.80). Differentiating (2.100) twice, with respect to r

$$\frac{\partial^2 \psi_t}{\partial r^2} + \frac{2}{r} \frac{\partial \psi_t}{\partial r} = (\zeta^2 - 2\zeta/r)\exp(-\zeta r) \tag{2.101}$$

and from expansion of the radial part of (2.80), and (2.99)

$$\mathscr{E} = \frac{\int_0^\infty \{\exp(-\zeta r)[-(h^2/8\pi^2\mu)(\zeta^2 - 2\zeta/r) - e^2/(4\pi\varepsilon_0)]\exp(-\zeta r)\}r^2 \, dr}{\int_0^\infty \exp(-2\zeta r)r^2 \, dr} \tag{2.102}$$

Proceeding from left to right, we have

$$\mathscr{E} = -(h^2\zeta^2/8\pi^2\mu)\int_0^\infty r^2 \exp(-2\zeta r) \, dr + (h^2\zeta/4\pi^2\mu)\int_0^\infty r \exp(-2\zeta r) \, dr$$
$$- e^2/(4\pi\varepsilon_0)\int_0^\infty 1/r \exp(-2\zeta r) \, dr \Big/ \int_0^\infty r^2 \exp(-2\zeta r) \, dr \tag{2.103}$$

Following Appendix 4, we find that the integrals are, in order, $1/8\zeta^3\Gamma(3)$, $1/4\zeta^2\Gamma(2)$, $1/4\zeta^2\Gamma(1)$ and $1/8\zeta^3\Gamma(3)$, whence

$$\mathscr{E} = \frac{\{-(h^2/32\pi^2\mu\zeta) + (h^2/16\pi^2\mu\zeta) - e^2/(16\pi\varepsilon_0\zeta^2)\}}{1/4\zeta^3}. \tag{2.104}$$

Thus

$$\mathscr{E} = h^2\zeta^2/8\pi^2\mu - e^2\zeta/4\pi\varepsilon_0. \tag{2.105}$$

We now differentiate $\mathscr{E}$ with respect to ζ and set the derivative to zero, so as to find the condition for minimum energy, namely

$$\zeta = \pi\mu e^2/h^2\varepsilon_0 \tag{2.106}$$

which is the reciprocal of the Bohr radius, a_0. Substituting (2.106) in (2.105) gives

$$E = -\mu e^4/8h^2\varepsilon_0^2 \tag{2.107}$$

which is equivalent to (2.81) with $n = 1$. This excellent result has been obtained because we made a particularly fortunate (or cunning) choice for the trial wave function, but it serves to give confidence in the variation method.

2.7.2   Linear combination of atomic orbitals

Instead of making guesses for the trial wave functions of molecules, we can compound known atomic wave functions by a technique known as the *linear combination of atomic orbitals* (LCAO). We have already separated the nuclear and electronic energies; now we shall treat a molecule in terms of atomic orbitals.

Let a molecular wave function Ψ be constructed as a sum of a *basis set* of atomic orbitals.

$$\Psi = c_1\psi_1 + c_2\psi_2 + \cdots + c_n\psi_n \tag{2.108}$$

where c_1 to c_n are variable parameters that will be adjusted so as to obtain a minimum energy configuration, that is, $\partial\mathscr{E}/\partial c_1 = 0$, $\partial\mathscr{E}/\partial c_2 = 0, \ldots, \partial\mathscr{E}/\partial c_n = 0$. The validity of the LCAO technique can be justified by the following argument. For any particular eigenfunction ψ_i, we can write

$$\mathscr{H}\psi_i = \varepsilon\psi_i \tag{2.109}$$

where ε is the eigenvalue of ψ_i. Let a linear combination be formed:

$$\Psi = \sum_i c_i\psi_i. \tag{2.110}$$

Then

$$\mathscr{H}\Psi = \sum_i c_i\mathscr{H}\psi_i = \varepsilon\sum_i c_i\psi_i = \varepsilon\Psi. \tag{2.111}$$

Thus, ε is also the eigenvalue of the linear combination Ψ.

If we substitute (2.108) in (2.99), restricting n to 2 for conciseness, we obtain, since[1] $\int \psi_1\mathscr{H}\psi_2 \, d\tau = \int \psi_2\mathscr{H}\psi_1 \, d\tau$ in this example,

$$\frac{c_1^2\int \psi_1\mathscr{H}\psi_1 \, d\tau + 2c_1c_2\int \psi_1\mathscr{H}\psi_2 \, d\tau + c_2^2\int \psi_2\mathscr{H}\psi_2 \, d\tau}{c_1^2\int \psi_1^2 \, d\tau + 2c_1c_2\int \psi_1\psi_2 \, d\tau + c_2^2\int \psi_2^2 \, d\tau}. \tag{2.112}$$

---

[1] By symmetry, at least for diatomic molecules.

Let

$$\int \psi_i \mathcal{H} \psi_j d\tau = H_{ij} \quad \text{and} \quad \int \psi_i \psi_j d\tau = S_{ij}. \tag{2.113}$$

Then

$$\mathcal{E} = \frac{c_1^2 H_{11} + 2c_1 c_2 H_{12} + c_2^2 H_{22}}{c_1^2 S_{11} + 2c_1 c_2 S_{12} + c_2^2 S_{22}}. \tag{2.114}$$

If, as is usual, ψ_i and ψ_j are separately normalized, then $\int \psi_i \psi_j \, d\tau \le 1$, the equality sign holding when $i = j$.

We next form $\partial \mathcal{E}/\partial c_1$ and $\partial \mathcal{E}/\partial c_2$, and equate them to zero; thus

$$\frac{\partial \mathcal{E}}{\partial c_1} =$$

$$\frac{(c_1^2 S_{11} + 2c_1 c_2 S_{12} + c_2^2 S_{22})(2c_1 H_{11} + 2c_2 H_{22}) - (c_1^2 H_{11} + 2c_1 c_2 H_{12} + c_2^2 H_{22})(2c_1 S_{11} + 2c_2 S_{12})}{(c_1^2 S_{11} + 2c_1 c_2 S_{12} + c_2^2 S_{22})^2}$$

$$\tag{2.115}$$

Equating (2.115) to zero and rearranging

$$(2c_1 H_{11} + 2c_2 H_{12}) = \frac{(c_1^2 H_{11} + 2c_1 c_2 H_{12} + c_2^2 H_{22})(2c_1 S_{11} + 2c_2 S_{12})}{c_1^2 S_{11} + 2c_1 c_2 S_{12} + c_2^2 S_{22}} \tag{2.116}$$

or, using (2.114), replacing $\mathcal{E}$ by E, we have

$$c_1 H_{22} + c_2 H_{12} = E(c_1 S_{11} + c_1 S_{12}). \tag{2.117}$$

Hence

$$c_1(H_{11} - ES_{11}) + c_2(H_{12} - ES_{12}) = 0. \tag{2.118}$$

By a similar argument

$$c_1(H_{12} - ES_{12}) + c_2(H_{22} - ES_{22}) = 0. \tag{2.119}$$

Equations (2.118) and (2.119) are the *secular equations* for the system; $\mathcal{E}$ has been replaced by E because the correct solution of these equations should lead to the experimental, true energy. Each of these equations can be solved for the ratio c_2/c_1:

$$\frac{c_2}{c_1} = -\frac{(H_{11} - ES_{11})}{(H_{12} - ES_{12})} = -\frac{(H_{12} - ES_{12})}{(H_{22} - ES_{22})} \tag{2.120}$$

Rearrangement of (2.120) leads to the secular equation

$$(H_{11} - ES_{11})(H_{22} - ES_{22}) - (H_{12} - ES_{12})^2 = 0 \tag{2.121}$$

which may be set in the determinantal form

$$\begin{vmatrix} H_{11} - ES_{11} & H_{12} - ES_{12} \\ H_{12} - ES_{12} & H_{22} - ES_{22} \end{vmatrix} = 0. \tag{2.122}$$

The left-hand side of (2.122) is called a *secular determinant*. The general requirement that a set of equations such as (2.118) and (2.119) has non-trivial solutions for c_1 and c_2 is that the corresponding secular determinant has a zero value.

2.7.2.1 Overlap integrals
The S_{ij} integrals are known as overlap integrals; they measure the extent to which orbitals overlap; typically, values of S range between 0.2 and 0.3. Fig. 2.25 shows

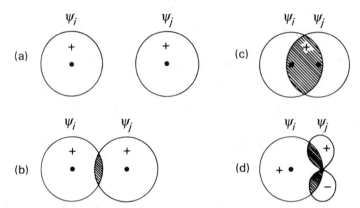

Fig. 2.25. Overlap of atomic orbitals ψ_i and ψ_j: (a) two separated 1s orbitals, S is zero; (b) two 1s orbitals, S is small; (c) two 1s orbitals, S is large; (d) overlap of 1s and 2p (axis of p orbital not along the molecular axis); S is zero because of the $\pm$ signs of the p-orbital.

differing degrees of overlap between 1s orbitals, and between 1s and 2p orbitals; S_{ij} is greater and leads to stronger bonding for the higher degree of overlap. The net overlap between 1s and 2p, in the orientation shown here, is zero, because of the opposing signs of the wave function in the two lobes of the p orbital. Pairs ij of orbitals for which S_{ij} is zero are termed *orthogonal* (see section 2.7.3).

2.7.2.2   Coulomb integrals

In studying diatomic molecules, one approximate procedure is to set $H_{11} = \alpha_1$, $H_{22} = \alpha_2$ and $H_{12} = \beta$; the α terms are called Coulomb integrals, and β is the resonance integral. A Coulomb integral measures the energy of an electron when it occupies its own orbital; in a homonuclear diatomic, $\alpha_1 = \alpha_2$. The resonance integral is negative in a bonding situation, but vanishes to zero when the two orbitals do not overlap.

2.7.3   Orthogonality

An important feature of well-designed atomic orbitals is that they are orthogonal. This property is desirable if we are to be able to discuss them separately, for otherwise we would find that one depended on another.

Mathematically, orthogonality is defined by

$$\int_{\substack{\text{all} \\ \text{space}}} \psi_i \psi_j \, d\tau = \delta_K \tag{2.123}$$

where δ_K is the Kronecker delta: for normalized orbitals δ_K is zero for $i \neq j$, and unity otherwise. As an example, consider the 1s and $2p_x$ orbitals from Table 2.4. We write the integral

$$\int_{\substack{\text{all} \\ \text{space}}} \psi_{1s} \psi_{2p_x} \, d\tau = N \int_0^\infty R_{1,0,0} R_{2,1,1} r^2 \, dr \int_0^\pi \sin \theta^2 \, d\theta \int_0^{2\pi} \cos \phi \, d\phi. \tag{2.124}$$

Without further elaboration, we can see that (2.124) is zero, because the integral in $\cos \phi$ is zero. In the case of the $2p_x$ orbital itself, however, we have

$$\int_{\substack{\text{all} \\ \text{space}}} \psi_{2p_x}^2 \, d\tau = (1/32\pi)(Z/a)_0^5 \int_0^\infty r^4 \exp(-Zr/a_0) \, dr \int_0^\pi \sin^3 \theta \, d\theta \int_0^{2\pi} \cos^2 \phi \, d\phi.$$

$$(2.125)$$

The integrals on the right-hand side of (2.125) are, respectively, $(a_0/Z)^5\Gamma(5)$, $4/3$ and π, so that the whole integral is unity.

2.8   SIMPLE MOLECULES—STRUCTURES

We are now poised to consider the application of quantum mechanics to molecules. There are two models that have been developed for this purpose, the valence-bond (VB) and the molecular orbital (MO) approximations. We shall consider first, and in more detail, the molecular orbital method, then look at the valence-bond technique, and finally draw some comparisons of the two procedures.

2.8.1   Molecular orbital method
We introduce this molecular model with reference first to the one-electron hydrogen molecule ion H_2^+, which plays a role for molecules similar to that of hydrogen for atoms. In the Born–Oppenheimer approximation, we have a single electron in the field of two stationary nuclei. Hence, with the notation already described, the Schrödinger equation may be written as

$$-\frac{h^2}{8\pi^2\mu}\nabla^2\Psi - \frac{e^2}{4\pi\varepsilon_0}\left(-\frac{1}{r_1} - \frac{1}{r_2}\right)\Psi = E\Psi \qquad (2.126)$$

which can be solved exactly. However, we shall use the LCAO MO technique for this problem.

The lowest configuration of a hydrogen atom is 1s, and we shall assume that the ground state MO for H_2^+ is similar. Let ψ_1 and ψ_2 represent the electron in the neighbourhoods of nuclei 1 and 2, respectively, that is, they are 1s hydrogen-like atomic orbitals, appropriately normalized. Then, following (2.108), we have

$$\Psi = c_1\psi_1 + c_2\psi_2. \qquad (2.127)$$

The two Ψ functions have equal weight in this species, by symmetry, and since probabilities are proportional to Ψ^2, we can say that $c_1^2 = c_2^2$. There are then two solutions, which may write as

$$\Psi_\pm = c_\pm(\psi_1 \pm \psi_2). \qquad (2.128)$$

Following through the variation method, and because of the identity of the normalized functions ψ_1 and ψ_2, we can write $H_{11} = H_{22} = \alpha$, $H_{12} = \beta$, $S_{11} = S_{22} = 1$ and $S_{12} = S$. Then we have for the secular equations

$$c_1(\alpha - E) + c_2(\beta - ES) = 0$$

and

$$c_1(\beta - ES) + c_2(\alpha - E) = 0$$

$$\left.\right\} \qquad (2.129)$$

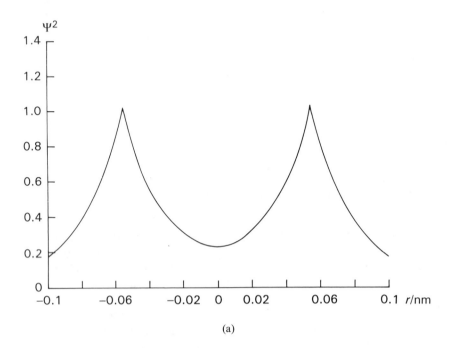

(a)

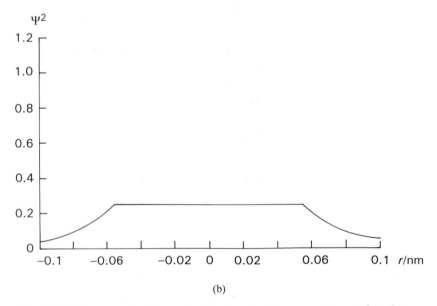

(b)

Fig. 2.26. Hydrogen molecule ion probability density: (a) superposition of $\psi_1^2 + \psi_2^2$ functions; (b) the function $2\psi_1\psi_2$—a constant for $(r_1 + r_2) \leq r_e$; (c) superposition (a) + (b), leading to a bonding $1s\sigma$ molecular orbital; (d) superposition of (a) − (b), leading to a $1s\sigma^*$ antibonding orbital.

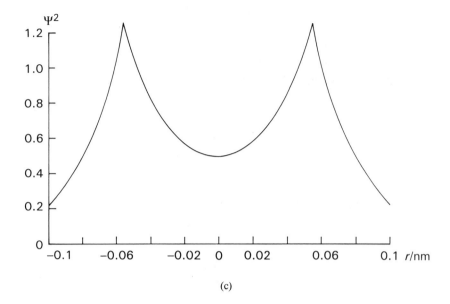

(c)

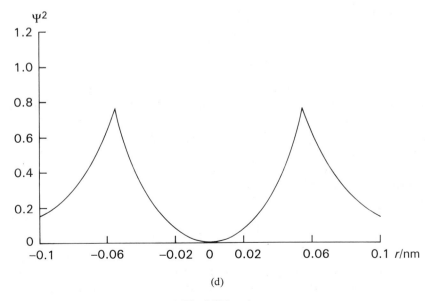

(d)

Fig. 2.26 (*cont.*)

for which the secular determinant becomes

$$\begin{vmatrix} \alpha - E & \beta - ES \\ \beta - ES & \alpha - E \end{vmatrix} = 0. \tag{2.130}$$

Expanding this determinant and solving the ensuing quadratic equation gives the two values

$$E = \frac{\alpha + \beta}{1 + S} \tag{2.131}$$

$$E = \frac{\alpha - \beta}{1 - S}. \tag{2.132}$$

Substituting these values of E in (2.120) shows that $c_1 = -c_2$, as we anticipated, and we obtain a value for c_+ by normalizing the molecular wave function Ψ according to (2.45). Thus

$$c_\pm^2 \int (\psi_1 \pm \psi_2)^2 \, d\tau = c_\pm^2 \int (\psi_1^2 + \psi_2^2 \pm 2\psi_1\psi_2) \, d\tau = 1$$

$$= c_\pm^2(1 + 1 \pm 2S) \tag{2.133}$$

whence c, the normalizing constant, becomes $1/\sqrt{(2 \pm 2S)}$ for the two wave functions in (2.128). The value for S in H_2^+ at its internuclear separation of approximately 0.11 nm is 0.57; this value is unusually large, as it is also in H_2. The evaluation of a simple overlap integral is given in Appendix 10.

2.8.1.1   Bonding and antibonding orbitals
From the foregoing discussion, we have two values for the parameter $c_\pm$ ($c_+ = 0.56$, and $c_- = 1.08$) and, hence, two wave functions for H_2^+:

$$\Psi_+ = 0.56(\psi_1 + \psi_2) \tag{2.134}$$

$$\Psi_- = 1.08 \, (\psi_1 - \psi_2) \tag{2.135}$$

with the corresponding two energies given by (2.131) and (2.132), respectively. Since α and β are both negative, (2.134), corresponding to (2.131), will have the lower (more negative) energy, and constitutes a *bonding* molecular orbital; on the other hand, (2.135) is an *antibonding* molecular orbital. Both of these orbitals have cylindrical symmetry about the internuclear axis, and are called σ *orbitals* and, more fully, $1s\sigma$ orbitals, since they have been formed from 1s atomic orbitals.

The probability density is proportional to the square of the wave function. For the $1s\sigma$ bonding orbital, we have

$$\Psi^2 = 0.56^2(\psi_1^2 + \psi_2^2 + 2\psi_1\psi_2)$$

$$\propto \{\exp(-2r_1/a_0) + \exp(-2r_2/a_0) + 2 \exp[-(r_1 + r_2)/a_0]\} \tag{2.136}$$

where r_1 and r_2 are the distances of the electron from nuclei 1 and 2, respectively. Fig. 2.26a shows the sum of the first two terms on the right-hand side of (2.136). It is the superposition of two atomic orbitals separated by a distance of 0.11 nm, the

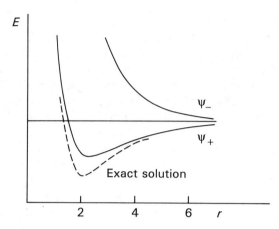

Fig. 2.27. Hydrogen molecule ion: variation of energy E with internuclear distance r for the bonding ψ_+ and antibonding ψ_- wave functions; the curve for the exact solution is shown for comparison.

equilibrium internuclear distance r_e in H_2^+; in the plot, $r_1 = |r + r_e/2|$ and $r_2 = |r - r_e/2|$. Fig. 2.26b shows the overlap density, which represents a build-up of the density over that shown in Fig. 2.26a, because of the constructive interference of the overlap of the two 1s atomic orbitals of positive amplitude ψ; it remains constant over the distance r_e. Fig. 2.26c is the sum of all three terms in (2.136), and shows clearly the enhancement of density, especially in the internuclear region.

The results of detailed calculations on this system give $r_e = 0.13$ nm (experimental, 0.11 nm) and $E = 1.8$ eV (experimental, 2.6 eV), which are acceptably close for a first approximation. Fig. 2.27 shows the variation of energy E with internuclear distance r for the two wave functions in (2.134), bonding, and (2.135), antibonding; the curve for the exact solution is given for comparison.

The energy minimum corresponds to the theoretical bond dissociation energy D_e (see Fig. 2.24). We may note, in passing, that D_e differs from the experimental dissociation energy D_0 by the magnitude of the zero-point energy of vibration.

If we repeat the calculation for the antibonding 1sσ* orbital, we need to change the sign in front of the third term on the right-hand side of (2.136). Figs. 2.26a, b still apply, but the subtraction of the third term constitutes a destructive interference, and leads to a reduction in the probability density compared to that of the sum of the first two terms. The result is shown in Fig. 2.26d, from which it is clear that the density falls to zero between the two nuclei. The source of cohesion between the nuclei has been lost, and comparison with Fig. 2.25a shows that the 1sσ* molecular orbital is less stable than the separated system of two nuclei and one electron.

2.8.2 Diatomic molecules

In this section, we shall study the application of molecular orbital theory to a selection of diatomic molecules, and introduce hybridization. Then, we shall go on to consider some polyatomic molecules, and in a later section we shall look at certain of these molecules again but by the valence-bond method, and draw comparisons between the two approaches.

2.8.2.1 *Homonuclear diatomics*

Much of the discussion on the hydrogen molecule ion can be carried over to other diatomic molecules. In the hydrogen molecule, a molecular orbital can be built from two ls atomic orbitals, as with the hydrogen molecule ion, and we obtain $1s\sigma$ and $1s\sigma^*$ molecular orbitals. We interject a word of caution here. A molecular orbital, like an atomic orbital, is a mathematical function. The implied coalescence of atomic orbitals to form molecular orbitals is a useful descriptive convenience rather than a physical reality.

The energies of the two $1s\sigma$ molecular orbitals for hydrogen are given approximately by (2.131) and (2.132) with multiplying factors of 2, since there are now two electrons, but other factors such as electron–electron repulsion need to be taken into account in detailed calculations. We shall not enter into this degree of sophistication, but just note here that at the internuclear distance of 0.074 nm in H_2 (smaller than that in H_2^+) the first approximation energy is 3.6 eV (experimental, 4.7467 eV). The best calculations have given an energy in exact agreement with the observed value[1]. The similarity between H_2^+ and H_2 can be brought out by use of the molecular orbital energy diagram, Fig. 2.28. In H_2^+, the $1s\sigma$ orbital is occupied by a single electron, whereas in H_2 the $1s\sigma$ orbital contains two electrons, following the aufbau principle, and with paired spins according to the Pauli exclusion principle.

It is now not difficult to see how He_2 does not form a stable molecule. The ground state configuration would be $(1s\sigma)^2(1s\sigma^*)^2$. Since, as we have seen (Fig. 2.27), an

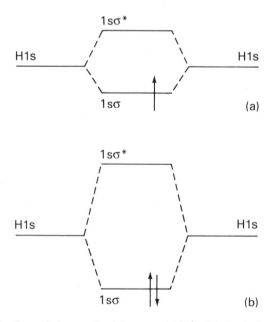

Fig. 2.28. Molecular orbital energy level diagrams: (a) H_2^+, with the single electron in a $1s\sigma$ MO; (b) H_2, with the $1s\sigma$ MO fully occupied by two electrons with paired spins.

[1] W. Kolos and L. Wolniewicz (1968) *J. Chem. Phys.* **49**, 404

antibonding orbital more than negates a bonding orbital in terms of energy, He_2 would be less stable than two separate helium atoms. The reader is invited to consider the feasibility of a molecule Li_2.

2.8.2.2 Symmetry of orbitals

Equation (2.128) was developed by LCAO theory without reference to the choice of atomic orbital. The combination of an orbital ψ_1 with another orbital ψ_2 led, through its development for the species H_2^+, to bonding and antibonding molecular orbitals. In general, the combination of n atomic orbitals leads to the same number n of molecular orbitals.

Homonuclear diatomic molecules exhibit symmetry; this symmetry can be seen also in their orbitals, thus permitting a useful classification of them. Thus,

(a) orbitals that are symmetrical about the internuclear axis are called σ molecular orbitals, Fig. 2.29a–d;
(b) orbitals that have a nodal plane containing the internuclear axis are called π molecular orbitals, Fig. 2.29e,f;
(c) orbitals may be described as even, g (German, $gerade$) or as odd, u (German $ungerade$) with respect to inversion across the centre of the internuclear axis, Fig. 2.29a–f. The subscripts g and u are called the $parity$ of the orbital.

It may be noted that the σ_g and π_u molecular orbitals are bonding, whereas the σ_u and π_g are antibonding, so that σ_u and σ^* represent the same type of orbital. The u and g descriptors cannot be applied to heteronuclear diatomics because these molecules are not centrosymmetric.

For species with $n = 2$, the valence electrons are 2s and 2p; the 1s form the $core$, without significant overlap. We can envisage 2s, 2s overlap to give sσ MOs of the type already discussed (Fig. 2.29a,b); $2p_z$, $2p_z$ overlap (z along the internuclear axis, by convention) to give pσ (Fig. 2.29c,d); $2p_x$ and/or $2p_y$, perpendicular to the internuclear axis, overlap to give pπ (Fig. 2.29e,f). Another type is 2s, 2p: 2s, $2p_z$ contributes to the pσ MO, with significant overlap given by $\int \psi(2s)\psi(2p)\,d\tau$, but 2s, $2p_x$ or 2s, $2p_y$ have zero overlap (see Fig. 2.25d).

Molecular orbital energy level diagrams for homonuclear diatomics up to and including nitrogen, and for oxygen and fluorine respectively, are shown in Figs 2.30 and 2.31. It may be noted that the 2pσ and 2pπ energies are reversed in order after nitrogen; for molecules in later horizontal periods the pσ is always below the pπ in energy. Hence, we write the complete configurations of N_2 and O_2 as follows:

$$N_2\ (1s\sigma)^2(1s\sigma^*)^2(2s\sigma)^2(2s\sigma^*)^2(2p\pi^4)(2p\sigma)^2$$
$$O_2\ (1s\sigma)^2(1s\sigma^*)^2(2s\sigma)^2(2s\sigma^*)^2(2p\sigma)^2(2p\pi)^4(2p_x\pi^*)^1(2p_y\pi^*)^1.$$

We separate the pπ^* MOs in oxygen so as to stress the single occupancy of these orbitals; it is this feature that is responsible for the paramagnetic property of molecular oxygen.

A net bonding parameter may be defined as

$$\kappa = (n_e - n_e^*)/2 \tag{2.137}$$

where n_e and $n_e^*$are, respectively, the numbers of electrons in bonding and antibonding molecular orbitals. The net bonding parameters are 3 and 2 for N_2 and O_2,

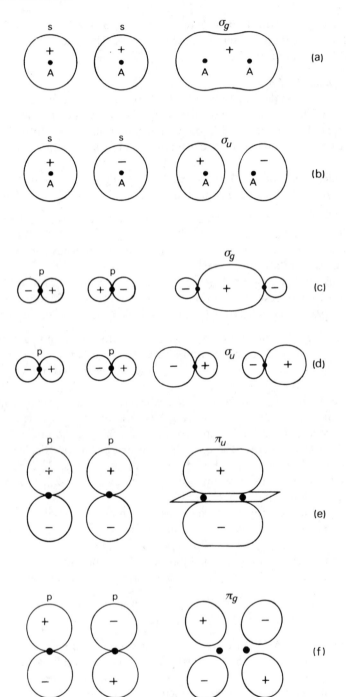

Fig. 2.29. Schematic diagrams of molecular orbitals of differing types and symmetries formed by linear combinations of atomic orbitals: the $s\sigma$ and $p\sigma$ MOs are g (bonding) and u (antibonding), whereas the $p\pi$ MOs are u (bonding) and g (antibonding). It should be noted that in homonuclear diatomics σ_u and σ^*, for example, are equivalent descriptions.

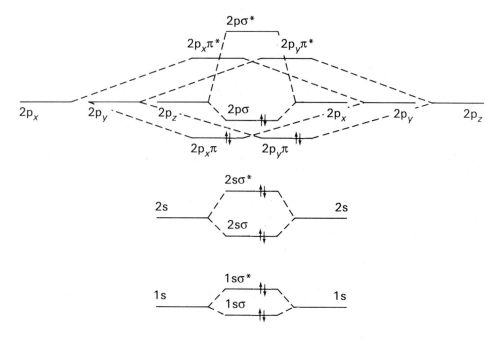

Fig. 2.30. Molecular orbital energy level diagram for N_2: the configuration is $(1s\sigma)^2(1s\sigma^*)^2(2s\sigma)^2(2s\sigma^*)^2(2p\pi)^4(2p\sigma)^2$, the arrows represent electrons occupying the MOs, and in N_2 they are all paired; the bond order parameter κ is 3, a classical triple bond.

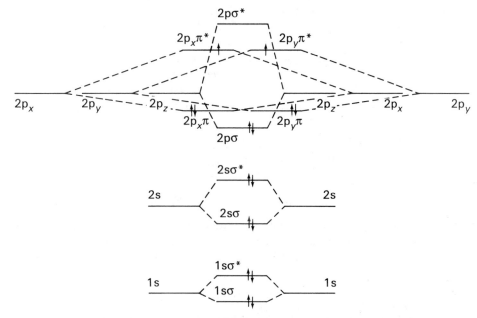

Fig. 2.31. Molecular orbital energy level diagram for O_2: the configuration is $(1s\sigma)^2(1s\sigma^*)^2(2s\sigma)^2(2s\sigma^*)^2(2p\sigma)^2(2p\pi)^4(2p\pi^*)^2$. Notice the reversal of the order of $2p\sigma$ and $2p\pi$ relative to N_2: the $2p\pi^*$ MOs are singly occupied, and the unpaired electrons lead to paramagnetism in O_2; κ is 2, a classical double bond.

respectively, corresponding to the classical formulations N≡N and O=O for these species.

2.8.2.3   *Heteronuclear diatomics*

When a diatomic molecule consists of different chemical species, the electron distribution is polarized, and there is an accumulation of negative charge δ_- on one species, the more electronegative, and a corresponding depletion δ_+ on the other; the bond is then said to be polar. In hydrogen fluoride, for example, the fluorine atom is the more electronegative, and carries a partial negative charge of approximately $-0.4e$; the hydrogen atom has a corresponding charge of $+0.4e$, and the molecule possesses a dipole moment.

The dipole moment is an important physical and theoretical property of a molecule. If a molecule is represented by partial numerical charges $\pm q$ separated by a distance $\imath$, its dipole moment μ is defined by

$$\mu = qe\imath \qquad (2.138)$$

where e is the elementary charge. The experimental value for a dipole moment[1] of the hydrogen fluoride molecule is 1.82 D and the bond length is 0.0927 nm. Hence q is 0.41.

The electron configuration of atomic fluorine is $(1s)^2(2s)^2(2p)^5$, and we can anticipate a bond between the H(1s) and F($2p_z$) AOs (Fig. 2.32). The LCAO MO may be written as

$$\Psi = c_H\psi(H, 1s) + c_F\psi(F, 2p). \qquad (2.139)$$

Because of the lack of symmetry between H and F, $c_H \neq \pm c_F$, but the values have been determined by the variation method. In the bonding orbital, $c_H^2/c_F^2 = 0.11/0.88$, and in the antibonding orbital the reciprocal of this ratio applies.

In the heteronuclear diatomics, we can continue to neglect, to a first approximation, the contributions of the 'core' AOs. Bonding overlap arises only for AOs of the same symmetry and of similar energy. Ionization energies can provide a useful guide to the orbitals that are likely to be involved in bonding; in the case of HF, $I(2s)$ is much greater than $I(2p)$, which itself is commensurate with $I(2s)$ for hydrogen. The wave functions for HF may be written as

$$\Psi_+ = 0.33\psi(H, 1s) + 0.94\psi(F, 2p) \qquad (2.140)$$

and

$$\Psi_- = 0.94\psi(H, 1s) - 0.33\psi(F, 2p). \qquad (2.141)$$

The electronic configuration of HF may be written as $[(1s\sigma)^2(2s\sigma)^2(2p\pi)^4](3p\sigma)^2$. The configuration in the brackets forms a 'core' that enters into the bonding to a relatively small extent, and the net bonding parameter may be reckoned in terms of only the $3p\sigma$ MO; thus, $\kappa = 1$.

2.8.2.4   *Hybridization*

The lithium hydride molecule provides a slightly new situation. The ionization energies for Li(2s) and Li(2p) are both commensurate with that for H(1s). Which of

---

[1] D, Debye; 1D = 3.3356×10^{-30} C m.

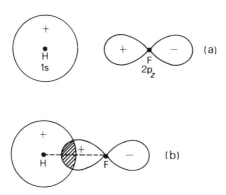

Fig. 2.32. Bond-forming AOs for HF: (a) H(1s) and F(2p$_z$), of similar energy and compatible symmetry; (b) overlap of the orbitals in bond formation through a 2pσ MO,

the AOs will lead to the best combination? The answer is that both of these lithium AOs are used, and we may write the bonding wave function as

$$\Psi = c_H\psi(\text{H, 1s}) + c_{\text{Li(s)}}\psi(\text{Li, 2s}) + c_{\text{Li(p)}}\psi(\text{Li, 2p}).$$ (2.142)

Application of the variation method leads to the result

$$\Psi = 0.69\psi(\text{H, 1s}) + 0.32\psi(\text{Li, 2s}) + 0.23\psi(\text{Li, 2p})$$ (2.143)

and we have the configuration $[(1s\sigma)^2](2s\sigma)^2$ for LiH, with participation of the Li 2p character in the 2sσ MO; $[(1s\sigma)^2]$ is the core of the two 1s electrons of lithium, with negligible bonding overlap because the energy of the Li 1s electron state is well below that of the H 1s level, and the net bonding parameter κ is 1.

In order to align this result with the idea of overlap between two atomic orbitals, we may introduce the concept of *hybridization*, or quantum mechanical 'mixing', of the 2s and 2p AOs on lithium. Thus, we may define a hybrid atomic orbital for lithium as

$$\psi(\text{Li, sp}) = \psi(\text{Li, 2s}) + \lambda\psi(\text{Li, 2p})$$ (2.144)

with λ approximately 0.72, which then overlaps with $\psi(\text{H, 1s})$.

We should note that hybridization is, again, something that we do to achieve a description of what exists, rather than a reflection of reality.

The ratio c_H^2/c_X^2, where X is another element, determines the extent to which electron density is transferred from hydrogen to the species X in a heteronuclear diatomic HX. Hybridization modifies the shapes of the participating AOs and, hence, those of the resulting MOs. Fig. 2.33 shows electron density contour maps for the heteronuclear hydrides from lithium to fluorine. It is clear that the MO for each species becomes 'fatter' around the non-hydrogen species as the series progresses from lithium to fluorine; in the same direction, electronegativity (section 2.11.4) increases, too.

2.9 POLYATOMIC MOLECULES

In this section, we shall consider some simple polyatomic, inorganic molecules, such as water and ammonia, and some of the hydrocarbons, the latter affording an opportunity to introduce Hückel MO theory in the ensuing section.

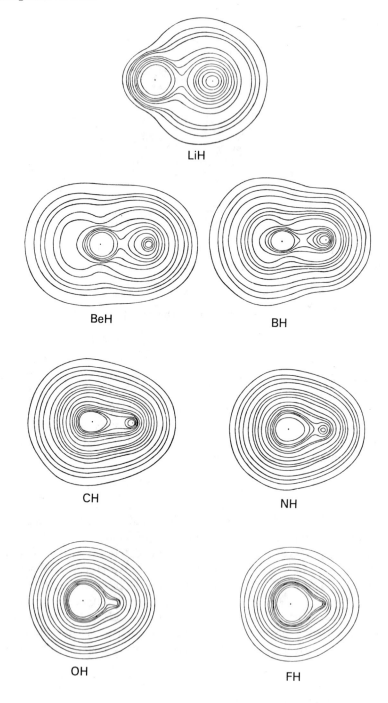

Fig. 2.33. Electron density contour diagrams for the first-row diatomic hydrides (*A*H); all maps are drawn to the same scale. The species *A* is on the left-hand side of each drawing, and its innermost contours have been omitted for the sake of clarity (after Bader, Keaveney and Cade, and reproduced by permission of the American Chemical Society).

Polyatomic molecules are built up in the same way as the diatomics but, in general, more orbitals are involved, and a greater range of shapes is possible.

2.9.1 H_2O molecule—directed bonds

In the water molecule, we know from experimental studies that the H—O bond length is 0.096 nm, and the H—O—H bond angle is 104.4°. Now the electronic configuration of oxygen is $(1s)^2(2s^2)(2p_z^2)(2p_x)^1(2p_y)^1$: we would expect divalency, with bonding between O(2p) and H(1s); and, from the use of the $2p_x$ and $2p_y$ orbitals, a bond angle of 90°. We could then say that the hydrogen atoms repel each other, thus opening out the bond angle to the observed value. But this approach would be rather simplistic.

Consider two similar p orbitals, say p_u and p_v, with their axes making an angle θ_{uv} with each other, and let another similar orbital p_w, in the plane of p_u and p_v, be directed orthogonally to p_u (Fig. 2.34). Then p_v can be resolved, like a vector, into components $p_u \cos \theta_{uv}$ *along* p_u and $p_w \sin \theta_{uv}$ *along* p_w. The overlap integral between p_u and p_v is given by

$$S_{uv} = \int \psi(p_u)\psi(p_v)\, d\tau \tag{2.145}$$

which may then be written as

$$S_{uv} = \int \psi(p_u)\{\psi(p_u)\cos \theta_{uv} + \psi(p_w)\sin \theta_{uv}\}\, d\tau \tag{2.146}$$

or

$$S_{uv} = \cos \theta_{uv} \int \psi^2(p_u)\, d\tau + \sin \theta_{uv} \int \psi(p_u)\psi(p_w)\, d\tau. \tag{2.147}$$

If the AOs are normalized, $\int \psi^2(p_u)\, d\tau = 1$, and $\int \psi(p_u)\psi(p_w)\, d\tau = 0$. Hence

$$S_{uv} = \cos \theta_{uv} \tag{2.148}$$

and is zero only when p_v coincides with p_w.

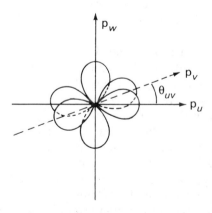

Fig. 2.34. Formation of hybrid orbitals: p_u and p_w are orthogonal, and p_v lies at an angle θ_{uv} to p_u, in the same plane as p_u and p_w; the components of p_v are $\cos \theta_{uv} \times$ a p orbital along p_u *plus* $\sin \theta_{uv} \times$ a p orbital along p_w.

Suppose next that, in the water molecule, the oxygen atom forms bonds with hydrogen by first using its 2s and 2p AOs to form a hybrid, which we may formulate as

$$\psi(sp) = \psi(p) + \lambda\psi(s) \tag{2.149}$$

implying p and s contributions to the electron density in the ratio of $1:\lambda^2$. We need two of these orbitals, which we shall require to be orthogonal; then with an obvious notation

$$\int \{\psi(p_u) + \lambda\psi(s)\}\{\psi(p_v) + \lambda\psi(s)\} \, d\tau = 0. \tag{2.150}$$

Expanding

$$\int \psi(p_u)\psi(p_v) \, d\tau + \lambda \int \psi(p_u)\psi(s) \, d\tau + \lambda \int \psi(p_v)\psi(s) \, d\tau + \lambda^2 \int \psi^2(s) \, d\tau = 0. \tag{2.151}$$

From previous results, the terms on the left-hand side of (2.151) are, successively, $\cos\theta_{uv}$, zero, zero and λ^2, whence

$$\cos\theta_{uv} = -\lambda^2. \tag{2.152}$$

Equation (2.152) shows that for $\lambda \neq 0$, θ_{uv} is greater than 90°. If we take the experimental value for the H—O—H bond angle as 104.4°, then it follows that λ is very close to 0.5; the hybrid has about 80% p and 20% s character ($\lambda^2/1 + \lambda^2$).

We have used part of the filled 2s orbital of oxygen in making each sp hybrid orbital. Energy has to be expended in opening the fully occupied 2s orbital, but it is more than compensated by the increased overlap of O(sp) with H(1s) compared to O(2p) and H(1s). The remaining 2s, $2p_x$ and $2p_y$ electron density, equivalent to a pair of electrons, occupy another orthogonal hybrid orbital. Their off-centre distribution together with the 'lone pair' $2p_z$ electrons, make the major contribution to the polar nature of the water molecule. This view is supported by the fact that the dipole moment of the water molecule is 1.8 D, whereas that of F_2O is only 0.2 D; the stronger polarity of the F—O bond, directed away from the lone pairs, reduces their contributions. Fig. 2.35 illustrates some of the features from our discussion of the water molecule.

Finally, we may note that if the 'remnant' 2s, $2p_x$ and $2p_y$ density is confined to another sp hybrid orbital, orthogonal to both bonding sp hybrids, then its axis bisects the H—O—H angle, and so lies at $(360 - 104.4)/2°$, or 127.8°, to the bonding hybrids; the non-bonding hybrid must have a 62% s and 38% p character. The total s content of the hybrids is 1.02 ($2 \times 0.20 + 0.62$), and that for the p is 1.98 ($2 \times 0.80 + 0.38$), very close to the formal values of 1 and 2, respectively.

Another way of looking at the water molecule is to assume a use of all 2s and 2p electrons of oxygen and form initially four equivalent sp^3 hybrid orbitals (see section 2.9.2) at 109.47° to one another. Two of them overlap with H 1s orbitals, and the remaining non-bonding hybrids, each containing two non-bonding electrons, repel each other so as to modify the angles from the tetrahedral value to those observed. In this way, the water molecule is treated as a modified tetrahedron, similar to the Bernal model for the water molecule in ice, where extensive hydrogen-bonding is present.

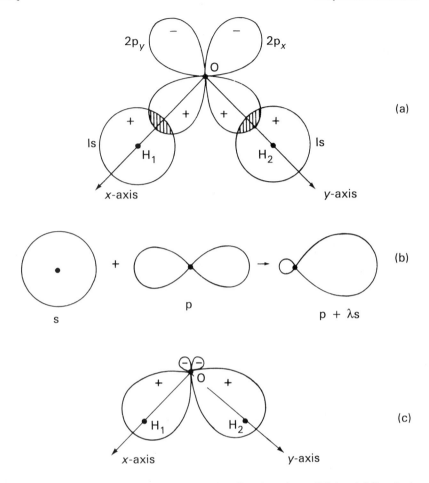

Fig. 2.35. Structure of the water molecule: (a) bonding through two H(1s) and O($2p_x$, $2p_y$) would lead to a bond angle of 90°; (b) formation of sp hybrid orbital, with the internuclear axis coincident with the axis of the 2p orbital; (c) bonding between H(1s) and O($2p + \lambda 2s$) hybrids leads to a bond angle of 104.4°, at the equilibrium distance. The two hybrid orbitals in H_2O are nearly independent of each other: if H_1 is replaced by, say, CH_3, the H_2—O bond electrons are not significantly altered, thus providing a basis for characteristic bond lengths and energies.

In this essentially qualitative approach, the water molecule was regarded as a regular tetrahedron with two charges of $+\frac{1}{2}$ and two of $-\frac{1}{2}$ at the four apices. The positive apices corresponded to the bonded hydrogen atoms and the negative apices to the non-bonding electrons. This model can be useful in considering the role of the water molecule in ice, hydrates and other structures containing water molecules.

2.9.2 Methane
The electron configuration for carbon is $(1s)^2(2s)^2(2p)^2$. It seems feasible to form four bonds with the 2s and 2p electrons, but if, in addition, they are to be equivalent, then we would expect that hybridization might be invoked. Since all four electrons are involved, the s:p hybrid character will be in the ratio of 1:3, and we call them sp^3

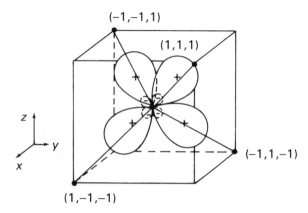

Fig. 2.36. Methane, CH_4: sp^3 hybrid orbitals, showing their directional relationship to the cube. The carbon atom is at the centre of the cube, and the relative coordinates of the corners of the cube that lie on the orbital axes (the threefold symmetry axes of the cube) are given with respect to an origin at carbon.

hybrid orbitals. Following the previous arguments, the hybrid wave function may be written as

$$\psi(sp^3) = \psi(p) + \lambda\psi(s). \tag{2.153}$$

From (2.152), if we are to achieve a tetrahedral angle between the directions of the sp^3 hybrids, then λ, given by $\sqrt{-\cos(109.47°)}$, is $1/\sqrt{3}$. The energy needed to promote carbon to the valence state has been calculated to be approximately 4.0 eV, whereas the average C—H bond energy is 4.25 eV (410 kJ mol^{-1}), so that bond formation through hybrid orbitals is energetically feasible. The relationship of the sp^3 hybrid directions to a cube is shown in Fig. 2.36; the hybrids lie along the body diagonals of the cube, the directions of the four threefold symmetry axes of the cube.

2.9.3   Delocalized systems
Under this heading, we shall consider some simple unsaturated hydrocarbons, conjugated hydrocarbons, and benzene as a representative of the aromatic system. We shall also use these compounds to introduce Hückel molecular orbital (HMO) theory and some features that follow from it.

2.9.3.1   *Ethene and ethyne*
By arguments similar to those given above, we can show that sp^2 bonds, as in ethene (C_2H_4), have, ideally, 120° valence angles. The C—C and C—H bonds are σ in type, and the $2p_z$ AO lies normal to the molecular plane. Two of these orbitals on adjacent carbon atoms overlap to form a π bond (Fig. 2.37; see also Fig. 2.29e). In ethyne (C_2H_2), the C—C and C—H bonds are again σ, but two π bonds are now formed from 2p AOs on adjacent carbon atoms (Fig. 2.38). The single, double and triple C—C bonds in ethane, ethene and ethyne form an order of increasing bond dissociation enthalpy:[1]

---

[1]Often called bond *energy* but, strictly, it is an enthalpy because it relates to the standard rate of 298.15 K and 1 atmosphere.

	C_2H_6	C_2H_4	C_2H_2
Average standard C—C bond dissociation enthalpy/kJ mol^{-1}	+344	+613	+831
Enthalpy of formation, $\Delta H_f^{\ominus}$/kJ mol^{-1}	−84.6	+52.3	+227

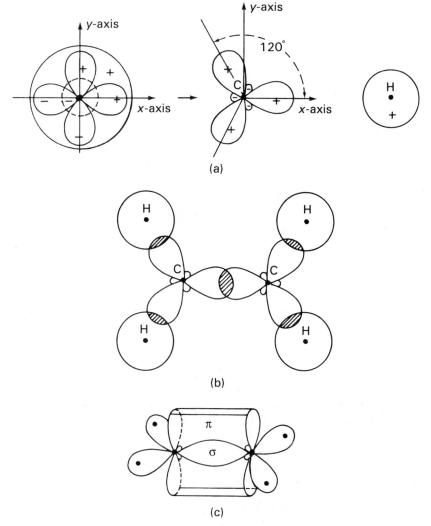

Fig. 2.37. Ethene, C_2H_4: (a) Formation of sp^2 hybrid bonds from carbon 2s, 2p$_x$ and 2p$_y$ AOs and hydrogen 1s AOs; $\psi(sp^2) = \psi(p) + \lambda\psi(s)$, and a value of $\lambda = 1/\sqrt{2}$ leads to angles of 120° between the hybrids. (b) Overlap of four sp^2 hybrids with four H(1s) AOs to form the molecule; the 2p$_z$ orbital is directed normal to the molecular plane. (c) σ MO formed from the sp^2 hybrids and H(1s) AOs, and π MO formed from the overlap of two adjacent 2p$_z$ AOs.

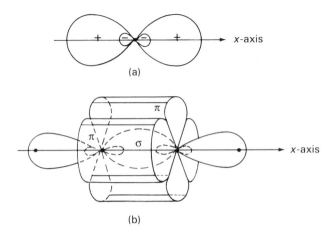

Fig. 2.38. Ethyne, C_2H_2. (a) Linear sp carbon hybrid orbital; $\psi(sp) = \psi(p) + \lambda\psi(s)$, with $\lambda = 1$. (b) The $2p_y$ and $2p_z$ AOs are mutually perpendicular with the x-axis, and the overlap from adjacent atoms leads to two π MOs, and full cylindrical symmetry is developed.

The π bond is the source of reactivity in unsaturated organic compounds: for example, the figures above show that two single C—C bonds are more stable with respect to the atoms than one double bond by ~ 75 kJ mol^{-1}, which is an energetic prerequisite for the ready reactions of unsaturated compounds. The stability of the corresponding compounds with respect to the formation from the elements lies in the reverse order, as the enthalpies of formation $\Delta H_f^{\ominus}$ indicate. A π-configuration in multiple bonds confers a degree of rigidity on any molecule in which it is present by inhibiting rotation about the σ bond.

2.10   Hückel molecular orbital theory

The Hückel MO theory treats only the π orbitals in a multiple bond system, and assumes that the σ orbitals form a sort of rigid molecular 'core', rather like the non-bonding electrons of fluorine in hydrogen fluoride. The carbon atoms are assumed to be identical, so that all Coulomb integrals α take the same value. Although of most interest in conjugated molecules, it is of value to develop the theory in terms of the simplest molecule with a multiple bond, ethene.

Let the two carbon atoms of ethene be identified by the subscripts 1 and 2; then we can write the wave function Ψ_π of the bonding π molecular orbital as

$$\Psi_\pi = c_1\psi_1 + c_2\psi_2 \tag{2.154}$$

where ψ_1 and ψ_2 represent 2p AOs of carbon. Following the technique of the variation method developed in section 2.7, we obtain the secular determinant

$$\begin{vmatrix} \alpha - E & \beta - ES \\ \beta - ES & \alpha - E \end{vmatrix} = 0. \tag{2.155}$$

In the Hückel approximation, three assumptions are made, namely, that all overlap integrals S (section 2.7.2.1) can be equated to zero, all resonance integrals (section

2.7.2.2) other than those between adjacent carbon atoms can be equated to zero, and all other resonance integrals have the same value, β. then (2.155) is reduced to

$$\begin{vmatrix} \alpha - E & \beta \\ \beta & \alpha - E \end{vmatrix} = 0 \qquad (2.156)$$

for which the roots are

$$E_{\pm} = \alpha \pm \beta. \qquad (2.157)$$

E_{+} corresponds to the bonding π MO (2.154) and is fully occupied, whereas E_{-} corresponds to an empty π^* antibonding MO. Both α and β are actually negative quantities, and β is used to form a measure of the delocalization energy. Since β is actually a negative quantity, the largest value of β correspond to the lowest-energy MO. The pair of MOs in (2.157) forms the *frontier molecular orbitals* of ethene. More generally, the frontier MOs are the highest energy occupied molecular orbital (HOMO) and the lowest energy unoccupied molecular orbital (LUMO). Frontier MOs are largely responsible for the chemical reactivity in unsaturated compounds.

An important molecule in HMO theory is buta-1,3-diene, C_4H_6; it is an example of a conjugated system, that is, a molecule containing, formally, alternating double and single bonds:

$$CH_2=CH-CH=CH_2.$$

However, when we consider the experimentally determined molecular geometry, we find the bond lengths/nm

$$C \quad \overset{0.135}{\rule{2cm}{0.4pt}} \quad C \quad \overset{0.146}{\rule{2cm}{0.4pt}} \quad C \quad \overset{0.135}{\rule{2cm}{0.4pt}} \quad C$$

from which it is clear that the double bonds are slightly longer than the double-bond value of 0.133 nm ethene, and the single bond is shorter than the standard value of 0.154 nm. We say that the electrons are partially delocalized over the molecule.

We can apply HMO theory to butadiene. Following the previous procedure, we may write a π MO wave function as

$$\Psi_{\pi} = c_1\psi_1 + c_2\psi_2 + c_3\psi_3 + c_4\psi_4. \qquad (2.158)$$

Making the same approximations as before leads to the secular determinant

$$\begin{vmatrix} \alpha - E & \beta & 0 & 0 \\ \beta & \alpha - E & \beta & 0 \\ 0 & \beta & \alpha - E & \beta \\ 0 & 0 & \beta & \alpha - E \end{vmatrix} = 0. \qquad (2.159)$$

To solve this determinant[1], it is convenient to let $(\alpha - E)/\beta = y$, whence (2.159) is transformed to:

$$\begin{vmatrix} y & 1 & 0 & 0 \\ 1 & y & 1 & 0 \\ 0 & 1 & y & 1 \\ 0 & 0 & 1 & y \end{vmatrix} = 0. \qquad (2.160)$$

[1] See, for example, A. C. Aitken (1964) *Determinants and Matrices*, Oliver & Boyd.

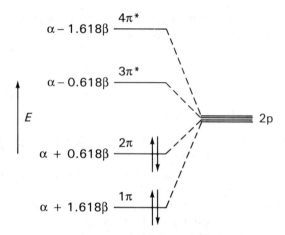

Fig. 2.39. HMO energy-level diagram for butadiene; the four π electrons occupy the two bonding MOs. The total energy is $4\alpha + 4.472\beta$, and the delocalization energy against ethene is 0.472β, or approximately -33 kJ mol^{-1}.

Expanding the determinant, by the method of cofactors (see section 2.10.2), gives

$$y^4 - 3y^2 + 1 = 0. \tag{2.161}$$

Solving, we obtain $y^2 = 2.618$ and 0.382. Since

$$y = (\alpha - E)/\beta \tag{2.162}$$

it follows that the energies of the four π MOs are

$$E = \alpha \pm 1.618\beta, \qquad \alpha \pm 0.618\beta. \tag{2.163}$$

Fig. 2.39 is an energy-level diagram for butadiene. The four bonding π electrons occupy the 1π and 2π MOs, and the frontier orbitals are 2π (bonding) and $3\pi^*$ (antibonding).

2.10.1   Delocalization energy
The total π binding energy E_π is the sum of the energy of each state multiplied by the number of electrons in that state. For ethene, E_π is $2\alpha + 2\beta$. In the case of butadiene, we have

$$E_\pi = 2(\alpha + 1.618\beta) + 2(\alpha + 0.618\beta) = 4\alpha + 4.472\beta. \tag{2.164}$$

If butadiene contained two ethenic double bonds, the π energy would be $4\alpha + 4\beta$: hence, butadiene is more stable by an amount 0.472β on account of the conjugation. This extra stability is called the *delocalization energy* D_π. Since β is approximately -70 kJ mol^{-1}, the delocalization energy in butadiene is about -33 kJ mol^{-1}.

2.10.2   Butadiene wave functions
To complete the wave functions of (2.158), we need the coefficients c_1 to c_4. Generally[1],

$$\frac{c_n}{c_1} = (-1)^{2n+1} \frac{(\text{cofactor})_n}{(\text{cofactor})_1}. \tag{2.165}$$

[1] (Cofactor)$_{ij} = (-1)^{i+j} | M_{ij} |$, where $| M_{ij} |$ is the *minor* obtained by striking out the ith row and jth column of the given determinant. In this application, $i = 1$; c_n is implicitly c_{1n} here.

Taking c_2/c_1 as a typical example, we have

$$\frac{c_2}{c_1} = (-1)\begin{vmatrix} 1 & 1 & 0 \\ 0 & y & 1 \\ 0 & 1 & y \end{vmatrix} \div \begin{vmatrix} y & 1 & 0 \\ 1 & y & 1 \\ 0 & 1 & y \end{vmatrix}$$

$$= (1 - y^2) \div y^2(y - 2). \tag{2.166}$$

The orthogonal MO wave functions in (2.158) can be normalized, through (2.45), by the term

$$N = \left\{ \sum_n (c_n/c_1)^2 \right\} \tag{2.167}$$

(we can equally well use c_n/c_1 as coefficients in place of c_n) and, from further properties of determinants, the nth coefficient is given by

$$c_n = \frac{(c_n/c_1)}{N}. \tag{2.168}$$

Proceeding in this way for the bonding MO of lowest energy, $\alpha + 1.618\beta$, $y = -1.618$. This value is inserted into equations such as (2.166), and solved for the ratios c_n/c_1, whence the c_n values are obtained by (2.168). The following results are obtained for the $\alpha + 1.618\beta$ MO:

n	c_n/c_1	$(c_n/c_1)^2$	c_n
1	1.0000	1.0000	0.3717
2	1.6180	2.6180	0.6015
3	1.6180	2.6180	0.6015
4	1.0000	1.0000	0.3717
			$\sum = 2.6900$

The final wave functions become

$$\Psi_1 = 0.3717\psi_1 + 0.6015\psi_2 + 0.6015\psi_3 + 0.3717\psi_4$$
$$\Psi_2 = 0.6015\psi_1 + 0.3717\psi_2 - 0.3717\psi_3 - 0.6015\psi_4 \tag{2.169}$$
$$\Psi_3 = 0.6015\psi_1 - 0.3717\psi_2 - 0.3717\psi_3 + 0.6015\psi_4$$
$$\Psi_4 = 0.3717\psi_1 - 0.6015\psi_2 + 0.6015\psi_3 - 0.3717\psi_4.$$

2.10.3 π bond order

The π electrons are not localized in pairs in conjugated hydrocarbons, so that classical double bonds do not exist in the HMO approximation. Each electron contributes to the bonding, and the relative π bonding between pairs of adjacent atoms ij can be related to the coefficients c_n by the π bond order $p_{i,j}$, sometimes called the mobile bond order, defined by

$$p_{i,j} = \sum_k \eta_k c_{ik} c_{jk}. \tag{2.170}$$

η_k is the number of electrons in the kth occupied π MO, i in c_{ik} refers to atom i in MO k, and j in c_{jk} refers to atom j in MO k, the sum being taken over all k occupied π MOs.

Applying this equation to the MO between atoms 1 and 2 of butadiene, noting that the antibonding orbitals are not occupied in this species, we have from (2.169):

$$P_{1,2} = 2 \times 0.3717 \times 0.6015 + 2 \times 0.3717 \times 0.6015 = 0.8943.$$

Continuing in this manner, we build up a picture for π bond order:

$$CH_2 \; \overline{\underset{}{0.894}} \; CH \; \overline{\underset{}{0.447}} \; CH \; \overline{\underset{}{0.894}} \; CH_2$$

If we assume a bond order of unity for the σ bonds, the total bond order $P_{i,j}$ becomes $(1 + p_{i,j})$, which is comparable with the net bonding parameter κ:

$$CH_2 \; \overline{\underset{}{1.894}} \; CH \; \overline{\underset{}{1.447}} \; CH \; \overline{\underset{}{1.894}} \; CH_2.$$

Bond order is an important parameter, as it correlates well with bond length. We can draw a graph of bond length d against bond order (Fig. 2.40), using the following standard values:

i, j	d/nm	$P_{i,j}$
C—C	0.154	1.00
C_{arom}—C_{arom}	0.140	1.667
C=C	0.134	2.00
C≡C	0.120	3.00

If we interpolate the calculated mobile bond orders for butadiene, we obtain the values of 0.136 nm and 0.144 nm for the bond lengths, in very good agreement with the experimental values.

2.10.4   Free valence

Another interesting parameter that derives from the HMO procedure is the free-valence parameter $\mathscr{F}_i$ for an atom. It is a measure of the degree to which the ith atom is bonded to its neighbours, and its value is large when this bonding is not strong: this result has a considerable bearing on the reactivity of a species.

The free-valence parameter is defined by

$$\mathscr{F}_i = M_i - \sum_k P_{i,k} \qquad (2.171)$$

where M_i is the maximum bonding power of the ith atom, and the sum is taken over the k occupied orbitals. We can show as follows that for a carbon atom, $M_C = 4.732$. Consider the molecule of trimethylenemethane

$$\begin{array}{c} H_2C \\[-2pt] \diagdown\!\!\!\diagdown \\ \qquad\quad C{=}CH_2. \\[-4pt] \diagup\!\!\!\diagup \\ H_2C \end{array}$$

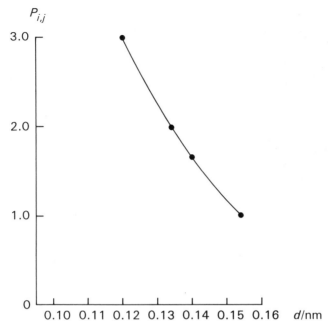

Fig. 2.40. Graph of total bond order $P_{i,j}$ against bond length d: the smooth curve has been drawn in terms of the standard bond lengths for the classical single, double and triple bonds (orders 1, 2 and 3) in ethane, ethene and ethyne, respectively, and benzene (1.667).

In this known but unstable species, the central carbon atom has three σ bonds and three π bonds to its neighbours. HMO calculations show readily that the mobile bond order is $1/\sqrt{3}$. Hence the total bond order is $3 + 3/\sqrt{3} = 4.732$. Now, continuing our analysis of butadiene, we have (σ and π bonds)

$$\mathscr{F}_1 = \mathscr{F}_4 = 4.732 - (2 + 1 + 0.894) = 0.838$$

and

$$\mathscr{F}_2 = \mathscr{F}_3 = 4.732 - (3 + 0.447 + 0.894) = 0.391$$

thus enhancing our picture of the butadiene molecule:

$\mathscr{F}$	0.838		0.391		0.391		0.838
$P_{i,j}$		1.894		1.447		1.894	
	CH_2	——	CH	——	CH	——	CH_2
d_{calc}/nm		0.136		0.144		0.136	
d_{obs}/nm		0.135		0.146		0.135	

$$D_\pi = 0.472\beta$$

It is not difficult to see how it is that bromine adds to butadiene most readily in the 1–4 positions; these positions have the higher free-valence parameter, thus indicating a greater readiness to enter into reaction than the 3–4 positions:

$$Br_2 + CH_2=CH-CH=CH_2 \rightarrow CH_2Br-CH=CH-CH_2Br$$

2.10.5 Aromatic systems

A complete delocalization of π valence electrons is achieved in benzene, C_6H_6. The six carbon atoms are linked to one another and to six hydrogen atoms by a series of sp^2 σ bonds, giving C—C—C and C—C—H bond angles of 120° (Fig. 2.41a). The remaining 2p AOs are directed normally to the molecular plane and overlap to give six MOs, including a 'double streamer' (Fig. 2.41b) all around the ring. Fig. 2.41c,d,e shows the three lowest-energy occupied MOs as seen in a direction normal to the molecular plane; Fig. 2.40c is another view of Fig. 2.41b.

The HMO treatment of benzene leads to the secular determinant

$$\begin{vmatrix} y & 1 & 0 & 0 & 0 & 1 \\ 1 & y & 1 & 0 & 0 & 0 \\ 0 & 1 & y & 1 & 0 & 0 \\ 0 & 0 & 1 & y & 1 & 0 \\ 0 & 0 & 0 & 1 & y & 1 \\ 1 & 0 & 0 & 0 & 1 & y \end{vmatrix} = 0 \qquad (2.172)$$

where y is again $(\alpha - E)/\beta$. Solution of this determinant gives

$$E = \alpha \pm 2\beta, \qquad \alpha \pm \beta, \qquad \alpha \pm \beta. \qquad (2.173)$$

The MO energy-level diagram is given in Fig. 2.42; twofold degeneracy exists for both the $\alpha + \beta$ and $\alpha - \beta$ energy states.

If we compare benzene with a hypothetical cyclohexatriene, with three ethenic double bonds, we find that the benzene delocalization energy is

$$D_\pi = 2(\alpha + 2\beta) + 4(\alpha + \beta) - 3(2\alpha + 2\beta) = 2\beta \qquad (2.174)$$

which is approximately -140 kJ mol^{-1}.

The well-known stability of the aromatic ring arises from three structural features:
(a) the ring is strain-free—the natural internal angle of a plane, regular hexagon is 120°;
(b) all the π electrons are in bonding orbitals, which gives rise to the large delocalization energy;
(c) the π electrons are completely delocalized, that is, they extend over the whole molecule.

Calculations of bond order show that the C—C bonds in benzene are equivalent, with a π bond order of 1.667. The free-valence parameter for each atom is 0.399, indicating only a moderate degree of reactivity; it may be compared with the value of 0.730 for ethene, 0.894 for butadiene and 1.15 for the unstable trimethylenemethane. It must be remembered that free valence is a static concept related to the nature of the reactants. The reactivity in terms of the rate of reaction depends also upon the energy of activation for the reaction in question.

2.10.6 Charge distributions

If a carbon atom forms three σ bonds, and is π bonded, it will remain neutral if there is an average of one electron in its π orbital. The charge distribution q_i may be taken as

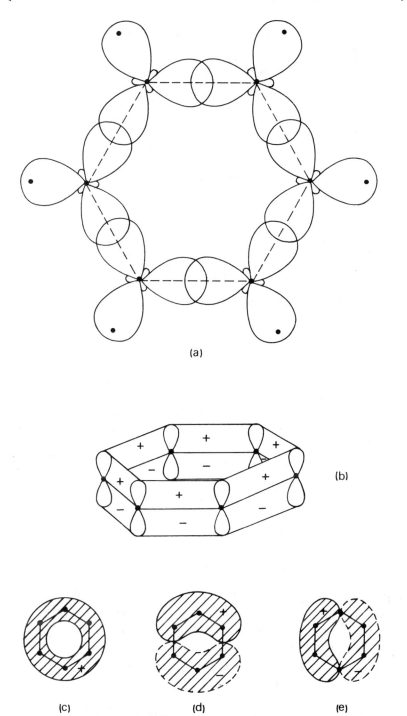

Fig. 2.41. Benzene, C_6H_6: (a) σ_{C-C} and σ_{C-H} bonds forming a regular, plane hexagon, with the $2p_z$ AO of each atom directed normally to the molecular plane; (b) double streamer π MO formed by the overlap of the $2p_z$ AOs; (c), (d), (e) the same double streamer and two degenerate π orbitals of next higher energy, respectively, as seen in a direction normal to the molecular plane.

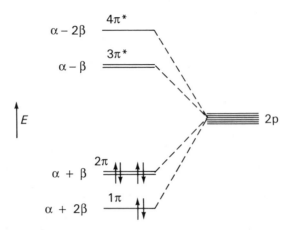

Fig. 2.42. HMO energy-level diagram for benzene; the MO of Fig. 2.41b,c is that of energy $\alpha + 2\beta$, and the two degenerate MOs for Fig. 2.41d,e are those of energy $\alpha + \beta$ each. The total energy E_π is $6\alpha + 8\beta$, and the delocalization energy D_π is 2β, or approximately -140 kJ mol^{-1}.

the deviation from neutrality, and is defined by

$$q_i = 1 - \sum_k \eta_k \, c_{ik}^2 \qquad (2.175)$$

where η_k is the number of electrons in the kth occupied MO Ψ_k, and c_i is the coefficient of ψ_i in Ψ_k; the sum is taken over the k occupied MOs. In butadiene, it is easy to show that the q values for all atoms are zero, as they are also for benzene. It is left as an exercise to the reader to show that, for benzene, the π bond order $p_{i,j}$ is $2/3$, the charge q_i is zero and the free-valence parameter $\mathscr{F}_i$ is 0.399, given the following final π molecular orbital wave functions:

$$\Psi_1 = 1/\sqrt{6}(\psi_1 + \psi_2 + \psi_3 + \psi_4 + \psi_5 + \psi_6)$$
$$\Psi_2 = 1/2(\psi_1 + \psi_2 - \psi_4 - \psi_5)$$
$$\Psi_3 = 1/\sqrt{12}(\psi_1 - \psi_2 - 2\psi_3 - \psi_4 + \psi_5 + 2\psi_6) \qquad (2.176)$$
$$\Psi_4 = 1/2(\psi_1 - \psi_2 + \psi_4 - \psi_5)$$
$$\Psi_5 = 1/\sqrt{12}(\psi_1 + \psi_2 - 2\psi_3 + \psi_4 + \psi_5 - 2\psi_6)$$
$$\Psi_6 = 1/\sqrt{6}(\psi_1 - \psi_2 + \psi_3 - \psi_4 + \psi_5 - \psi_6).$$

2.10.6.1   Aniline and pyridine
Two other interesting examples that show how the HMO method can help to explain reactivity are afforded by aniline and pyridine. Calculations show that these molecules have the following charge distributions:

Aniline reacts with bromine water to give 2,4,6-tribromoaniline. The active brominat-
ing agent in bromine water is the Br^+ species, which seeks preferentially the negative
2, 4 and 6 centres. In a similar way the nitration of pyridine, the active agent being the
nitronium ion NO_2^+, forms 3-nitropyridine. On the other hand, a nucleophilic reagent
seeks the 2 position:

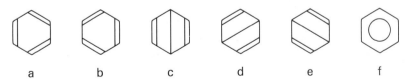

2.11   VALENCE-BOND APPROXIMATION

The molecular orbital technique that we have been discussing was preceded in time
by the valence-bond model for molecular structure. The valence-bond (VB) approach
is of importance in certain applications, and some of its concepts, such as 'resonance',
are useful. However, it has been overtaken by the developments in MO theory, which
has proved to be the more versatile technique for studying theoretically the structures
of molecules. In this section, we shall look briefly at the valence-bond model. Subsequently, we shall consider ways in which the VB and MO methods differ in their
approach.

2.11.1   Valence-bond model
The valence-bond model focuses its attention on the electron-pair bond, which was the
central feature of Lewis's earlier hypothesis of covalency. A VB wave function is made
up of all structures of a given molecule that display pairing of the electrons. The two
Kekulé structures (a) and (b) for benzene show two modes of electron pairing; three
more forms are given by the Dewar structures (c), (d) and (e), and ionic structures can
also be written; (f) is just a representation of benzene often used to stress the absence
of formal double bonds in this molecule, and not a possible electron-paired formulation.

In total, structures such as (a)–(e) are referred to as *canonical forms*, and the
complete wave function is obtained by their superposition, with the structures of
lower energy making the larger contributions. This combination of canonical forms is
referred to as *resonance*. It is important to note that, in this context, resonance refers
to the complete structure: it does not imply a process of alternation between the
several canonical forms. We shall pursue the VB model first with reference to the
hydrogen molecule.

2.11.2   Hydrogen molecule
Consider again the graph of Fig. 2.24. On the extreme right-hand side of the graph, at
large values of r, the separated atoms exist, and they retain their individual character.

As they are brought closer together, they enter the bond-forming region of the curve. We can imagine that they retain much of their individual character, but at bond-forming distances, the electrons may change positions in the AOs. A bond is formed by the overlap of the AOs, and we must write down a possible wave function to represent this situation.

In the case of the hydrogen molecule, there is only one canonical structure H—H, and we can write $\psi_1(1)\psi_2(2)$ to represent the canonical form where electrons 1 and 2 are on atoms 1 and 2, respectively, and $\psi_1(2)\psi_2(1)$ for the equally likely situation where the electron attachments are reversed. The superposition, or resonance, of these wave functions is given as

$$\Psi = c_1\psi_1(1)\psi_2(2) + c_2\psi_1(2)\psi_2(1) \qquad (2.177)$$

where $c_1^2 = c_2^2$, because the electrons are equivalent. Then we can write

$$\Psi_\pm = \psi_1(1)\psi_2(2) \pm \psi_1(2)\psi_2(1). \qquad (2.178)$$

Heitler and London (1927) used this wave function to obtain two energies E_+ and E_- for the hydrogen molecule. Their analysis gave r_e as 0.087 nm (experimental, 0.074 nm) and E as 3.14 eV (experimental, 4.74 eV). A situation similar to that given in Fig. 2.27 was obtained and, again, Ψ_+ is the bonding wave function.

We must consider also the spin factor, as well as the space factor, in order to explain the two signs in (2.178), and we will look at this matter in a little detail in this section, although we have implicitly used the result earlier.

2.11.2.1 Spin factor
The function Ψ_+ is symmetrical with respect to the interchange of electrons. Now the Pauli principle (from which the Pauli exclusion principle derives) requires the total wave function for electrons to be antisymmetrical under interchange of electrons. A function $f(x)$ is antisymmetrical if on replacing x by $-x$, $f(-x)$ changes to $-f(x)$. Its application to electrons is reasonable; electrons are indistinguishable, and so their interchange should not change the calculated properties of the system, such as electron density, or energy.

It follows that the antisymmetry of Ψ_+ must reside in the spin. Furthermore, the $+$ sign leads to a single energy state (singlet), whereas the $-$ sign leads to a triplet energy state (triplet), as explained through Table 2.7.

Table 2.7. Total VB wave function for H_2

Space terms	Spin terms (a)	Spin terms (b)
$\psi_1(1)\psi_2(2) + \psi_1(2)\psi_2(1)$	$\alpha(1)\beta(2) - \alpha(2)\beta(1)$	$\alpha(1)\alpha(2)$ $\beta(1)\beta(2)$ $\alpha(1)\beta(2) + \alpha(2)\beta(1)$
$\psi_1(1)\psi_2(2) - \psi_1(2)\psi_2(1)$	$\alpha(1)\alpha(2)$ $\beta(1)\beta(2)$ $\alpha(1)\beta(2) + \alpha(2)\beta(1)$	$\alpha(1)\beta(2) - \alpha(2)\beta(1)$

We can begin our analysis of spin by considering that electron 1 has spin α, and that electron 2 has spin β; the spin wave function will be designated $\alpha(1)\beta(2)$. However, we must also consider $\alpha(2)\beta(1)$, and since the two are degenerate, the linear combinations $\alpha(1)\beta(2) \pm \alpha(2)\beta(1)$ are permissible functions. The complete space and spin functions are summarized in Table 2.7.

If we interchange electrons 1 and 2 throughout the space and spin functions, we find that those involving spin terms (a) are multiplied by -1 (antisymmetric), whereas those involving spin terms (b) are unchanged (symmetric). Heisenberg showed (1926) that only antisymmetrical wave functions for electrons explained the spectra of the so-called *ortho*helium (triplet) and *para*helium (singlet) species of this element. Thus, we reject spin terms (b), and the bonding wave function becomes

$$\Psi_+ = [\psi_1(1)\psi_2(2) + \psi_1(2)\psi_2(1)][\alpha(1)\beta(2) - \alpha(2)\beta(1)]. \tag{2.179}$$

The spin term corresponds to a total spin of zero; the spins are paired. The same considerations of spin apply to the MO wave functions that we have considered, although we have delayed the discussion until this point.

To conclude our study of the VB model of hydrogen, we note that, implicitly, it does not permit contributions from ionic structures, yet there is a significant probability that both electrons can be found on any one of the two atoms. Hence the wave function can be improved by including $H^+ \, H^-$ and $H^- \, H^+$ with the canonical form. The ionic contribution may be written as

$$\psi_{ion} = \psi_1(1)\psi_1(2) + \psi_2(1)\psi_2(2) \tag{2.180}$$

to take account of the two possible situations. Then we have a bonding wave function Ψ_+ given by

$$\Psi_+ = \psi_{cov} + \lambda\psi_{ion} \tag{2.181}$$

where ψ_{cov} is given by (2.179). The value of λ that gives a minimum energy for the structures included here is about 0.25, which corresponds to an approximately 6% ionic contribution to the bond. In this context, we speak of ionic–covalent resonance, and in heteronuclear molecules, ionic terms become, not surprisingly, more significant.

2.11.3   Hydrogen fluoride

As an example of a heteronuclear species, we will consider again the molecule of hydrogen fluoride. The VB picture is given in terms of the overlap of the hydrogen 1s AO with the $2p_z$ AO of fluorine, with the molecular axis along z. We anticipate the importance of ionic contributions in hydrogen fluoride, and write the total structures as H—F, $H^+ \, F^-$ and $H^- \, F^+$. The bond-forming VB wave function can then be given as

$$\Psi_+ = \psi_{cov} + \lambda_1\psi_{ion_1} + \lambda_2\psi_{ion_2}. \tag{2.182}$$

Not surprisingly, calculation shows that λ_2, corresponding to $H^- \, F^+$, is negligible, because of the high value for the ionization energy of fluorine, and we have

$$\Psi_+ = \psi_{cov} + \lambda\psi_{ion}. \tag{2.183}$$

Table 2.8. Fractional ionic character in the hydrogen halides

	HF	HCl	HBr	HI
$10^{30}\mu/\text{C m}$	6.07	3.44	2.77	1.50
$10^{10}d/\text{m}$	0.927	1.27	1.41	1.61
q	0.41	0.17	0.12	0.058
q_χ	0.40	0.20	0.12	0.089
λ	0.83	0.45	0.37	0.25

λ is determined by minimizing the Rayleigh ratio (2.99), and (2.183) is a formulation of ionic–covalent resonance. We used earlier (2.138) a value of 0.41 for the partial charge in the hydrogen fluoride molecule. The partial charge is related to the fraction of ionic character in the bond, and we have the relationship

$$\lambda^2 = q/(1 + q) \qquad \text{or} \qquad q = \lambda^2/(1 + \lambda^2). \tag{2.184}$$

We summarize results for the hydrogen halides in Table 2.8.

2.11.4   Electronegativity

It is convenient to introduce the (relative) electronegativity χ at this stage. It is a measure of the power of an atom to attract electrons in compound formation: the greater the value of χ, the greater is the attraction of electrons. Pauling's scale of electronegativities (1932) was drawn up by relating the difference in electronegativity of two species $\Delta\chi_{AB}$, given by $|\chi_A - \chi_B|$, for the compound A–B to the excess of the bond energy of AB over the geometric mean of the bond energies for A_2 and B_2 in the diatomic species A_2 and B_2. The electronegativity is then a sort of resonance energy for the compound. Thus for hydrogen and fluorine, we have $E(\text{HF}) = 565$, $E(\text{H}_2) = 436$ and $E(\text{F}_2) = 155$, kJ mol^{-1}. Hence, $\Delta\chi_{\text{HF}} = 0.102 \{565/\text{kJ mol}^{-1} - \sqrt{(436/\text{kJ mol}^{-1} \times 155/\text{kJ mol}^{-1})}\}^{1/2} = 1.78$, and this is the difference between the Pauling electronegativities of hydrogen and fluorine. (Pauling's original proportionality factor was 0.208, because he used energies in kcal mol^{-1}; $0.208/\sqrt{4.184} = 0.102$.)

Mulliken (1934) introduced a so-called absolute value for χ by defining it as $(I + E)/2$; here I is the ionization energy and E the electron affinity of the atom *in its valence state*. The valence state of an atom is that state wherein it is part of a molecule, so detailed calculation is required to obtain the Mulliken values. The more recent Allred–Rochow scale sets χ equal to $(0.359\, Z_{\text{eff}}/r^2) + 0.744$, where r is the covalent radius in Å. These values are closer to the Pauling values than are those of Mulliken. Since electronegativity is at best a qualitative parameter, one scale seems adequate, and Table 2.9 lists the Pauling electronegativities for some species.

The relationship between electronegativity and ionic character has been characterized, and the best equation is that given by Hannay and Smith, and is most reliable for q less than about 0.5:

$$q = 0.16\,|\Delta\chi| + 0.035\,|\Delta\chi|^2. \tag{2.185}$$

The results of applying this equation to the hydrogen halides are the q_χ values in Table 2.8; it will be seen that the agreement is acceptable. A graph may be drawn to relate

Table 2.9. Electronegativities of some species

H						
2.20						
Li	Be	B	C	N	O	F
0.98	1.57	2.04	2.55	3.04	3.44	3.98
Na	Mg	Al	Si	P	S	Cl
0.93	1.31	1.61	1.90	2.19	2.58	3.16
K	Ca	Ga	Ge	As	Se	Br
0.82	1.00	1.81	2.01	2.18	2.55	2.96
Rb	Sr	In	Sn	Sb	Te	I
0.82	0.95	1.78	1.96	2.05	2.10	2.66
Cs	Ba	Tl	Pb	Bi		
0.79	0.89	2.33	2.02	2.02		

fractional ionic character q to the difference in electronegativities $|\Delta\chi|$, based on the data in Table 2.8, but it adds little more to (2.185).

2.12   MO AND VB MODELS COMPARED

The MO and VB models have in common that they seek to determine a structure by finding the minimum energy configuration for the combination of its component atoms. They both include the concept of overlap density, and its accumulation in the internuclear region being responsible for bonding (see Figs. 2.26a–d).

The valence-bond method leans strongly towards the chemical concept of the electron-pair bond. It is preferable to the MO model at internuclear distances greater than the bonding value. However, the number of canonical structures increases significantly with increase in the number of atoms in the molecule (over 1000 for naphthalene), and is one reason why the valence-bond model has recieved less development than has the molecular orbital method.

The molecular orbital technique is much closer to that developed for atoms, with molecular orbitals that encompass the whole molecule replacing the atomic orbitals of the component atoms. The formation of MOs through the LCAO procedure is relatively straightforward, and lends itself readily to computational methods.

Electron correlation, which is the tendency of electrons to repel one another, is not specifically included in either model. Its effect is over-emphasized by the VB treatment, while the MO description does not take it sufficiently into account. It appears explicitly in more detailed theoretical calculations, the discussion of which lies outside the scope of this book.[1]

[1] See for example, J. N. Murrell, S. F. A. Kettle and J. M. Tedder (1985) *The Chemical Bond* Wiley; R. L. DeKock and H. B. Gray (1980) *Chemical Structure and Bonding* Benjamin; C. A. Coulson (1979) revised by R. McWeeney, *Valence* OUP.

2.12.1 VB and MO models with homonuclear diatomics

We can compare the two models also by noting how they treat simple homonuclear diatomic molecules, and we shall look at some examples here.

Dihydrogen, H_2

VB: A quantum mechanical combination, or resonance hybrid, of the canonical form H—H, with H^+H^- and H^-H^+ is postulated, with equal weights for the two ionic forms so as to preserve apolarity in the molecule. Resonance leads to a single structure that is more stable than any one of the individual forms.

MO: Two spin-paired electrons; a $(1s\sigma)^2$ bonding MO; $\kappa = 1$.

Dihelium, He_2

VB: No electrons are available for pairing, so no canonical forms can be described. There is no resonance energy available to overcome the repulsion between the closed shells of the helium atoms, and no bond is formed.

MO: He_2 would be written $(1s\sigma)^2(1s\sigma^*)^2$ which is overall an antibonding configuration, and He_2 would not be formed; $\kappa = 0$.

Dilithium, Li_2

VB: Li—Li, Li^+Li^- and Li^-Li^+ are all appropriate forms for resonance. The 1s electrons are localized in AOs, and the 2s electrons form a bond.

MO: $(1s\sigma)^2(1s\sigma^*)^2(2s\sigma)^2$; the bond is $(2s\sigma)^2$; $\kappa = 1$.

Dinitrogen, N_2

VB: Three unpaired 2p electrons with parallel spins reside on each nitrogen atom. One σ and two π bonds are formed

MO: $(1s\sigma)^2(1s\sigma^*)^2(2s\sigma)^2(2s\sigma^*)^2(2p\pi)^4(2p\sigma)^2$; one σ and two π bonds (Fig. 2.30), the classical triple bond ($\kappa = 3$).

Dioxygen, O_2

VB: A simple double bond is precluded by the fact that oxygen is paramagnetic, an effect that depends on the presence of unpaired electrons. The VB model is given in terms of one σ bond and two 3-electron bonds.

MO: $(1s\sigma)^2(1s\sigma^*)^2(2s\sigma)^2(2s\sigma^*)^2(2p\sigma)^2(2p\pi)^4(2p\pi^*)^2$: since the electrons in the two $2p\pi^*$ MOs are unpaired (Fig. 2.31), the source of paramagnetism becomes clear; $\kappa = 2$.

Difluorine, F_2

VB: A single electron-pair bond F—F is formed. The ionic structures are not important because of the high ionization energy of fluorine: $F \rightarrow F^+$, 1681 kJ mol^{-1}; cf. Li, 520 kJ mol^{-1} and H, 1312 kJ mol^{-1}.

MO: $(1s\sigma)^2(1s\sigma^*)^2(2s\ )^2(2s\sigma^*)^2(2p\sigma)^2(2p\pi)^4(2p\pi^*)^4$ (Fig. 2.31, with two more electrons in the $2p\pi^*$ MOs); a stable molecule, with $\kappa = 1$.

In this short section, we have shown that the VB and MO models both give simple and satisfactory accounts of homonuclear diatomic molecules. Sometimes, as in the MO description of O_2, one picture is clearer than the other.

2.12.2   Electron density

Important distinctions between the VB and MO models arise in their expression of electron density. From the space part of (2.179) the VB electron density for the hydrogen molecule, given by the square of the wave function, is

$$\Psi^2_{+\,VB} = \{\psi_1(1)\psi_2(2) + \psi_1(2)\psi_2(1)\}^2$$

$$= \psi_1(1)^2\psi_2(2)^2 + \psi_1(2)^2\psi_2(1)^2 + 2\psi_1(1)\psi_2(2)\psi_1(2)\psi_2(1). \qquad (2.186)$$

We may compare this with the MO wave function, again for the *two electrons* in the hydrogen molecule, by expanding (2.127), taking $c_1 = c_2$:

$$\Psi^2_{+\,MO} = \{[\psi_1(1) + \psi_2(1)][\psi_1(2) + \psi_2(2)]\}^2$$

$$= \psi_1(1)^2\psi_1(2)^2 + \psi_1(1)^2\psi_2(2)^2 + \psi_2(1)^2\psi_1(2)^2 + \psi_2(1)^2\psi_2(2)^2$$

$$+ 4\psi_1(1)\psi_2(2)\psi_1(2)\psi_2(1)$$

$$+ 2\psi_1(1)^2\psi_1(2)\psi_2(2) + 2\psi_1(1)\psi_1(2)^2\psi_2(2) + 2\psi_1(1)\psi_2(2)^2\psi_2(1)$$

$$+ 2\psi_2(1)^2\psi_1(2)\psi_2(2). \qquad (2.187)$$

The significant differences are the presence in (2.187) of terms such as $\psi_1(1)^2\psi_1(2)^2$ and $\psi_1(1)^2\psi_1(2)\psi_2(2)$. They represent ionic contributions to the bond that are specifically excluded by the initial valence-bond formulation, but which needed to be introduced, as in (2.180) and (2.181), to obtain the best wave function for the molecule.

2.13   LIGAND-FIELD THEORY

Ligand-field theory is used to describe the structures of transition-metal compounds in which bonding takes place through d orbitals. The series of transition metals from scandium to zinc is characterized by the filling of the 3d orbitals from d^1 to d^{10} (see Table 2.5). The d atomic orbitals were described in section 2.6.1, and their wave functions listed in Table 2.4. Fig. 2.43 illustrates $|\psi|^2$ surfaces (AOs) for the d wave functions.

Ligand-field theory is essentially a molecular orbital theory in which the MOs are built from the atomic orbitals of ligands and the d orbitals of a central atom, which has a very symmetrical (often cubic) environment. A typical transition-metal com-pound (sometimes called transition-metal complex, or coordination compound) is hexamminecobalt(III) chloride, $[Co(NH_3)_6]Cl_3$. The compound has octahedral sym-metry (point-group symbol O_h, or $m3m$ in crystallographic notation), and the environment of the cation is shown in Fig. 2.44; we shall use it as an example of the theory.

From Figs 2.43 and 2.44, we can see that the d_{z^2} and $d_{x^2-y^2}$ orbitals are energetically less favourable for the approach of the lone-pair nitrogen atom in the ammonia molecule than are the other three d orbitals. The result is that the d orbitals of the metal atom are split into a group of two, the e_g orbitals, and a group of three, the t_{2g} orbitals. (This notation is relevant to the study of group theory[1], and we employ its conventional use here.) The difference between the e_g and t_{2g} levels is called Δ, or $10Dq$, the ligand-field-splitting energy parameter, which is also the HOMO–LUMO

---

[1]See, for example, F. A. Cotton (1990) *Chemical Applications of Group Theory*, Wiley; M. F. C. Ladd (1989) *Symmetry of Molecules and Crystals*, Horwood.

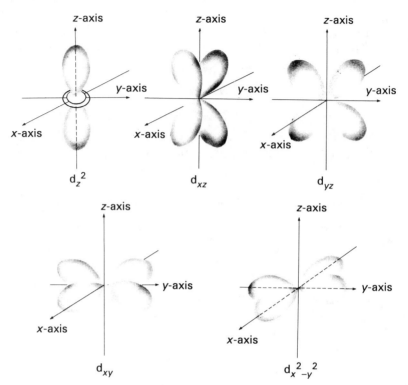

Fig. 2.43. Atomic d orbitals; surfaces of constant $|\psi|^2$. The notation is straightforward: the , lobes in d_{z^2} are directed along the z-axis; the lobes in $d_{x^2-y^2}$ are directed along the x- and the y-axes; and the lobes in d_{xz}, d_{yz} and d_{xy} lie in the corresponding planes.

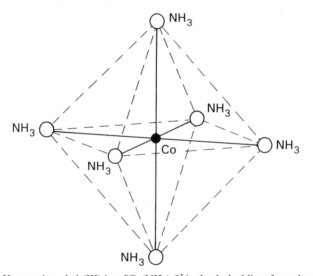

Fig. 2.44. Hexamminecobalt(III) ion, $[Co(NH_3)_6]^{3+}$; the dashed lines form the edges of an octahedron, symmetry O_h, or $m3m$. The d orbitals of Fig. 2.43 can be set into the same octahedral type of environment. These orbitals, and their linear combinations, are often referred to as *symmetry adapted* orbitals; they are related by, and make use of in forming wave functions, the symmetry of the environment.

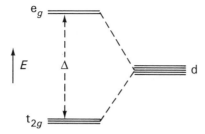

Fig. 2.45. MO energy-level diagram for the five degenerate d orbitals of a transition metal that are split under the influence of the field of the ligands into a group of three degenerate t_{2g} levels and a group of two degenerate e_g levels; the energy difference between t_{2g} and e_g is Δ (sometimes called 10 Dq).

energy difference. Fig. 2.45 illustrates the situation for a d orbital compound in the field of a ligand.

In our example compound, the approach of the six NH_3 ligands to the central cobalt ion is energetically favourable because of the opposite polarities of the ligands and the ion. The aufbau principle tells us that the t_{2g} MOs will be occupied preferentially by the six d electrons of cobalt, with paired spins, *provided* that the magnitude of Δ is large enough. The twelve electrons from the six lone pairs and the six d electrons of cobalt are fed into bonding MOs that are termed[1], in order of increasing energy, a_{1g}, t_{1u} and e_g. Six bonds are thus formed with the ammonia ligands, and the remaining six electrons occupy non-bonding t_{2g} MOs on the metal. Since Δ has been assumed to be large in this example, and the strength of the field of the ligand molecules not high, the t_{2g} non-bonding MOs are occupied by the six electrons in pairs. This fact is of significance, for it means that the hexammine-cobalt(III) ion will be diamagnetic; the configuration of cobalt in the compound is given as $(t_{2g})^6$, and this type of compound is often called a *strong-field* or *low-spin* complex. The t_{2g} level is $2\Delta/5$ in energy below the average energy, while the e_g level is $3\Delta/5$ above. Hence the t_{2g} level has an energy -0.4Δ, and the e_g level an energy $+0.6\Delta$. The net energy of a $t_{2g}^p e_g^q$ configuration, relative to the average energy is then $(0.6q - 0.4p)\Delta$, and is termed the *ligand-field stabilization energy* (LFSE).

The magnitude of the energy gap Δ is very important. In a weak-field situation, with a ligand such as Br^- or SCN^-, the LFSE parameter may be only about one-fifth of the above value. Then it becomes commensurate with the value of the repulsion energy between paired electrons. Thus, Hund's rule would indicate that the config-uration with orbitals singly occupied with electrons having parallel spins is then to be preferred. Thus, for the $[Co(SCN)_6]^{3-}$ ion, the six non-bonding electrons will enter the t_{2g} and e_{2g} orbitals, with one t_{2g} orbital fully occupied and four with unpaired electrons. Thus, this complex would be a high-spin compound, and strongly paramag-netic. The MOs that we have described are summarized in Fig. 2.46.

Table 2.10 indicates the relationship between numbers of unpaired electrons and the theoretical and observed magnetic moments for a selection of transition-metal species. The total magnetic moment may arise from either electron spin or a combination of spin and orbital motion.

---

[1]Again, a conventional notation; see, for example, F. A. Cotton and G. Wilkinson (1988) *Comprehensive Inorganic Chemistry*, Wiley

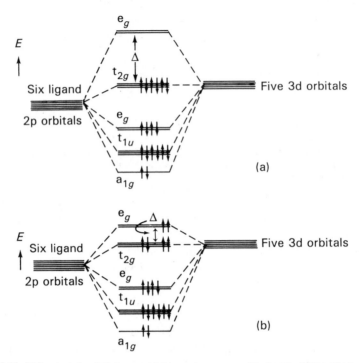

Fig. 2.46. MO energy-level diagrams. (a) Low-spin case, as with the $[Co(NH_3)_6]^{3+}$ ion; the bonding orbitals a_{1g}, t_{1u} and e_g are fully occupied by paired electrons, as are the non-bonding t_{2g} MOs, and the species is $(t_{2g})^6$, diamagnetic. (b) High-spin case: the splitting energy parameter Δ is commensurate with the repulsion energy of paired electrons, and Hund's rule leads to singly occupied non-bonding e_g MOs; the species is $(t_{2g})^4(e_g)^2$, and has paramagnetic properties. In both cases, other non-bonding MOs exist, but are unoccupied and of little interest for the ground state configurations.

Table 2.10   Theoretical and experimental high-spin magnetic moments

Species	Number of unpaired e⁻	μ_B (spin)[a]	μ_B(spin + orbital)	μ_B(exprt)
V(IV)	1	1.73	3.00	1.7–1.8
Cu(II)	1	1.73	3.00	1.7–2.2
V(III)	2	2.83	4.47	2.6–2.8
Ni(II)	2	2.83	4.47	2.8–4.0
Cr(III)	3	3.87	5.20	~ 3.8
Co(II)	3	3.87	5.20	4.1–5.2
Fe(II)	4	4.90	5.48	5.1–5.5
Co(III)	4	4.90	5.48	~ 5.4
Mn(II)[b]	5	5.92	5.92	~ 5.9
Fe(III)	5	5.92	5.92	~ 5.9

[a]The μ_B values are in Bohr magneton units ($1\mu_B = eh/4\pi m_e = 9.274 \times 10^{-24}$ J T⁻¹).
[b]The Mn(II) and Fe(III) species have no orbital component of magnetic moment.

Another common type of complex is tetrahedral, as in the $[Zn(NH_3)_4]^{2+}$ ion (symmetry T_d, or $\overline{4}3m$), for example. We shall consider this type briefly by means of a problem, and the reader is directed to more detailed works for further discussion of coordination compounds.[1]

2.14   APPARENTLY ABNORMAL VALENCE

Certain classes of compounds appear to have abnormal valences (they are not really abnormal, otherwise they would not occur) in some of their compounds. For example, silicon in the $[SiF_6]^{2-}$ ion appears to have an 'expanded octet' of twelve valence electrons. In other compounds, such as the boron hydrides, there appear to be insufficient valence electrons for the number of bonds formed. These topics are neatly explained by molecular orbital theory.

Silicon has the electronic configuration $(Ar)(3s)^2(3p)^2$, but the 3d orbitals are not energetically unavailable. Hence, this atom may make use of 3d orbitals in bond formation. However, the involvement of d orbitals is not essential. There are sufficient AOs from the Si and the six F atoms to form six occupied, bonding MOs, and the ability to pack the requisite number of atoms around the central atom, without undue repulsion effects, may also be significant.

In the second class of these compounds, we can consider the so-called electron deficient compound B_2H_6. The eight atoms provide a total of 14 valence shell molecular orbitals, four from each boron, excluding the use of $(1s)^2$, and one from each hydrogen atom. There can then be seven bonding and seven antibonding MOs. The twelve valence electrons then fill the six lower bonding MOs in accordance with the aufbau principle, which results in a bonding situation. In the actual structure, there are two BHB MOs that give rise to two 3-centre, 2-electron bridging bonds. In the three-centre two-electron bridging system, a bonding orbital encompasses three atoms, BHB, and the bridge utilizes two electrons in binding two pairs of atoms, BH and HB.

```
    H       H       H
     \     / \     /
      B       B
     /     \ /     \
    H       H       H
```

The MO theory provides neat explanations for compounds such as these, which were insuperable obstacles to the Lewis electron-pair hypothesis, and not easily explained by the valence-bond model.

2.15   VALENCE-SHELL ELECTRON-PAIR REPULSION THEORY

This model, generally known as VSEPR theory, predicts the shapes of simple polyatomic molecules in terms of repulsion between pairs of electrons, particularly lone pairs, on the component atoms. We recall that the term 'lone pair' refers to electrons not engaged in bonding. The water molecule (see section 2.9.1) has two such lone pairs.

---

[1] See, for example, D. F. Shriver, P. W. Atkins and C. H. Langford (1990) *Inorganic Chemistry*, OUP; F. A. Cotton and G. Wilkinson (1988) *Comprehensive Inorganic Chemistry*, Wiley; F. Basolo and R. Johnson (1987) *Coordination Chemistry*, Science Reviews, Northwood.

The VSEPR model is based on the Lewis bonding-pair/lone-pair model, and its basic tenet is that a molecule adopts that shape which minimizes the repulsions between pairs of electrons, that is, it places the electron pairs as far apart as possible. For any molecule of the type MX_n each electron pair may be represented by a point on a sphere, centre M, and the positions of the n X species that minimize the repulsive forces give the structure of MX_n. The basic arrangements are as follow:

Number of electron pairs	Configuration
2	Linear
3	Trigonal planar
4	Tetrahedral
5	Trigonal bipyramidal
6	Octahedral

Fig. 2.47 shows the tetrahedral arrangement for four electron pairs. An alternative regular arrangement would be a square, but the electron pairs are then closer together.

The model may be given an added realism by allowing each electron pair to take up the space of a sphere, to represent a fully occupied atomic orbital. This refinement has the advantage of showing regions, or domains, of space that may not be overlapped by other similar domains. Whereas the point-charge model can only adopt the order for repulsion of pairs:

lone pair/lone pair $>$ lone pair/bonding pair $>$ bonding pair/bonding pair,

the domain model can incorporate size factors for the domains.

The general formula for molecules may be recast as MX_nL_m, where now L_m refers to m lone pairs. We may note that:

(a) a bonding domain involves the M and X valence electrons, and both are attracted to the nuclei, whereas the non-bonding L domain belongs only to M;
(b) classical double and triple bonds involved two and three shared pairs respectively, and the domain sizes increase in the order, single bond $<$ double bond $<$ triple bond;
(c) the electronegativities determine the extent to which electrons may be transferred from M to X.

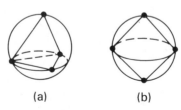

(a)   (b)

Fig. 2.47. VSEPR point-charge model for a species MX_4: (a) the preferred tetrahedral arrangement; (b) square planar arrangement. The distances apart of adjacent points in the tetrahedral and square modes are in the ratio of $r\sqrt{(8/3)}:r\sqrt{2} = 1.15:1$, where r is the radius of the sphere.

The lone pair domain L is a larger sphere than a bonding domain X and, because it belongs to M, it will tend to be closer to M than are the X species. If we consider NH_3 (MX_3L) and H_2O (MX_2L_2), the total of four domains should lead formally to a tetrahedral arrangement. However, if the L domain is larger than that for X, and more strongly attracted to M, it follows that the X–M–X angle must be less than the tetrahedral value. In fact for NH_3 it is 107°, and for H_2O, 104.4°, as we have discussed already.

If we compare SF_2 and SCl_2, both of the type MX_2L_2, we see that they are like the water molecule. Hence, they will be again distorted tetrahedral arrangements of M, X and L. Furthermore, because fluorine is more electronegative than chlorine, it will draw relatively more charge from the sulphur atom, and this in turn leads to a smaller bond angle in SF_2 (SF_2, 98°; SCl_2, 102°). On the point-charge VSEPR model, we would say that the type MX_2 is formally linear, but that the lone pairs on M would tend to repel the X species and so produce a bent molecule. The result is similar, except that it is not so easy to comment on the relative sizes of the X–M–X angle.

Sulphur dioxide is a species with multiple bonds and the formal structure shown below. Thus, it may be classed as an MX_2L species; larger domains are assumed for multiple bonds, and lead to a bond angle larger than the 120° of the trigonal planar

arrangement. In the case of SO_3, the symmetry is restored, and the trigonal planar shape is obtained. In carbon dioxide, although the multiple bonds produce larger bonding domains, there are no lone pairs to disturb the symmetry of the linear arrangement for this molecule that corresponds to minimum repulsion.

VSEPR is a simple method for predicting the shapes of small molecules containing a central atom, and is particularly applicable to main-group elements. It is not so successful with transition-metal compounds because the central atom domain does not retain the implicitly spherical shape that we have used so far. However, the d^0, d^5 and d^{10} configurations do respond to the simple treatment, and $TiCl_4$ (d^0) is tetrahedral, $[CoF_6]^{2-}$ (d^5) is octahedral, and $[Ag(NH_3)_2]^+$ is linear. The subject of VSEPR is treated in most good discussions on inorganic chemistry.[1]

2.16   COVALENT STATES OF MATTER

In this chapter, we have studied the structures of molecules in which the covalent bond was the predominant form of bonding. We encounter chemical species normally in one or other of the three states of matter. In gases and liquids, although the molecules are formed through covalent forces, the binding of these molecules arises through weaker intermolecular bonds. These states of matter will be treated in the next chapter, after a study of intermolecular forces, and we include here some of the solids that may be termed covalent.

[1]See, for example, R. J. Gillespie (March, 1992) *Chemical Society Reviews*; (Chemical Society, London) D. F. Shriver, P. W. Atkins and C. H. Langford (1990) *Inorganic Chemistry*, OUP.

2.16.1   Covalent solids

Covalent solids exhibit a three-dimensional disposition of covalent bonds of the nature that we have discussed already. Thus, there are very few solids that fall into this class. The best example is the diamond form of carbon, which we have illustrated in Fig. 1.2. In this structure, each carbon atom is bonded to four others through sp^3 orbitals, in a characteristic regular tetrahedral configuration. Other elements of group 14 of the periodic table exhibit the diamond structure type. Tin is dimorphous in having also a metallic structure type (see Fig. 5.17a). As we consider the elements of this periodic group, we can trace a continuous change in bond type from covalent carbon to metallic lead. This change is paralleled, for example, by a gradation in the electrical resistivity ρ_T of the elements at a given temperature T.

	C(diamond)	Si	Ge	Sn(grey)	Sn(white)	Pb
ρ_T/ohm m	5×10^{12}	2×10^3	5×10^{-1}	1×10^{-5}	1×10^{-7}	2×10^{-7}

Other solids that are essentially covalent include silicon carbide (SiC), boron nitride $(BN)_n$, α-quartz (SiO_2, Fig. 1.10a) and β-quartz (Fig. 2.48).

The two common forms of zinc sulphide (Würtzite, α-ZnS and Blende, β-ZnS), although tetrahedral in structure, contain a high degree of ionic character, and are considered under that heading. A dilemma will always arise in the classification of compounds: what has been attempted in this book is just one way of dealing with the problem.

2.16.2   Structural and physical properties of the covalent bond

Covalent bonds exist between an atom and, generally, a small number of neighbours. The bond is strongly directional, and we have discussed how this feature arises.

Covalent solids form strong, hard crystals, with low compressibility and expansivity, but high melting-point temperatures. In these properties, they are similar to ionic crystals but, in strong contrast, they are electrical insulators in both the solid and liquid states. Covalent solids are chemically unreactive, and insoluble in all usual solvents.

We omit from this discussion the numerous compounds of organic chemistry. While the molecules themselves are bound by covalent forces, the linkage of the molecules is through weaker intermolecular forces, and so organic solids and fluids will be considered in that context.

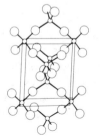 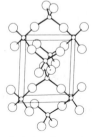

Fig. 2.48. Stereoview of the unit cell of β-quartz; circles in increasing order of size are Si and O. The tetrahedral SiO_4 structural unit is present in this structure, as well as in the α polymorph, and in silica glass (see Fig. 1.10).

2.17   SEMI-EMPIRICAL AND *AB INITIO* MOLECULAR ORBITAL THEORIES

We conclude this chapter with a very brief mention of two higher-level approximations in molecular orbital theory.

In sections 2.7 and 2.8, we examined the LCAO approximation for diatomic molecules, and used the minimization of the energy to give the optimum values for the coefficients of the molecular wave functions. Problems remain in obtaining satisfactory values for the Coulomb and resonance integrals. One method, to which we have already alluded, employs values of the ionization energies of the atoms. For example, if we write a molecular wave function for a heteronuclear diatomic molecule AB as

$$\Psi = c_A \psi(A) + c_B \psi(B) \tag{2.188}$$

and if we assume negligible overlap, then S_{AB} may be equated to zero and, with $S_{AA} = S_{BB} = 1$, the secular determinant may be written as

$$\begin{vmatrix} \alpha_A - E & \beta \\ \beta & \alpha_A - E \end{vmatrix} = 0 \tag{2.189}$$

This determinant is readily expanded to the quadratic equation

$$(\alpha_A - E)(\alpha_B - E) - \beta^2 = 0 \tag{2.190}$$

for which the solutions are

$$E = \{(\alpha_A + \alpha_B) \pm (\alpha_A - \alpha_B)(1 + [2\beta/(\alpha_A - \alpha_B)]^2)^{1/2}\}/2 \tag{2.191}$$

where the Coulomb integrals may be equated approximately to the negatives of the ionization energies. We will apply this procedure to the hydrogen chloride molecule, so as to find the energy of the σ H—Cl bond. We write the wave function as

$$\Psi = c_H \psi(H) + c_{Cl} \psi(Cl) \tag{2.192}$$

We take $\alpha_H = -I_H = -13.60$ eV and $\alpha_{Cl} = -I_{Cl} = -13.01$ eV; a value for β is approximately -2.0 eV. Using (2.191), the values for E_σ are -15.33 eV and -11.29 eV. From (2.120) and the fact that $c_H^2 + c_{Cl}^2 = 1$, we obtain the wave funxctions

$$\Psi_+ = 0.76\psi(H) + 0.65\psi(Cl) \tag{2.193}$$
$$\Psi_- = 0.65\psi(H) - 0.76\psi(Cl) \tag{2.194}$$

the former, bonding orbital corresponding to the energy -15.33 eV. It may be noted that the bonding orbital is richer in the s function than in the p, but the converse is true for the antibonding orbital.

This type of procedure that utilizes physical data in the quantum mechanical equations is included in the term semi-empirical MO theory, and further details of it are addressed by specialized texts.[1] The reader might like to carry out the above procedure for hydrogen fluoride, with $\alpha_F = -17.42$ eV, keeping β at -2.0 eV, and compare the amounts of s and p wave functions between the two diatomic molecules.

In *ab initio* molecular orbital theory, which is the subject of much current research and practice, both the Coulomb and resonance integrals are calculated from first principles. This high-level approximation leads to results of much greater precision,

---

[1] See, for example, J. N. Murrell and A. J. Harget (1972) *Semi-empirical Self-consistent Molecular Orbital Theory of Molecules*, Wiley.

but the calculations are lengthy and require considerable computational time. The expression of this method lies outside the remit of the present book, and the reader is referred to specialized texts on this subject.[1]

PROBLEMS FOR CHAPTER 2

2.1 Construct Lewis diagrams for (a) ethyl ethanoate, (b) methylamine.

2.2 Calculate the energy and momentum of an electron travelling at $1/10$ of the speed of light *in vacuo*, if the energy is wholly kinetic.

2.3 At what wavelength is the energy density for a black-body radiator a maximum at 500 K?

2.4 Determine the energy density of light in the wavelength range 780 nm to 800 nm in a black-body cavity at 1500°C. Is it important to use the Planck equation rather than that of Rayleigh–Jeans for this calculation?

2.5 Millikan exposed a freshly cut surface of sodium metal, in a vacuum, to monochromatic radiation from a quartz-mercury arc source. The photoelectrons emitted were collected in an oxidized copper Faraday cylinder. The current obtained for different values of an applied potential difference V was measured with an electrometer, and in different experiments differing wavelengths λ were employed.

The deflexion θ recorded by the electrometer was proportional to the photoelectric current, and θ increased as V was made more negative at the metal. A field was set up in the space between the dissimilar sodium and oxidized copper metals, and a contact potential difference existed between them; in Millikan's experiments, it acted from sodium to copper, that is, sodium was positive with respect to the Faraday cylinder. The following results were obtained for three wavelengths:

$\lambda = 546.1$ nm		$\lambda = 365.0$ nm		$\lambda = 312.6$ nm	
V/V	θ/deg	V/V	θ/deg	V/V	θ/deg
−2.257	28	−1.157	67.5	−0.5812	52
−2.205	14	−1.105	36	−0.5288	29
−2.152	7	−1.0525	19	−0.4765	12
−2.100	3	−1.0002	11	−0.4242	5.7
		−0.9478	4	−0.3718	2.5

(a) For each wavelength, plot θ against the independent variable V, and estimate the *minimum* applied voltage V_0 that prevents the fastest moving photoelectrons from reaching the Faraday cylinder.

(b) An electron moving through a potential difference V acquires an energy of eV eV. This energy is equivalent to the kinetic energy $\frac{1}{2} m_e v^2$, where m_e and v are, respectively, the mass and speed of the electron. Show that the equation $V_0 e = kv - \phi$ represents the variation of V_0 with frequency v. Find the value of the constant k in this context. What is k?

2.6 X-rays of wavelength 100 pm, incident upon a material, are scattered with a Compton angle of 45°. What is the wavelength of the scattered radiation?

2.7 In a certain electron diffraction experiment, electrons of a wavelength 0.5 nm were found to be desirable. What would be the velocity of such electrons?

[1] See, for example, R. McWeeny (1979) *Coulson's Valence*, OUP; J. A. Pople and D. A. Beveridge (1970) *Approximate Molecular Orbital Theory*, McGraw-Hill; J. Almlöf, K. Faegri Jr and K. Korsell (1982) *J. Computational Chemistry*, **3**, 385.

2.8 The Balmer series in the spectrum of atomic hydrogen was analysed originally in terms of wavelength λ, through the equation $\lambda = K[n^2/(n^2 - 4)]$, where K is a constant and n is an integer greater than 4. Show that this equation is equivalent to (2.29), and find K in terms of R_H. What is the energy associated with the spectral line in the Balmer series, nearest to the red end of the spectrum?

2.9 A proton and an electron, taken as point charges, are held at a distance from each other not greater than 10^{-15} m by Coulombic forces. Is this system feasible in the light of the uncertainty principle?

2.10 Consider a 'particle in a box' with a box length of 20 nm. Calculate the probability that the particle will lie between (a) 5 nm and 15 nm, (b) 9 nm and 11 nm for (i) the ground state and (ii) the first harmonic state of the function.

2.11 What is the smallest value of the kinetic energy for an electron in a cubical box of side 10^{-15} m? Give the energy in both J and Hartree energy E_H.

2.12 Calculate the probability of finding a hydrogen 1s electron in the volume bounded by $r = 1.10a_0$ to $1.11a_0$, $\theta = 0.20\pi$ to 0.21π, $\phi = 0.60\pi$ to 0.61π. The $\psi(1s)$ wave function may be assumed to be constant over the small volume considered. Use the value of the wave function given in Table 2.4.

2.13 Write the electron configurations for (a) N, (b) Al, (c) Cl^-,(d) K.

2.14 Calculate the ground state energy of the species He^+.

2.15 Set up the Schrödinger equation for the helium atom as fully as possible. (It is not recommended that a solution of the equation be attempted.)

2.16 By using the orthogonality criterion determine whether or no and under what restraints, if any, the linear combination of hydrogen-like atomic orbitals (a) 1s + 2s, (b) 1s + 2p will lead to bonding molecular orbitals.

2.17 The overlap integral for two hydrogen-like 1s AOs is given by $S_1 = \exp(-\rho)$ $\{1 + \rho + \rho^2/3\}$ and between 1s and 2p it is $S_2 = \rho\exp(-\rho)\{1 + \rho + \rho^2/3\}$, where ρ here is r/a_0, r being the internuclear distance. Plot both S_1 and S_2 for $r = 0$ to $3a_0$. Determine the value of the overlap integral for (a) H_2^+ ($r_e = 0.11$ nm), (b) H_2 ($r_e = 0.074$ nm), (c) HF ($r_e = 0.093$ nm). (d) At what value of r is the 1s, 2p overlap a maximum? Confirm this result by differentiation of the appropriate overlap integral function.

2.18 What are the ground state electronic configurations for (a) Be_2, (b) C_2? Which of these species is likely to be more stable than the corresponding separate atoms?

2.19 Draw molecular orbital energy level diagrams for (a) NO, (b) CN, and give the ground state electronic configuration and the net bonding parameter for each species. Which of the species is likely to be more stable as either a singly negative or singly positive ion? Give reasons.

2.20 The dipole moment of the water molecule is 1.8 D. Given that O—H = 0.096 nm and H—O—H = 104.4°, calculate the partial charges on oxygen and hydrogen.

2.21 Show that sp^2 hybridization of normalized AOs leads to σ bond angles of 120°.

2.22 Show that an sp^x hybrid from normalized AOs has a normalized wave function given by

$$\psi(sp^x) = (1 + x)^{-1/2}\{\psi(s) + x^{1/2}\psi(p)\}.$$

2.23 (a) Use the result from problem 2.22 to show that an sp^3 hybrid orbital has a normalized wave function given by

$$\psi(sp^3) = \tfrac{1}{2}\{\psi(s) + \sqrt{3}\psi(p)\}.$$

(b) Refer to Fig. 2.36. For the corner 1, 1, 1, the p orbital may be resolved into components p_x, p_y and p_z. Show that

$$\psi(sp^3) = \tfrac{1}{2}\{s + p_x + p_y + p_z\}$$

using s to represent the $\psi(s)$ AO, and so on. Formulate the other three sp^3 wave functions in a similar manner, and show that they are mutually orthogonal.

2.24 Determine and solve the secular determinant for cyclobutadiene; remember that atoms 1 and 4 are now adjacent. Determine the energies of the first four MOs and hence find the delocalization energy for cyclobutadiene. Which orbitals constitute the frontier MOs in this molecule?

2.25 The polymethene dye

may be treated as the 'box' $-\ddot{N}-C{=}C-C{=}C-C{=}\overset{+}{N}-$

where the mean bond length is 140 pm. The box contains $2N + 2$ π electrons (two from each double bond and two from the neutral nitrogen atom), where N is the number of double bonds; these electrons occupy the first $N + 1$ MOs. The colour of the dye arises from the transition of an electron between the $N + 1$ and $N + 2$ orbitals. Show that the wavelength of the transition is given by

$$\lambda/m = 3.297 + 10^{12}a^2/(2N + 3)$$

where a is the length of the box. Calculate λ, and state the colour of the dye with respect to white light.

2.26 Write down and comment on the VB and MO descriptions for Ne_2 and LiH.

2.27 Construct VB wave functions for the HF molecule, assuming the use of H(1s) and $F(2p_z)$ normalized AOs, as (a) wholly covalent, (b) wholly ionic, (c) a resonance hybrid of types (a) and (b).

2.28 Use the HMO approximation to obtain the complete wavefunctions, MO energies, delocalization energy, bond orders, free-valence parameters and charge distributions for bicyclobutadiene:

2.29 Estimate the fractional ionic character in the VB wave function for the hydrogen sulphide molecule; its dipole moment is 1.1 D, the S—H bond length 0.134 nm and the H—S—H bond angle 92°.

2.30 For the transition-metal ions $(3d)^1$ to $(3d)^9$ write down the d electron configuration and number of unpaired electrons in (a) a weak ligand field, (b) a strong ligand field.

2.31 $[Zn(NH_3)_4]^{2+}$ has regular tetrahedral symmetry (point group T_d, or $\overline{4}3m$). The d orbitals are split into e_g and t_{2g} as before; for similar central ion and ligands, Δ (tetrahedral) is approximately $\tfrac{1}{2}\Delta$ (octahedral). By considering how a regular tetrahedron is related in symmetry to an octahedron (both can be placed inside a cube so that corresponding symmetry axes coincide), and with the aid of Fig. 2.43, show what d orbital splitting is to be expected for tetrahedral complexes. Would the zinc compound be diamagnetic or paramagnetic?

2.32 Use VSEPR theory to predict probable structures for (a) F_2O, (b) NF_3, (c) $SiCl_4$, (d) SF_4.

3

Bonding between molecules

This chapter is divided into three sections, gases, liquids and solids, they being the states of matter encompassed by the plan of this book. We have made some introductory remarks on each of these topics in the first chapter, and we now develop them further.

3.1 GASES

The gaseous state is the most straightforward to study. The simplicity arises from the fact that the molecules of a gas are, normally, independent of one another, with negligible forces of interaction. We referred to the ideal gas in Chapter 1, and to the equation of state (1.7) that relates the p, V and T variables of a gas in bulk. In this section, we shall consider some aspects of the kinetic theory of gases and properties that derive from it. We shall see how the approximations in the kinetic theory are revealed when studying real gases, and we shall consider the nature of the interactions between species in the gas phase.

3.1.1 Kinetic theory of gases
The kinetic theory of gases is based on three main postulates:
  (a) a gas consists of a large number of molecules (or atoms) in continuous random thermal motion, that is, motion related to the translational (kinetic) energy of the molecules at a given temperature, and which is independent of their orientation—an isotropy of motion;
  (b) the gas molecules interact only by elastic collisions, that is, where the total kinetic energy of the colliding species is conserved;
  (c) the gas molecules are of negligible size, that is, their diameters are vanishingly small in relation to the distances of travel of molecules between collisions.
From these assumptions, a number of important properties of gases can be derived.

3.1.1.1 Pressure exerted by a gas
Consider a single molecule of mass m in a rectangular box of sides a, b and c, and let its velocity v be resolvable into components v_x, v_y and v_z parallel to a, b and c,

respectively. Let the molecule travel in the $+x$-direction and make an elastic collision with the wall bc. The momentum of the molecule in the x-direction is changed from mv_x to $-mv_x$, so that the total momentum transferred to the wall is $2m|v_x|$, because the momentum of the wall–molecule system is conserved. The molecule will rebound and strike the opposite wall bc', and so on. Collisions with the wall bc will occur every $2a/|v_x|$ seconds. Hence, the rate of change of momentum is $2m|v_x|/2a/|v_x|$, or mv_x^2/a.

From Newton's second law, the rate of change of momentum is the force on the wall bc and, since pressure is force per unit area, we have the pressure on the wall bc, due to the molecule, given by

$$p_x = \frac{mv_x^2/a}{bc} = \frac{mv_x^2}{V} \tag{3.1}$$

where V is the volume of the rectangular box. Similar expressions can be generated for p_y and p_z, and they will not, in general, be equal.

If we now increase the number of molecules in the box to N then, because it is a postulate that they do not interact, their contribution to the pressure will be additive. Hence, for the wall bc, the total pressure p will be given by the sum over all N molecules:

$$p = \frac{m}{V} \sum_{i=1}^{N} v_{x_i}^2 \tag{3.2}$$

and similarly for the other walls of the box. The mean square velocity $\overline{v_x^2}$ is given by

$$\overline{v_x^2} = \frac{1}{N} \sum_{i=1}^{N} v_{x_i}^2. \tag{3.3}$$

Hence

$$p = N \, m \, \overline{v_x^2}/V. \tag{3.4}$$

When N is large, the impacts with the wall become a steady isotropic pressure. Since no direction is preferred, it follows that $\overline{v_x^2} = \overline{v_y^2} = \overline{v_z^2}$, and the mean square speed $\overline{v^2}$ is then

$$\overline{v^2} = \overline{v_x^2} + \overline{v_y^2} + \overline{v_z^2} \tag{3.5}$$

so that (3.4) becomes

$$pV = \tfrac{1}{3} N \, m \, \overline{v^2}. \tag{3.6}$$

This equation assumes implicitly that there is no attraction between the molecules and the wall of their containing vessel, which we need to justify.

Let there be $\mathcal{N}_x$ molecules of a perfect gas per unit volume at a distance x from a wall, and let it be assumed that the molecules are attracted by the wall. If U_x is the attractive energy at a distance x from the wall, the force of attraction by the wall F_x on the molecule is given by

$$F_x = -\frac{dU_x}{dx} \tag{3.7}$$

and that on the wall is, by Newton's laws, dU_x/dx. Then the attractive pressure p_a on the wall is given by

$$p_a = \int_0^\infty \mathscr{N}_x \, (dU_x/dx) \, dx \qquad (3.8)$$

where the limit of ∞ indicates a position in the gas removed from any effect of the wall. If we assume that the $\mathscr{N}_x$ molecule density follows the Boltzmann distribution (1.4), we may write

$$\mathscr{N}_x = \mathscr{N}_0 \exp(-U_x/k_B T). \qquad (3.9)$$

We have from (3.9)

$$\frac{d\mathscr{N}_x}{dx} = \mathscr{N}_0 \exp(-U_x/k_B T)(-1/k_B T)\frac{dU_x}{dx} = \frac{-\mathscr{N}_x}{k_B T}\frac{dU_x}{dx} \qquad (3.10)$$

whence, with (3.8)

$$p_a = -k_B T \int_0^\infty (d\mathscr{N}_x/dx) \, dx = -k_B T(\mathscr{N}_\infty - \mathscr{N}_0) \qquad (3.11)$$

In (3.11), $\mathscr{N}_0$ is the number density at the wall, which relates to the pressure at the wall p_0, given by $\mathscr{N}_0 k_B T$, whereas $\mathscr{N}_\infty$ refers to molecules in the bulk, which would identify with $\mathscr{N}_w$, the number density at the wall in the absence of attraction, with a corresponding pressure p_w given by $\mathscr{N}_w k_B T$. The actual pressure p due to collision at the wall is the difference between the ideal value p_0 and that of attraction p_a, that is,

$$p = p_0 - p_a = \mathscr{N}_0 k_B T + k_B T(\mathscr{N}_\infty - \mathscr{N}_0) = \mathscr{N}_\infty k_B T = \mathscr{N}_w k_B T = p_w. \qquad (3.12)$$

Thus, the actual pressure, taking attraction into account, identifies with the pressure assuming no attraction. This result applies also to imperfect gases.

3.1.1.2 *Equipartition*
The classical distribution of energy in a system of molecules was considered in section 1.3.1. Since the average translational kinetic energy $\bar{\varepsilon}$ is equal to $\frac{1}{2} m\overline{v^2}$, it follows from (1.7) and (3.6) that

$$pV = \tfrac{2}{3} N \bar{\varepsilon} = n\mathscr{R}T. \qquad (3.13)$$

Since, from section 1.3.2, $Lk_B = \mathscr{R}$, and with $nL = N$, it follows that

$$\bar{\varepsilon} = \tfrac{3}{2} k_B T \qquad (3.14)$$

or $\frac{3}{2}\mathscr{R}T$ per mole. In view of (3.5), we see that each degree of translational freedom of a molecule has an average energy $\frac{1}{2} k_B T$ associated with it (see also Appendix 16).

3.1.1.3 *Mixtures of ideal gases*
Let a gas contain N_i mole of gases i ($i = 1, 2, 3, \dots$). At equilibrium, we have for the energies

$$\bar{\varepsilon}_1 = \bar{\varepsilon}_2 = \bar{\varepsilon}_3 = \cdots = \bar{\varepsilon} \qquad (3.15)$$

and

$$N = \sum_i N_i. \tag{3.16}$$

Each gaseous component follows the equation of state (1.7), and from (3.13)

$$p_i V = \tfrac{2}{3} N_i \, \bar{\varepsilon}. \tag{3.17}$$

Hence

$$V \sum_i p_i = \tfrac{2}{3} \bar{\varepsilon} \sum_i N_i \tag{3.18}$$

or

$$pV = \tfrac{2}{3} N \, \bar{\varepsilon} \tag{3.19}$$

where p is the total pressure. Thus, we have Dalton's law of partial pressures, namely:

$$p = \sum_i p_i. \tag{3.20}$$

We can reach further results in an interesting way through the Maxwell–Boltzmann distribution of velocities.

3.1.1.4   Maxwell–Boltzmann distribution of velocities

The calculation of the pressure of a gas (section 3.1.1.1) implied a steady-state distribution of molecular velocities in a gas under equilibrium conditions. The steady-state description refers to average properties, individual variations being a natural consequence of intermolecular collisions.

Let the probability that a molecule has a velocity with components in the range v_x to $v_x + \mathrm{d}v_x$, v_y to $v_y + \mathrm{d}v_y$ and v_z to $v_z + \mathrm{d}v_z$ be $\Phi(v_x, v_y, v_z) \, \mathrm{d}v_x \, \mathrm{d}v_y \, \mathrm{d}v_z$. Since the three components of velocity lie along Cartesian axes, each is independent of the other two. In other words, we can write

$$\Phi(v_x, v_y, v_z) \, \mathrm{d}v_x \, \mathrm{d}v_y \, \mathrm{d}v_z = \phi(v_x)\phi(v_y)\phi(v_z) \, \mathrm{d}v_x \, \mathrm{d}v_y \, \mathrm{d}v_z \tag{3.21}$$

where $\phi(v_i)$ ($i = x, y, z$) is the individual probability that the ith velocity component lies in the range v_i to $v_i + \mathrm{d}v_i$. The kinetic theory postulate of isotropy means that the velocity distribution is independent of orientation, so that $\Phi(v_x, v_y, v_z)$ depends only on the speed v and can be replaced by $\Phi(v)$, where

$$v^2 = v_x^2 + v_y^2 + v_z^2 \tag{3.22}$$

whence

$$\Phi(v) = \Phi(v_x, v_y, v_z) = \phi(v_x)\phi(v_y)\phi(v_z). \tag{3.23}$$

For the one-dimensional case along x, partial differentiation with respect to v_x leads to

$$\frac{\partial \Phi(v)}{\partial v_x} = \Phi'(v) \frac{\partial v}{\partial v_x} = \phi'(v_x)\phi(v_y)\phi(v_z). \tag{3.24}$$

From (3.22), $\partial v / \partial v_x = v_x/v$, and (3.24) becomes

$$\Phi'(v)/v = \phi(v_y)\phi(v_z)\phi'(v_x)/v_x. \tag{3.25}$$

Dividing both sides of (3.25) by $\Phi(v)$, and using (3.23), we obtain

$$\Phi'(v)/[v\ \Phi(v)] = \phi'(v_x)/[(v_x\ \phi(v_x)]. \tag{3.26}$$

Since the left-hand side of (3.26) is a function of v alone, and the right-hand side a function of just v_x, each side must separately be equal to a constant, say $-\alpha$. Hence,

$$\phi'(v_x)/[v_x\ \phi(v_x)] = \frac{d\phi(v_x)}{dv_x}\,[v_x\ \phi(v_x)]^{-1} = -\alpha \tag{3.27}$$

whence

$$d \ln \phi(v_x) = -\alpha\, v_x\, dv_x \tag{3.28}$$

Integrating (3.28) gives

$$\phi(v_x) = \phi(0)\, \exp(-\alpha v_x^2/2) \tag{3.29}$$

where $\ln \phi(0)$ is the integration constant. It may be noted that the separation constant in (3.27) was made negative so that $\phi(v_x)$ should not increase indefinitely. In normalizing $\phi(v_x)$, as follows, the limits of $\pm \infty$ are used for convenience of integration: the probability of $|v_x|$ being in excess of the speed of light is vanishingly small.

$$\int_{\text{all } v_x} \phi(v_x)\, dv_x = \phi(0) \int_{-\infty}^{\infty} \exp(-\alpha v_x^2/2)\, dv_x = 1. \tag{3.30}$$

Since v_x^2 is symmetrical about $v_x = 0$, we can write (3.30) as

$$2\, \phi(0) \int_{0}^{\infty} \exp(-\alpha v_x^2/2)\, dv_x = 1. \tag{3.31}$$

Integrals of this type are solved easily by the use of the gamma function (Appendix 4). Following the procedure as before (Chapter 2), we obtain

$$(2/\alpha)^{1/2}\, \phi(0) \int_{0}^{\infty} t^{-1/2} \exp(-t)\, dt = 1 \tag{3.32}$$

where $t = \alpha v_x^2/2$. The value of the integral is $\Gamma(\tfrac{1}{2})$, or $\sqrt{\pi}$. Hence the normalization constant $\phi(0)$ is $\sqrt{(\alpha/2\pi)}$, and we obtain

$$\phi(v_x) = \sqrt{(\alpha/2\pi)} \exp(-\alpha v_x^2/2). \tag{3.33}$$

From the foregoing, we have that $\bar{\varepsilon} = \tfrac{1}{2}\, m\overline{v^2} = \tfrac{3}{2}\, k_B T$, so that

$$\overline{v_x^2} = \tfrac{1}{3}\, \overline{v^2} = k_B T/m. \tag{3.34}$$

We can now determine α from the distribution function, as follows. The average value of a parameter X with a distribution function $\phi(X)$ is given generally by

$$\bar{X} = \frac{\int X\, \phi(X)\, dX}{\int \phi(X)\, dX}. \tag{3.35}$$

Where the distribution function itself is normalized to unity between the limits of integration, the average value $\bar{X}$ is given by the numerator in (3.35). Hence, we may write

$$\overline{v_x^2} = \int_{-\infty}^{\infty} v_x^2\, \phi(v_x)\, dv_x \tag{3.36}$$

or

$$\overline{v_x^2} = \sqrt{(2\alpha/\pi)} \int_0^\infty v_x^2 \exp(-\alpha v_x^2/2) \, dv_x. \tag{3.37}$$

Solving the integral as before leads to $2/(\alpha\sqrt{\pi})\,\Gamma(\tfrac{3}{2})$, or $1/\alpha$, so that from (3.34) it follows that $\alpha = m/k_B T$, and (3.33) becomes

$$\phi(v_x) = (m/2\pi k_B T)^{1/2} \exp(-mv_x^2/2k_B T). \tag{3.38}$$

As an example of the use of (3.38), consider metallic sodium heated at 700 K in an oven with a small hole in one wall. What would be the average velocity of emergent sodium atoms?
    The emergent stream is unidirectional; hence; from (3.35) and (3.38)

$$\overline{v_x} = \frac{\displaystyle\int_0^\infty v_x \exp(-mv_x^2/2k_B T) \, dv_x}{\displaystyle\int_0^\infty \exp(-mv_x^2/2k_B T) \, dv_x}.$$

Solving the integrals as before, we obtain

$$\overline{v_x} = \frac{(k_B T/m)}{(\pi k_B T/2m)^{1/2}} = (2k_B T/\pi m)^{1/2}$$

which, for the given example, is $\{(2 \times 1.3807 \times 10^{-23}\text{ J K}^{-1} \times 700\text{ K})/(\pi \times 22.99 \times 1.6605 \times 10^{-27}\text{ kg})\}^{1/2}$, or 401.5 m s^{-1}.

Results similar to (3.38) can be obtained for $\phi(v_y)$ and $\phi(v_z)$; hence from (3.22) and (3.23)

$$\Phi(v) = (m/2\pi k_B T)^{3/2} \exp(-mv^2/2k_B T). \tag{3.39}$$

The normalization constant for the three-dimensional distribution may be confirmed from

$$\int_{-\infty}^{\infty} \int_{-\infty}^{\infty} \int_{-\infty}^{\infty} \Phi(v_x, v_y, v_z) \, dv_x \, dv_y \, dv_z = 1. \tag{3.40}$$

The exponential term in (3.39) is a form of the Boltzmann equation (1.4), with an average molecular energy equal to $mv^2/2$. The distribution of velocities along a single dimension is illustrated in Fig. 3.1. It is a normalized Gaussian distribution, with a mean velocity $\overline{v_x}$ of zero and a standard deviation of $\sqrt{(k_B T/m)}$.
    At the higher temperature, the distribution becomes broader, and more molecules attain higher speeds, and energies, which is an important factor in reaction kinetics. The derivations of (3.38) and (3.39) neglected molecular interactions, consistent with the isotropy postulate of the kinetic theory. More rigorous calculations by statistical mechanics lead to similar results, and the conclusions that we have drawn remain valid.

3.1.1.5 Maxwell–Boltzmann distribution of speeds
In order to obtain a distribution of *speeds*, we can express velocity in terms of magnitude and direction conveniently by a transformation from the Cartesian *x*, *y*

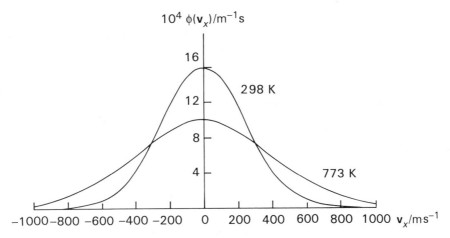

Fig. 3.1. One-dimensional distribution of velocities for argon gas at 298 K and 773 K: $\phi(v_x) = (m/2\pi k_B T)^{1/2} \exp(-mv_x^2/2k_B T)$; each distribution is a normalized Gaussian with a mean $\bar{v}_x$ of zero and a standard deviation of $\sqrt{(k_B T/m)}$.

and z coordinates to spherical polar coordinates v, θ and Ω, where

$$v_x = v \sin \theta \cos \Omega$$

$$v_y = v \sin \theta \sin \Omega \qquad (3.41)$$

$$v_z = v \cos \theta$$

following Appendix 5, but with the distance coordinate r in polar (linear) space replaced by the velocity coordinate v in polar (velocity) space, and the ϕ coordinate replaced by Ω to avoid a confusing notation in this context.

The probability that a molecule has a speed between v and $v + dv$ is $\Phi(v, \theta, \Omega)\,d\tau$, where $d\tau$ is an element of volume in velocity space given by $v^2 \sin \theta \, dv \, d\theta \, d\Omega$. The value of $\Phi(v, \theta, \Omega)$ is independent of orientation, so we can integrate over θ and Ω in order to obtain the probability for any region of space between the limits v and $v + dv$. Hence, from (3.23), (3.38), (3.22) and Appendix 5, we may write

$$\Phi(v)\,dv = \int_0^\pi \sin \theta \, d\theta \int_0^{2\pi} d\Omega \ (m/2\pi k_B T)^{3/2} v^2 \exp(-mv^2/2k_B T)\,dv \quad (3.42)$$

which, on evaluating the integrals, may be written as

$$\Phi(v) = 4\pi (m/2\pi k_B T)^{3/2} v^2 \exp(-mv^2/2k_B T). \qquad (3.43)$$

This equation may be obtained also from (3.39) by noting that the probability $\Phi(v)\,dv$ of a speed lying between v and $v + dv$ is $4\pi v^2 \ \Phi(v)\,dv$, where $4\pi v^2 \, dv$ is the volume of the spherical shell defined by radii v and $v + dv$.

In Fig. 3.2, $\Phi(v)$ is plotted as a function of v at 298 K and 773 K, and may be compared with Fig. 3.1. Whereas $\Phi(v)$ is proportional to both v^2 and $\exp(-mv^2/2k_B T)$, $\phi(v_x)$ is proportional only to $\exp(-mv_x^2/2k_B T)$. Although velocity components around zero are the most probable, there are fewer combinations of them so that $\Phi(v)$ is actually small near $v = 0$. The most probable speed is that at which $\Phi(v)$ is a

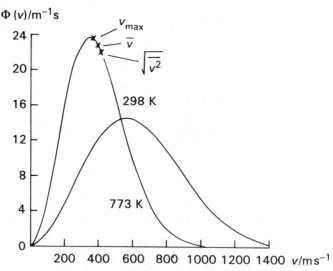

Fig. 3.2. Three-dimensional distribution of speeds for argon gas at 298 K and 773 K: $\Phi(v) = 4\pi(m/2\pi k_B T)^{1/2} \, v^2 \exp(-mv^2/2k_B T)$; the position of v_{max}, $\bar{v}$ and $\sqrt{\overline{v^2}}$ are shown (x) for the 298 K curve. At the higher temperature, a larger proportion of higher speeds (and energies) exists.

maximum. Thus, v_{max} may be found readily by differentiating (3.43) with respect to and setting the derivative to zero, whence

$$v_{max} = \sqrt{(2k_B T/m)}. \tag{3.44}$$

The root mean square speed $\sqrt{\overline{v^2}}$ derived from (3.22) and (3.34) is $\sqrt{(3k_B T/m)}$; this result can be derived directly from (3.43) and (3.35). These two measures of speed, together with the average $\bar{v}$, are indicated on Fig. 3.2 for the 298 K curve; it should be noted that $\bar{v}$ is not the same as $\sqrt{\overline{v^2}}$.

As an application of (3.43), we can extend the previous example to determine the mean speed of sodium atoms in the oven at 700 K. The distribution $\Phi(v)$ is normalized to unity between the limits of 0 and ∞. Hence, from (3.35) and (3.43)

$$\bar{v} = 4\pi(m/2\pi k_B T)^{3/2} \int_0^\infty v^3 \exp(-mv^2/2k_B T)\, dv.$$

Proceeding as before, the integral becomes $\frac{1}{2}(2k_B T/m)^2$, so that the mean speed is $\sqrt{(8k_B T/\pi m)}$ which for sodium, evaluates to 803 m s^{-1}.

3.1.2   Imperfect gases

The kinetic theory applies, strictly, to the ideal gas. All real gases deviate from the model of the ideal gas, to an extent dependent upon the external conditions. Fig. 3.3 shows the graph of $pV_m/\mathscr{R}T$ against p for nitrogen: we may refer to $(pV_m/\mathscr{R}T)$ as the compression factor $\mathscr{Z}$, and $\mathscr{Z}$ is clearly greater at low temperatures and high pressures —conditions that bring the gas molecules closer together. In fact, the equation of state for the ideal gas may be regarded as a limiting case of the general *virial equation*

$$pV_m = \mathscr{R}T\{1 + B(T)/V_m + C(T)/V_m^2 + \cdots\} \tag{3.45}$$

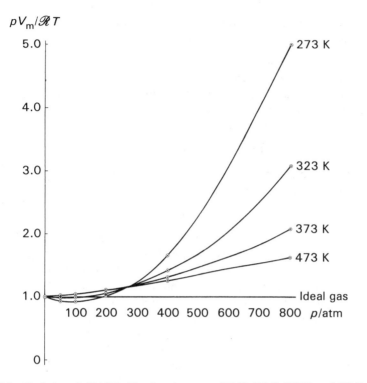

Fig. 3.3. Variation of $pV_m/\mathscr{R}T$ with p for nitrogen at 273 K, 323 K, 373 K, and 473 K; the horizontal line indicates ideal behaviour.

so that we can identify $\mathscr{Z}$ with the braced part of (3.45). Up to certain pressures, $C(T)$ and higher terms are negligible, and the second virial coefficients $B(T)$ for argon and nitrogen have the following values:

$$10^6 \, B(T)/\text{m}^3 \, \text{mol}^{-1}$$

T/K	273	323	373	412	873
Ar	− 21.7	−	− 4.2	0.0	11.9
N_2	− 10.5	0.0	6.2	−	21.7

From (3.45), by neglecting the third and higher virial coefficients and putting $V_m = \mathscr{R}T/p$, we can derive

$$\left(\frac{\partial \mathscr{Z}}{\partial p}\right)_T = \frac{B(T)}{\mathscr{R}T}. \tag{3.46}$$

At a particular temperature, the curve of Fig. 3.3 has a slope of zero up to a relatively high pressure. This temperature is called the Boyle temperature T_B because Boyle's law is valid over that pressure range. Now, $(\partial \mathscr{Z}/\partial p)_T$ must be zero at the Boyle temperature. Hence, from (3.46), $B(T)$ must be negative below T_B and positive above it; for nitrogen, T_B is 323 K. It follows that to explain the curves of Fig. 3.3 above about 100 atm pressure, the third virial coefficient, at least, would be needed.

3.1.2.1   Van der Waals' equation of state

Several equations of state have been proposed in order to represent the behaviour of real gases under widely varying conditions of temperature and pressure. One of the more satisfactory is the van der Waals' equation of state:

$$(p + an^2/V^2)(V - nb) = n\mathcal{R}T \tag{3.47}$$

where a and b are constants for a given gas. This equation can be written as a power series in V. Thus, for one mole of gas

$$V_m^3 - (\mathcal{R}T/p + b)V_m^2 + (a/p)V_m - (ab/p) = 0. \tag{3.48}$$

At a temperature known as the critical temperature T_c, (3. 48) has only one real root, the critical molar volume $V_{m,c}$; the corresponding pressure is the critical pressure p_c. At a temperature greater than its critical temperature a gas cannot be liquefied, no matter how great the applied pressure. The increase of pressure will increase the density of gas, but will not cause liquefaction. The liquid–gas interface disappears at the critical temperature, and the demarcation between these two states of matter is then indistinct. Fig. 3.4 shows the pVT diagram for argon, in projection on the pV plane; isotherms for the system at 80, 100, 150 and 200 K are shown. For argon, $T_c = 150.7$ K, $p_c = 48$ atm and the triple point temperature is 83 K.

3.1.2.2   Comparing gases

The critical constants of a gas are characteristic of it, and they lead to a method whereby gases can be compared. The p,V,T variables are first converted to reduced values:

$$p_r = \frac{p}{p_c} \qquad V_r = \frac{V_m}{V_{m,c}} \qquad T_r = \frac{T}{T_c}. \tag{3.49}$$

Fig. 3.5 shows a plot of $\mathcal{Z}$ ($pV_m/\mathcal{R}T$) against p, for three common gases. If, instead of plotting against p, we plot against p_r, a single curve, of similar form, would be obtained which suffices for a wide range of gases . The *principle of corresponding states* expresses the observation that real gases at the same reduced molar volume and reduced temperature exert (approximately) the same reduced pressure. This principle is followed most closely for gases composed of spherical, non-polar molecules.

Equation (3.48) can be expressed as

$$(V_m - \alpha_1)(V_m - \alpha_2)(V_m - \alpha_3) = 0 \tag{3.50}$$

where α_1, α_2 and α_3 are parameters that have the dimensions of molar volume. At the critical point, $\alpha_1 = \alpha_2 = \alpha_2 = V_{m,c}$, and

$$(V_m - V_{m,c})^3 = 0 \tag{3.51}$$

or

$$V_m^3 - 3V_{m,c} V_m^2 - 3V_{m,c}^2 V_m + V_{m,c}^3 = 0. \tag{3.52}$$

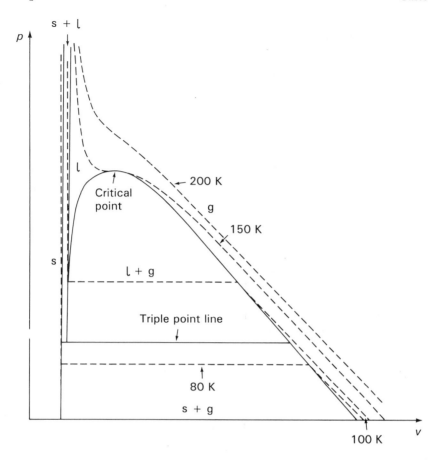

Fig. 3.4. pVT phase diagram for argon, in projection on to the pV plane; the isotherms are shown at 80, 100, 150 and 200 K.

Setting p, V and T in (3.48) to p_c, $V_{m,c}$ and T_c, respectively, and then comparing coefficients between (3.48) and (3.52), we obtain

$$V_{m,c} = (\mathscr{R}T_c/p_c + b)/3 = (a/3p_c)^{1/2} = (ab/p_c)^{1/3} \qquad (3.53)$$

from which we obtain

$$p_c = \frac{a}{27b^2} \qquad V_{m,c} = 3b \qquad T_c = \frac{8a}{27\mathscr{R}b}. \qquad (3.54)$$

Hence, $\mathscr{Z}_c = p_c V_{m,c}/\mathscr{R}T_c = 0.375$. Some values of the van der Waals' constants and critical constants are given in Table 3.1.

The van der Waals' equation of state is a good model for real gases, but it can fail under conditions of very high pressure or very low temperature. Other equations of state have been formulated for such circumstances, or the virial equation (3.45) may be more appropriate. The near constancy of $\mathscr{Z}_c$, although different from the predicted value of 0.375, is an indication of good self-consistency of the model used over three gases of very different physical properties.

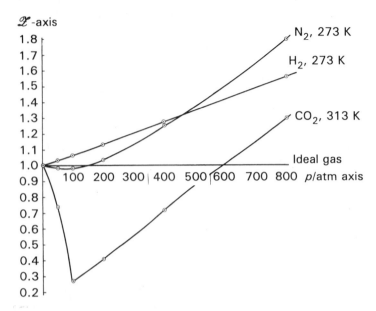

Fig. 3.5.   Variation of compression factor $\mathscr{Z}$ ($pV_m/\mathscr{R}T$) with p for nitrogen, hydrogen and carbon dioxide; the horizontal line represents ideality.

Table 3.1. Van der Waals' constants and critical constants of three gases

	Ar	N_2	CO_2
a/dm^6 atm mol^{-2}	1.345	1.390	3.592
$10^2b/dm^3$ mol^{-1}	3.219	3.913	4.267
p_c/atm	48.0	33.5	72.8
$V_{m,c}/cm^3$ mol^{-1}	75.2	90.1	94.0
T_c/K	150.7	126.2	304.2
T_B/K	411.5	327.2	714.8
$\mathscr{Z}_c$	0.292	0.291	0.274

3.1.3   Intermolecular attraction

The departure from ideal behaviour of real gases under certain conditions indicates a degree of failure in the approximations of the kinetic theory. Molecules do have a finite volume, which is significant in circumstances where the molecules are brought close together, and they do exert small forces upon one another. However, whereas attractive forces are effective at a relatively long range, repulsive forces become important only at intermolecular distances near to or less than equilibrium values. The effects of molecular interaction are revealed, for example, by the compression factor $\mathscr{Z}$ (Fig. 3.5), and we can explain them in terms of the van der Waals' equation of state.

3.1.3.1   *Molecule volume effect*

The finite volume of the molecules in a gas reduces the effective molar volume V to ($V_m - b$), and the effect of b is greater the closer the molecules are brought together.

The constant b in (3.47) is evidently related to the volume occupied by the gas molecules. More precisely, it is an *excluded* volume, as the following analysis shows.

Let the molecules of a gas be spherical and of diameter d. Two molecules cannot approach more closely than twice $d/2$, or twice the sum of their van der Waals' radii. Fig. 3.6 shows the situation of two molecules in closest contact: the spherical volume of space in which the molecules cannot move freely is shaded, and the radius of this sphere is d. Thus, the volume excluded per pair of molecules is $4\pi d^3/3$, and that per single molecule is $4\pi d^3/6$. The actual volume of a single molecule is $4\pi(d/2)^3/3$, or $\pi d^3/6$. Hence, the excluded volume b for a single molecule is four times its own volume. It should be noted that the values of b (and a) are normally quoted in molar terms.

3.1.3.2   Collision frequency

A collision occurs when two molecules come within the distance d of each other; d is called the *collision diameter* of the species. A given molecule sweeps out a cylinder of area πd^2 within which collisions occur; this area is called the *collision cross-section* σ. If the molecules, of mean speed $\bar{v}$, travel for a time t in the cylinder, and there are $\mathcal{N}$ molecules per unit volume, then the number of molecules with their centres inside the volume swept out is $\sigma \bar{v} t \mathcal{N}$. The collision frequency z is the number of collisions per unit time, and is equal to $\sigma \bar{v} \mathcal{N}$. So far we have considered only the motion of the given molecule, the others implicitly being assumed to be stationery. The average *relative* speed is $\sqrt{2}\bar{v}$; hence, z is more properly given by

$$z = \sqrt{2}\sigma \bar{v} N/V \qquad (3.55)$$

where $N/V = \mathcal{N}$, N being the number of molecules in a volume V, and $\bar{v}$ is $(8k_B T/\pi m)^{1/2}$.

The $\sqrt{2}$ factor can be seen to arise in the following way. Two molecules, each of average speed $\bar{v}$, moving in the same direction have a relative speed of zero. If they are moving away from each other along the same line then their relative speed would be $2\bar{v}$. An average situation between these two extremes would be motion at 90° to each

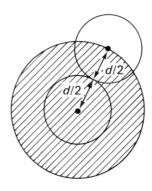

Fig. 3.6.   Excluded volume in a gas: the van der Waals' radius of the species is $d/2$, and the centres of two species cannot approach more closely than twice $d/2$. Hence, species in proximity cannot move freely in a sphere of radius less than d.

Table 3.2. Collision diameter, collision cross-section and van der Waals' constant b for three gases

	Ar	N_2	CO_2
d/nm	0.33	0.34	0.38
σ/nm^2	0.34	0.36	0.45
$10^2 b$/dm^3 mol^{-1}	4.49	5.10	6.83

other, in which case the relative speed would be $\sqrt{2}\bar{v}$. If we require the total number of collisions per unit volume per unit time Z, then we have

$$Z = \tfrac{1}{2} zN/V = \sigma\bar{v}(N/V)^2/\sqrt{2} \qquad (3.56)$$

where the factor of $\tfrac{1}{2}$ arises because collisions between any two given molecules must be counted once only.

The collision diameter d of a gas may be calculated from its viscosity (section 3.1.6); some values of d and the collision cross-section σ are given in Table 3.2, together with the van der Waals' constant b calculated from them.

The correspondence of the values of b with those in Table 3.1 indicates the approximate nature of b (similar to that of a).

3.1.3.3   Mean free path

A molecule of average speed $\bar{v}$ and collision frequency z spends a time $1/z$ between collisions and so travels a distance $\bar{v}/z$. The *mean free path* l of the molecule (between collisions) is given by

$$l = \bar{v}/z = 1/(\sqrt{2}\sigma N/V). \qquad (3.57)$$

We illustrate the results (3.55) to (3.57) by continuing the study of sodium vapour, heated at 700 K in an oven of volume 1 dm^3; the collision diameter of sodium is 0.37 nm and its vapour pressure is 1 mmHg at the temperature of the experiment. We shall calculate (a) the number of collisions of a single sodium atom in 1 s; (b) the total number of collisions in 1 s; (c) the mean free path of sodium atoms in the oven at 700 K.

Since $N/V = nL/V = p/k_B T$, $z = \sqrt{2}\sigma\bar{v}p/k_B T$, where σ is πd^2 and $\bar{v}$, from the example problem in section 3.1.1.5, is 803 m s^{-1}. Thus,

$$\text{(a) } z = \frac{\sqrt{2} \times \pi \times 0.37^2 \times 10^{-18}\ \text{nm}^2 \times 803\ \text{m s}^{-1} \times (1/760) \times 101325\ \text{N m}^{-2}}{1.3807 \times 10^{-23}\ \text{J K}^{-1} \times 700\ \text{K}}$$

$$= 6.74 \times 10^6\ \text{s}^{-1}$$

$$\text{(b) } Z_{NaNa} = \tfrac{1}{2} zN/V = \tfrac{1}{2} zp/k_B T = \frac{\tfrac{1}{2} \times 6.74 \times 10^6\ \text{s}^{-1} \times (1/760) \times 101325\ \text{N m}^{-2}}{1.3807 \times 10^{-23}\ \text{J K}^{-1} \times 700\ \text{K}}$$

$$= 4.65 \times 10^{28}\ \text{s}^{-1}\ \text{m}^{-3}$$

(c) $l = \bar{v}/z = 803\ \text{m s}^{-1}/(6.74 \times 10^6\ \text{s}^{-1}) = 1.19 \times 10^{-4}\ \text{m}$

We can see that l is approximately $3 \times 10^5 d$, so we would expect the kinetic theory to be reliable in the several example problems with sodium vapour that we have studied.

3.1.4   Origin of intermolecular forces

We touched briefly on the origin of one form of van der Waals' forces in section 1.2.3, and indicated there an induced dipole–induced dipole interaction in order to explain the attraction between atoms of argon. Van der Waals' interactions relate to interactions between closed-shell species and may be divided into several classes, such as ion (monopole)–dipole, dipole–dipole, dipole–induced dipole and induced dipole–induced dipole. There are also interactions involving quadrupoles and, in general, multipoles, but they are usually of much smaller magnitudes.

3.1.4.1   *Electric moments*

The first electric moment of a charge distribution is the dipole moment, which consists of numerical point (partial) charges $\pm\ qe$ separated by a distance $\imath$. We introduced this topic in section 2.8.2.3; here, we note further that the dipole moment is a vector quantity, directed along $\imath$, from the positive to the negative point charge.

The second moment is the quadrupole moment, and consists of four point charges with an overall charge of zero and, thus, a zero dipole moment. Carbon dioxide has a zero dipole moment but a significant quadrupole moment Θ equal to -14×10^{-40} C m^2. In this species, the two positive partial charges are resident on the carbon atom.

Generally, the moment of a 2^n-pole consists of each point charge multiplied by the nth power of its distance from the centroid of the charges, summed vectorially over all the charges in the distribution.

3.1.4.2   *Polarizability*

In an electric field, the electron distribution of an atom or molecule becomes distorted. As a result, a dipole moment is created (in addition to one that may be present permanently) because the positions of the centroids of positive and negative charge are altered. The species is said to be polarized, and the induced dipole moment μ is proportional to the strength $\mathscr{E}$ of the applied electric field, and given by

$$\mu = \alpha\mathscr{E} \tag{3.58}$$

where α is the *polarizability* of the species.

Since μ and $\mathscr{E}$ are, strictly, vector quantities, α is a second-order tensor, with components α_{ij} ($i, j = 1, 2, 3$). If we choose examples wherein isotropy can reasonably be assumed (spherical or near-spherical species), $\alpha_{ii} = \alpha$ and α_{ij} ($i \neq j$) $= 0$. For a molecule such as benzene, α would be very anisotropic, and an average polarizability may be approximated by $\frac{1}{3}\sum_i \alpha_{ii}$. Because of the shape of the benzene molecule, the electrons are much more readily distorted (polarized) along directions in the plane of the ring than they are in a direction normal to it.

The hydrogen chloride molecule has a polarizability α of 2.93×10^{-4} F m^2. By dividing a value of α by $4\pi\varepsilon_0$, where ε_0 is the permittivity of a vacuum (8.854×10^{-12} F m^{-1}), we obtain the quantity α', equal to 2.63×10^{-24} cm^3, which has the dimensions of volume, and serves to emphasize the relationship between polarizability and the size of the species. A *volume polarizability* α' may be defined therefore as

$$\alpha' = \alpha/4\pi\varepsilon_0. \tag{3.59}$$

Generally, the larger the species, the more easily it can be deformed by an external electric field, as the figures in Table 3.3 show.

Table 3.3. Selected radii and volume polarizabilities

	He	Ar	F$^-$	Cl$^-$	Br$^-$	I$^-$
r/nm	0.09	0.191	0.136	0.181	0.195	0.216
$10^{30}\,\alpha'$/m^3	0.20	1.7	0.90	3.1	4.3	6.6

Evidently, in the larger species the nuclear charge holds the electrons less strongly, so that external fields have greater effect on the electrons than in small species with few electrons (cp section 2.6.5). The effect may be enhanced by an anionic charge.

3.1.4.3   Ion–dipole interaction

Ion–ion interactions will be considered in the next chapter. Here, we consider first the case of the interaction between an ion of charge $+Q_2$ where, for convenience, Q is a numerical charge q multiplied by the electron charge e, and a dipole of moment μ_1 (Fig. 3.7a). The potential energy of interaction is based on the Coulombic formula $-Q_1Q_2/4\pi\varepsilon_0 r$ and, in this example, is given by two pairwise additive terms:

$$V_{i,d} = \frac{1}{4\pi\varepsilon_0}\left\{\frac{Q_1Q_2}{AC} - \frac{Q_1Q_2}{BC}\right\}. \tag{3.60}$$

Now, for AC, we have

$$AC = (r^2 + \imath r\cos\theta + \imath^2/4r^2)^{1/2} = r\{1 + (\imath\cos\theta/r + \imath^2/4r^2)\}^{1/2} \tag{3.61}$$

Similarly for BC:

$$BC = r\{1 + (-\imath\cos\theta/r + \imath^2/4r^2)\} \tag{3.62}$$

whence

$$V_{i,d} = -(Q_1Q_2/4\pi\varepsilon_0 r)\{[1 + (-\imath\cos\theta/r + \imath^2/4r^2)]^{-1/2}$$
$$- [1 + (\imath\cos\theta/r + \imath^2/4r^2]^{-1/2}\}. \tag{3.63}$$

Since, in general, r is significantly larger than $\imath$, we can expand by the binomial theorem to terms no larger than $(\imath/r)^2$ and, remembering that $\mu_1 = Q_1\imath$, we obtain

$$V_{i,d} = \frac{-\mu_1 Q_2\cos\theta}{4\pi\varepsilon_0 r^2}(1 - 3\imath^2/8r^2). \tag{3.64}$$

Further, since $\imath^2/r^2 \ll 1$, we may write

$$V_{i,d} = \frac{-\mu_1 Q_2\cos\theta}{4\pi\varepsilon_0 r^2}. \tag{3.65}$$

The maximum attraction arises when Q_2 is collinear with the dipole axis ($\theta = 0$), when

$$V_{i,d} = \frac{-\mu_1 Q_2}{4\pi\varepsilon_0 r^2}. \tag{3.66}$$

3.1.4.4   Dipole–dipole interaction

The dipoles of two polar molecules attract each other, and a force is set up between them. In a gas, or generally a fluid, the molecules are able to rotate, and the field of one dipole tends to orientate the dipole of a neighbouring molecule. Furthermore, the

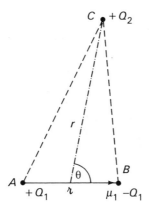

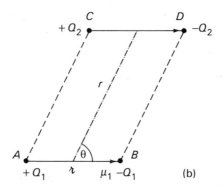

(b)

Fig. 3.7. Dipolar interactions: (a) dipole AB of length $\imath$, and point charge Q_2 distant r from the centre of the dipole; (b) dipoles AB and CD, each of length $\imath$, lying parallel, with r as the distance between their centres. (In this and ensuing figures, Q represents the numerical charge q multiplied by the electron charge e.)

attractive forces dominate, being of longer range than the repulsive forces, and a net attractive potential exists.

Consider first two molecules with permanent dipole moments of magnitudes μ_1 and μ_2, separated by a distance r, fixed in orientation and lying in one and the same plane (not the most general case), as shown in Fig. 3.7b. For the purpose of calculation, we can ensure that the dipoles have the same length by considering Q_2 modified as necessary with respect to Q_1. Proceeding as before, their potential energy $V_{d,d}$ is given by

$$V_{d,d} = \frac{1}{4\pi\varepsilon_0}\{2Q_1Q_2/d - Q_1Q_2/AC - Q_1Q_2/BC\}$$

$$= \frac{Q_1Q_2}{4\pi\varepsilon_0}\{2/d - 1/AC - 1/BC\}$$

$$= \frac{Q_1Q_2}{4\pi\varepsilon_0 r}\{2 - [1 + (\imath^2/r^2 + 2\imath\cos\theta/r)]^{-1/2}$$

$$- [1 + (\imath^2/r^2 - 2\imath\cos\theta/r)]^{-1/2}\}. \qquad (3.67)$$

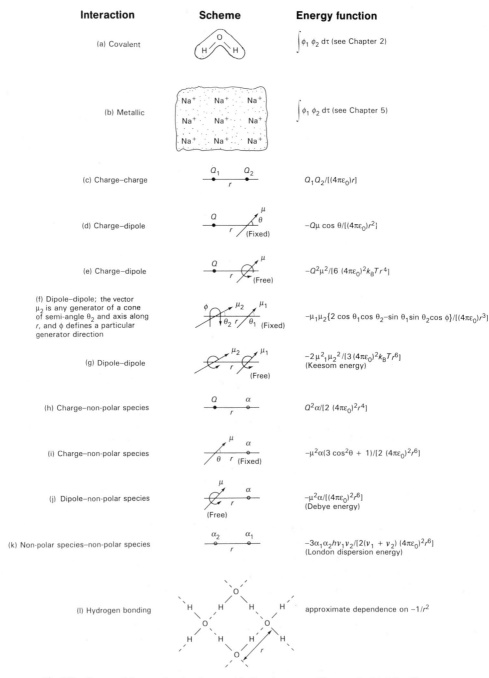

Interaction	Scheme	Energy function

(a) Covalent — $\int \phi_1 \phi_2 \, d\tau$ (see Chapter 2)

(b) Metallic — $\int \phi_1 \phi_2 \, d\tau$ (see Chapter 5)

(c) Charge–charge — $Q_1 Q_2 / [(4\pi\varepsilon_0)r]$

(d) Charge–dipole (Fixed) — $-Q\mu \cos\theta / [(4\pi\varepsilon_0)r^2]$

(e) Charge–dipole (Free) — $-Q^2\mu^2 / [6 \, (4\pi\varepsilon_0)^2 k_B T r^4]$

(f) Dipole–dipole; the vector μ_2 is any generator of a cone of semi-angle θ_2 and axis along r, and ϕ defines a particular generator direction (Fixed) — $-\mu_1\mu_2\{2 \cos\theta_1 \cos\theta_2 - \sin\theta_1 \sin\theta_2 \cos\phi\} / [(4\pi\varepsilon_0)r^3]$

(g) Dipole–dipole (Free) — $-2\mu^2_1\mu_2^2 / [3 \, (4\pi\varepsilon_0)^2 k_B T r^6]$ (Keesom energy)

(h) Charge–non-polar species — $Q^2\alpha / [2 \, (4\pi\varepsilon_0)^2 r^4]$

(i) Charge–non-polar species (Fixed) — $-\mu^2\alpha(3 \cos^2\theta + 1) / [2 \, (4\pi\varepsilon_0)^2 r^6]$

(j) Dipole–non-polar species (Free) — $-\mu^2\alpha / [(4\pi\varepsilon_0)^2 r^6]$ (Debye energy)

(k) Non-polar species–non-polar species — $-3\alpha_1\alpha_2 h\nu_1\nu_2 / [2(\nu_1 + \nu_2) \, (4\pi\varepsilon_0)^2 r^6]$ (London dispersion energy)

(l) Hydrogen bonding — approximate dependence on $-1/r^2$

Fig. 3.8. Types of intermolecular forces. (a) Covalent; see Chapter 2. (b) Metallic; see Chapter 5. (c) Charge–charge; r is the distance of separation of the centres of the charges, (d) Charge–fixed dipole; r is the distance between the charge and the centre of the dipole axis. (e) Charge–free dipole; the inverse dependence upon temperature shows that the tendency towards orientation of the dipole decreases as T increases. (f) Fixed dipole–fixed dipole; ϕ is the angle of relative rotation of μ_1 to μ_2 with respect to the dipole axis; for $\theta_1 = \theta_2$ and $\phi = 0$, the formulation becomes identical with (3.68). (g) Free dipole–free dipole; also known as the Keesom energy; temperature dependent. (h) Charge–non-polar species; α is the polarizability

Making assumptions similar to those in the ion–dipole interaction, we obtain

$$V_{d,d} = \frac{-\mu_1\mu_2(3\cos^2\theta - 1)}{4\pi\varepsilon_0 r^3}. \tag{3.68}$$

The general expression is (see Fig. 3.8)

$$V_{d,d} = \frac{-\mu_1\mu_2(2\cos\theta_1\cos\theta_2 - \sin\theta_1\sin\theta_2\cos\phi)}{4\pi\varepsilon_0 r^3}. \tag{3.69}$$

For the situation considered above $\theta_1 = \theta_2$, and $\phi = 0$, whence (3.68) is obtained. Maximum interaction occurs in either a collinear, head-to-tail orientation ($\theta_1 = \theta_2 = 0$), or an antiparallel orientation ($\theta_1 = 0$, $\theta_2 = \pi$); in these cases, we have

$$V_{d,d} = \frac{\mp 2\mu_1\mu_2}{4\pi\varepsilon_0 r^3}. \tag{3.70}$$

In a fluid, the molecules can rotate, and we can write an average potential energy in the form

$$\bar{V}_{d,d} = \frac{-\mu_1\mu_2}{4\pi\varepsilon_0 r^3}\,\overline{f(\theta,\phi)}\,\overline{P(\theta,\phi)}. \tag{3.71}$$

If we take a constant (zero) value of ϕ, in accord with Fig. 3.7b, $f(\theta,\phi)$ is the expression $(3\cos^2\theta - 1)$ developed above; for $P(\theta,\phi)$ we assume a Boltzmann distribution depending on θ, namely, $\exp(-V_\theta/k_BT)$, where V is given by (3.68). Expanding the exponential to two terms because $V_\theta \ll k_BT$, ϕ may be approximated by $(1 - V_\theta/k_BT)$. Hence

$$\bar{V}_{d,d} = \frac{-\mu_1\mu_2}{4\pi\varepsilon_0 r^3}\overline{(3\cos^2\theta - 1)(1 - \bar{V}_\theta/k_BT)}$$

$$= \frac{-\mu_1\mu_2}{4\pi\varepsilon_0 r^3}\{\overline{(3\cos^2\theta - 1)} - (-\mu_1\mu_2/(4\pi\varepsilon_0 r^3 k_BT)\overline{(3\cos^2\theta - 1)^2}\}. \tag{3.72}$$

We can easily show through (3.35) and (3.41) that the average value of $\cos^2\theta$ is 1/3, whence $(3\overline{\cos^2\theta} - 1) = 0$. In a similar manner $\overline{(3\cos^2\theta - 1)^2}$ evaluates to 4/5, but more detailed analysis, for the general case, shows that this factor is more correctly 2/3. Hence, we write

$$\bar{V}_{d,d} = \frac{-2\mu_1^2\mu_2^2}{3\,(4\pi\varepsilon_0)^2\,k_BT\,r^6}. \tag{3.73}$$

In this expression, we may note both the r^{-6} factor and the dependence upon temperature. The former shows that the dipole–dipole energy will in general be less in

---

of the non-polar species. (i) Fixed dipole–non-polar species; α is the polarizability of the non-polar species. (j) Free dipole–non-polar species; also known as the Debye energy. (k) Non-polar species–non-polar species; also known as the London dispersion energy. (l) Hydrogen-bonding; there is an approximate dependence on $-1/r^2$, where r is the closest non-bonded distance between two non-hydrogen atoms; thus, it is common to refer to the strength of a hydrogen bond by this length: 0.25 nm for O—H $\cdots$ O would be termed 'strong' and 0.29 nm for the same liaison 'weak'.

magnitude than that for ion–dipole interactions. The inverse dependence upon temperature means that the average dipole–dipole energy in a fluid decreases in magnitude as the temperature increases, because the tendency towards orientation of the dipoles is then randomized by the increased thermal motion of the molecules. For molecules of dipole moment approximately 1 D separated by 0.30 nm, (3.73) gives an energy of -1.3 kJ mol^{-1}; at 0.40 nm separation the energy is only -0.2 kJ mol^{-1}.

3.1.4.5   Dipole–induced dipole interaction
A polar molecule of dipole moment μ_1 can induce a moment μ_2 in a neighbouring molecule, whether or no that molecule is itself polar, so that the induced dipole interactions depend on the polarizability α_2 of the second species. The general calculation is lengthy, but it may be shown[1] that the field $\mathscr{E}_1$ due to a dipole of moment μ_1 at a point distant r from it, where the line from the point to the centre of the dipole makes an angle θ with the dipole axis, is given by

$$\mathscr{E}_1 = \frac{\mu_1(3\cos^2\theta + 1)^{1/2}}{(4\pi\varepsilon_0 r^3)}. \tag{3.74}$$

In inducing a dipole, energy is taken up by the molecule in displacing its positive and negative charges. This energy is Coulombic, given by

$$w = \int_0^{\iota} \mathscr{E}_1 Q_2 \, d\iota = \int_0^{\iota} \frac{\mu_2 Q_2}{\alpha_2} \, d\iota = \frac{Q_2^2}{\alpha_2} \int_0^{\iota} \iota \, d\iota. \tag{3.75}$$

where ι is the difference between the centres of positive and negative charge. Integrating, we find

$$w = \tfrac{1}{2} Q_2^2 \, \iota^2/\alpha_2 = \tfrac{1}{2} \mu_2^2/\alpha_2 = \tfrac{1}{2} \alpha_2 \mathscr{E}_1^2. \tag{3.76}$$

Since this work is the negative of the required interaction energy, we have from (3.74) and (3.76)

$$V_{d,id} = \frac{-\tfrac{1}{2}\alpha_2 \mu_1^2 (3\cos^2\theta + 1)}{(4\pi\varepsilon_0)^2 r^6}. \tag{3.77}$$

or

$$V_{d,id} = \frac{-\tfrac{1}{2}\alpha_2' \mu_1^2 (3\cos^2\theta + 1)}{4\pi\varepsilon_0 r^6}. \tag{3.78}$$

where α_2' is the volume polarizability, given by (3.59). The maximum interaction would arise for $\theta = 0$ (in-line and head-to-tail), but because $\cos^2\theta$ is $\tfrac{1}{3}$, the average energy is

$$\bar{V}_{d,id} = \frac{-\alpha_2' \mu_1^2}{4\pi\varepsilon_0 r^6}. \tag{3.79}$$

A hydrogen chloride molecule of dipole moment approximately 1 D interacting with a methane molecule of polarizability 2.89×10^{-40} F m^2, has an interaction energy, given by (3.79), of -0.2 kJ mol^{-1} at $r = 0.30$ nm, and of -0.04 kJ mol^{-1} at $r = 0.40$ nm.

---

[1]See, for example, J. N. Israelachvili (1992) *Intermolecular and Surface Forces*, Academic Press: E. A. Moelwyn-Hughes (1961) *Physical Chemistry*, Pergamon Press.

Induction energies are negative because the induction process always tends to induce moments in the direction of the induction field; there is no antiparallel induction.

3.1.4.6 Induced dipole–induced dipole interaction

The form for the induced dipole–induced dipole interaction (dispersion energy) was deduced first by London; it is quantum mechanical in origin, but the following simplified picture may be given. For a non-polar system such as argon, there must exist at any instant in a given species, because of the fluctuations in electron density, a transient dipole of moment, say μ_1, given by

$$\mu_1 = ea_0 \tag{3.80}$$

where a_0 is the Bohr radius. The field arising from this dipole will polarize a neighbouring species and produce an interaction that is analogous to the dipole–induced dipole interaction. Hence, from (3.79), with $\alpha'_2 = \alpha_2/4\pi\varepsilon_0$

$$V_{id,id} = \frac{-\mu_1^2\alpha_2}{(4\pi\varepsilon_0)^2r^6} = \frac{-(ea_0)^2\alpha_2}{(4\pi\varepsilon_0)^2r^6}. \tag{3.81}$$

As the first transient dipole changes its value and orientation, so the second (induced) dipole will follow it. It is because of this correlation that the multitude of transients do not average to zero, but give an overall attraction, which is another way of looking at the overall negative energetic result of induction. Now the polarizability α at a distance r from a polarizing influence is given by

$$\alpha = 4\pi\varepsilon_0 r^3. \tag{3.82}$$

In our analysis, (3.82) becomes

$$\alpha_2 = 4\pi\varepsilon_0 a_0^3 \tag{3.83}$$

and a_0 may be shown to be that distance at which the Coulomb energy $e^2/4\pi\varepsilon_0 a_0$ is equal to $2h\nu$, that is

$$a_0 = \frac{e^2}{8\pi\varepsilon_0 h\nu}. \tag{3.84}$$

For a Bohr atom $h\nu = 2.1799 \times 10^{-18}$ J ($\frac{1}{2}E_H$; see section 2.6), which is the first ionization energy. Combining (3.81), (3.83) and (3.84) gives

$$V_{id,id} = \frac{-\alpha_2^2 h\nu}{(4\pi\varepsilon_0)^2r^6}. \tag{3.85}$$

London's 1937 formula for the dispersion interaction between two identical atoms of polarizability α gives the more precise result

$$V_{id,id} = \frac{-\frac{3}{4}\alpha^2 h\nu}{(4\pi\varepsilon_0)^2r^6}. \tag{3.86}$$

and between dissimilar atoms

$$V_{id,id} = \frac{-\frac{3}{2}\alpha_1\alpha_2 h\nu_1\nu_2}{(\nu_1 + \nu_2)(4\pi\varepsilon_0)^2r^6}. \tag{3.87}$$

Using ionization energies, (3.87) becomes

$$V_{\text{id,id}} = \frac{-\frac{3}{2}\alpha_1\alpha_2 I_1 I_2}{(I_1 + I_2)(4\pi\varepsilon_0)^2 r^6} \tag{3.88}$$

and with volume polarizabilities

$$V_{\text{id,id}} = \frac{-\frac{3}{2}\alpha'_1\alpha'_2 I_1 I_2}{(I_1 + I_2)r^6}. \tag{3.89}$$

Again, the inverse sixth power of distance is in evidence. We may apply (3.89) to a non-polar substance, such as methane. We take α' as 2.6×10^{-24} cm^3, I as 12.6 eV, and the distance of closest approach as 0.37 nm, whence $V_{\text{id,id}}$ $(CH_4) = -8.5$ kJ mol^{-1}. A check on this value is given by the cohesive energy of the solid, which may be approximated by the sum of the enthalpies of melting and of vaporization; the resulting value of 8.9 kJ mol^{-1} is in good agreement with our calculation.

In any practical situation a range of intermolecular forces will operate. In a substance such as solid argon, the dispersion energy is of paramount importance. It is operative in other substances too, but it may be overshadowed in magnitude by other forces that are present. The range of possible intermolecular forces is shown diagrammatically by Fig. 3.8, together with their formulations, most of which we have derived in the foregoing sections.

3.1.4.7   Pairwise additivity

In calculations such as that leading to (3.66), we assumed correctly a pairwise additivity of the separate terms in the expression for the interactional energy. The procedure is correct for Coulombic and dispersion energies, but induction energies are not of exactly the same nature. In Fig. 3.9 we illustrate the distortion of an atom by a neighbouring charge. A single charge distorts the atom in one direction; a change in the position of the charge by 180° across the atom-Q axis reverses the direction of the distortion. The presence of both charges at one and the same time leads to a symmetrical distortion with a zero resultant field, so that the total effect is not pairwise additive. This situation leads to a quadrupole moment, as in carbon dioxide, and is important in the study of polar materials.

3.1.5   Intermolecular potentials

At large distances of separation r, two gas molecules do not interact, and their joint potential energy is effectively zero. As they approach each other, an attraction develops between them which increases continuously as r decreases. However, at small distances, strong repulsion between the electrons (and nuclei) of adjacent species causes the potential energy to rise very steeply. The superposition of these two potentials is a curve of the form of Fig. 3.10c (see also Fig. 2.24), with a minimum energy at the equilibrium distance r_e. A potential energy function must be able to represent this behaviour in a way that allows experimental properties to be determined theoretically.

Some potential energy functions that behave with varying degrees of success are illustrated in Fig. 3.10: they are, in order, the hard-sphere model, the square-well model and the Lennard-Jones (n–m) model. In the hard-sphere model (Fig. 3.10a), the potential energy is zero for r greater than an effective (hard-sphere) diameter δ, and

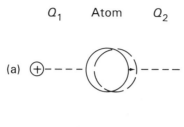

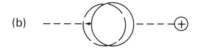

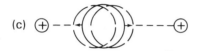

Fig. 3.9. Non-additivity of induction interactions. The total energy is not equal to the sum of the Q_1-Q_2, Q_1-atom and Q_2-atom energies: (a),(b) reversing the position of the charge from Q_1 to Q_2 reverses the direction of the distortion; (c) with both charges Q_1 and Q_2, a symmetrical distortion with zero resultant field obtains. If the atom were replaced by another point charge Q_3, the total energy would be exactly pairwise additive.

rises to infinity for $r < \delta$; $r = \delta$ represents a discontinuity in the function. This potential can provide reasonably satisfactory explanations of flow properties such as viscosity and thermal conductivity, because these properties are strongly dependent upon the repulsion forces between species. The hard-sphere model cannot treat phase equilibria, because a minimum energy is needed in order to represent equilibrium conditions. The square-well function (Fig. 3.10b) is an approximation that can treat equilibrium although it, too, possesses discontinuities.

A better and more widely used energy function is the Lennard-Jones n–m potential, particularly the 12–6 form:

$$V(r) = 4\varepsilon_L[(\delta/r)^{12} - (\delta/r)^6] \tag{3.90}$$

which can be seen from Fig. 3.10c to incorporate the features of both (a) and (b), but without discontinuity. The repulsion potential, given by the term $(\delta/r)^{12}$ in (3.90), may be better represented by an exponential function, so that we obtain the so-called 'exp–6' potential

$$V(r) = A \exp(-Br) - Cr^6 \tag{3.91}$$

where A and B are constants of the repulsion function, and C is the constant of the attraction function. The exponential function is in evidence in the overlap integrals discussed in Chapter 2 and so has a sound theoretical basis, whereas the r^{-12} function is empirical, chosen to reproduce the rapid increase in potential energy as r falls below δ. The value of 12 for the repulsion function has no other significance, but it is mathematically useful for n to be twice m.

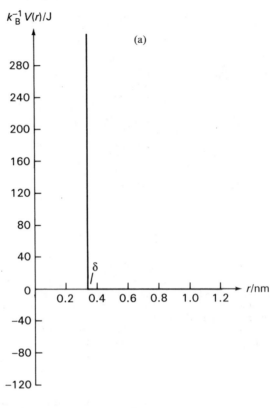

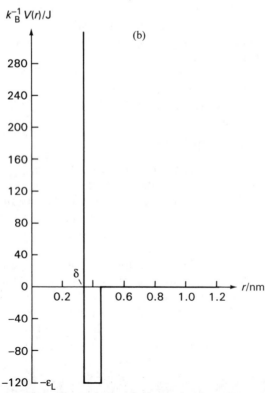

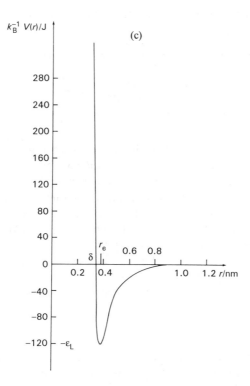

Fig. 3.10.   Intermolecular potential models. (a) Hard-sphere model; discontinuity at r equal to the hard-sphere diameter δ. (b) Square-well model; a minimum in V is represented in this function but, again, discontinuities are present, (c) Lennard-Jones 12–6 model; $V(r) = 0$ at $r = \delta$, and $-\varepsilon_L$ at $r = r_e$, where $r_e = 2^{1/6}\delta$. This diagram has been drawn to a correct relative scale, and the others set to match it.

The graph of (3.91) is very similar to that in Fig. 3.10c, except that it rises even more steeply for $r < \delta$ than does the Lennard-Jones function. In our discussion, we shall use the Lennard-Jones 12–6 function.

3.1.5.1   Lennard-Jones 12–6 potential function

If we differentiate (3.90) with respect to r and set the derivative to zero for $r = r_e$, we find that $r = 2^{1/6}\delta$. Inserting this value in (3.90) shows that $V(r)$ is equal to $-\varepsilon_L$ at the equilibrium distance r_e. Some values for the parameters ε_L and δ are listed in Table 3.4; ε is usually given in the form ε_L/k_B.

Approximate values for these parameters are given by $\varepsilon_L/k_B = 0.775T_c$, and $\delta = 0.1(V_{m.c}/3)^{1/3}$, where the values of T_c and $V_{m.c}$ are as in Table 3.1; the values in parentheses in Table 3.4 were obtained in this way. The values of δ are very similar to the corresponding collision diameters d (Table 3.2). This is not unreasonable: the essential difference is that whereas the values of d are obtained through experimental measurements, such as on viscosity, δ (and ε_L) are chosen as best-fit parameters for (3.90).

Table 3.4. Parameters for the Lennard-Jones 12–6 potential function

	$\varepsilon_L/k_B/K$	δ/nm
Ar	124 (117)	0.34 (0.29)
N_2	92 (98)	0.37 (0.31)
CO_2	190 (235)	0.40 (0.32)

For argon and nitrogen, ε_L has the values 1.0 kJ mol^{-1} and 3.7 kJ mol^{-1}, respectively. The corresponding classical thermal energies at 298 K are 3.7 kJ mol^{-1} ($\sim \frac{3}{2}\mathcal{R}T$) and 6.2 kJ mol^{-1} ($\sim \frac{5}{2}\mathcal{R}T$); the larger value for nitrogen arises because of extra (rotational and vibrational) degrees of freedom as well as the translational energy. It is clear that, at normal temperatures, the thermal energy is four or more times greater than the intermolecular attractive energy, so that the kinetic theory postulates are sensibly upheld by these results.

Experimental measurements, by Scoles and coworkers, of the elastic cross-section in beams of argon gas gave the value of 147 ± 3 K for ε_L/k_B and 0.37 ± 0.02 nm for r_e in the Lennard-Jones potential; the corresponding energy is $(2.03 \pm 0.04) \times 10^{-21}$ J, or 1.22 ± 0.02 kJ mol^{-1}.

3.1.6   Gas viscosity

We consider finally in this section on gases one of the important transport properties of a gas, namely, its viscosity. All fluids exert viscous forces on bodies moving through them.

Consider a body moving in air, such as a car travelling along a road, in the y-direction with a velocity v_y (Fig. 3.11a). The thin layer of air adjacent to the body B will be pulled along with it, and a molecule in this layer would acquire the steady-state average velocity v_y. In the x-direction, perpendicular to y, layers of air normal to it will have drift velocities that decrease as the distance along x from the body increases. In other words, a velocity gradient dv_y/dx is set up along x. The stress that acts across any plane normal to x is not a simple pressure, but contains a shearing component that tends to equalize the velocities at different points in the gas. If the effect is not too large turbulence will not occur, and the gas obeys Newton's laws; the gas is said then to be a Newtonian fluid.

The shearing stress (viscous force per unit area) tending to retard motion is proportional to the velocity gradient, and if F_y is the y-component of the viscous force per unit area, we have

$$F_y = \eta \frac{dv_y}{dx} \tag{3.92}$$

where η is the viscosity coefficient, which depends upon the temperature and the pressure.

Let us consider an imaginary layer S perpendicular to the x-direction at an arbitrary origin $x = 0$ (Fig. 3.11b), and evaluate the momentum carried by molecules across S down the velocity gradient, in the direction of $+x$. Molecules traversing S from left to right arrive with more momentum than the molecules that they meet,

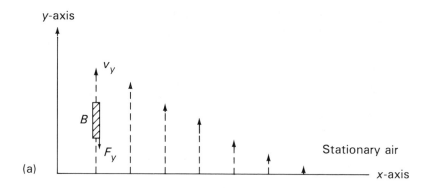

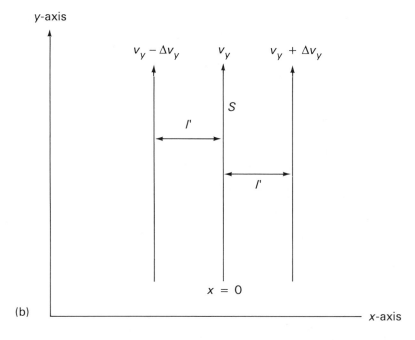

Fig. 3.11.   Viscosity of a gas. (a) A body B travelling with a velocity v_y in the y-direction sets up a velocity gradient in the x-direction. The viscous force F_y, acting on unit area of the body, is given by $F_y = \eta(dv/dx)$; it acts in the $-y$-direction because the velocity gradient is negative along $+x$. (b) An imaginary plane S at $x = 0$: molecules moving from left to right across the plane carry y-momentum $m(v_y - v_y)$ along $+x$, and molecules moving from right to left carry y-momentum $-m(v_y + v_y)$ along $+x$, so the momentum flux $J_{x(y)}$ is $m(v_y - v_y) - m(v_y + v_y)$, or $-2mv_y$.

whereas molecules traversing S from right to left carry a deficiency of momentum. Thus, there is a net transfer of y-momentum in the $+x$-direction.

Molecules travelling from a faster layer to a slower layer transport momentum $m(v_y - l' dv_y/dx)$ to the layer S, whereas those travelling from a slower layer to a faster layer transport momentum $m(v_y + l' dv_y/dx)$ to the layer S; the parameter l' is the effective mean free path in the x-direction. The net transfer is, therefore, $-2ml' dv_y/dx$ per molecule.

In a time t, a molecule of velocity v_y will strike plane S if it lies within a distance $v_y t$ of it. Hence, all molecules in a volume $Av_y t$ with velocities along y will strike S in the time t; A is an area of surface in S. The average total number of collisions N_c is, thus

$$N_c = (N/V) At \int_0^\infty v_y \, \phi(v_y) \, dv_y. \tag{3.93}$$

From (3.33), and integrating as before, we obtain

$$N_c = (N/V) At (k_B T/2\pi m)^{1/2}. \tag{3.94}$$

Therefore, the number of collisions with S per unit area per unit time (Z_s) is $(N/V)(k_B T/2\pi m)^{1/2}$. Since we have seen that $\bar{v} = (8k_B T/\pi m)^{1/2}$, we have

$$Z_s = \tfrac{1}{4} (N/V) \, \bar{v}. \tag{3.95}$$

The y-momentum flux along x, $J_{x(y)}$, is the rate of flow of momentum per unit area per unit time. Hence, the total momentum flux is given by

$$J_{x(y)} = -2Z_s \, ml' \frac{dv_y}{dx} = \tfrac{1}{2}(N/V) \, \bar{v} \, ml' \frac{dv_y}{dx}. \tag{3.96}$$

Since momentum flux is the rate of change of momentum per unit area, we have, in general, $J = d(mv)/dt = m \, dv/dt = ma$, where a is an acceleration. Thus, we can identify J with the viscous force per unit area. Alternatively, we can say that if momentum $J_{x(y)}$ is transferred away in the $+x$-direction, then by Newton's second law an equal and opposite force must be applied in order to maintain forward motion. Hence,

$$J_{x(y)} = = -F_y = -\eta \frac{dv_y}{dx} \tag{3 97}$$

whence

$$\eta = \tfrac{1}{2} m (N/V) l' \bar{v}. \tag{3.98}$$

The mean free path l' in the x-direction is related to the true mean free path l, and effusive flow analysis has shown that l' is $2l/3$, so that

$$\eta = \tfrac{1}{3} m (N/V) l \bar{v}. \tag{3.99}$$

A more detailed treatment, taking into account the fact that molecules in elastic collision retain about 40% of their original direction and that the mean free path is not the same for molecules travelling with different speeds, leads to the factor of $\tfrac{1}{3}$ in (3.99) being replaced by 0.499.

Gas viscosity is independent of pressure over a very wide range, but it increases with temperature because $\bar{v}$ is proportional to $\sqrt{T}$, and momentum is transferred more rapidly across a given area. This behaviour is opposite to that in liquids, because the viscosity of a liquid is strongly dependent upon intermolecular forces: increasing the thermal energy of a system where intermolecular forces are important results in a breakdown of these interactions.

We have considered certain aspects of ideal and real gases, together with some of their properties, and to what extent these properties are dependent upon intermolecular forces. Some of the results that we have obtained in this chapter will be transferred to our study of liquids in the ensuing section.

3.2   LIQUIDS

A characteristic of liquids (and solids) is the presence of significant cohesive forces between the species, atoms or molecules. In gases, these interactions can be neglected, except under extreme conditions of temperature and pressure, as we have discussed.

In the context of this chapter, the term 'liquid' will be used to indicate a pure substance. The presence of a second substance would bring us into the realm of solutions, which we have already indicated lies outside the present scope of the book.

Liquids may consist of single atomic species, such as argon, or of molecules, such as nitrogen or water. These substances are often called *classical liquids* because, for them, quantum effects are negligible, and their behaviour can be studied through the equations of classical physics. However, there are certain liquids, among which helium is best known, that exhibit abnormal behaviour in their flow and conduction properties. They are termed *quantum liquids*, or *superliquids*, since quantum mechanical principles must be invoked to explain them fully. Our main concern will be with classical liquids, but we shall refer briefly to superliquids in section 3.2.10.

3.2.1   Liquid–gas equilibrium

It is interesting to move into the discussion of liquids by considering the equilibrium between a gas and the corresponding liquid.

An individual molecule in a gas possesses a cohesive self-energy $\mathcal{U}_s$ which is the sum of the interactional energies with its neighbours. The self-energy is related to the potential (attractive) energy discussed in the first part of this chapter. We may write the potential energy between a pair of molecules (we shall use 'molecule' for any species) as $-C/r^m$ $(m > 3)$ for $r > d$, and infinity for $r < d$, where d is the collision diameter of the species. The energy $\mathcal{U}_s$ may be evaluated as the sum of the pair potential over all neighbours within the distance r.

Let the number density (assumed to be uniform) in the gas be $\mathcal{N}$. Then the number of molecules in a region of space between r and $r + dr$ is $4\pi\mathcal{N}r^2\,dr$. Hence,

$$\mathcal{U}_s = -4\pi\,\mathcal{N}\,C\int_d^{\infty}\frac{1}{r^{m-2}}\,dr$$

$$= \frac{-4\pi\,\mathcal{N}\,C}{(m-3)d^{m-3}}. \tag{3.100}$$

This equation can be written also as

$$\mathcal{U}_s = -\alpha\,\mathcal{N} \tag{3.101}$$

where $\alpha = 4\pi C/[(m-3)d^{m-3}]$. The entropy s of a gas is proportional to the logarithm of the free volume $v-\beta$ available to each molecule:

$$s/k_B = \ln(v-\beta)^{-1} = \ln[\mathcal{N}/(1-\beta\mathcal{N})] \tag{3.102}$$

where v is the volume occupied by a molecule.

The chemical potential μ_g for the gas is given by

$$\mu_g = -\alpha \mathcal{N} + k_B T \ln[\mathcal{N}/(1 - \beta \mathcal{N})]. \tag{3.103}$$

Since we know from thermodynamics the general result that $(\partial \mu / \partial p)_T (= v) = 1/\mathcal{N}$, it follows that

$$\left(\frac{\partial p}{\partial \mathcal{N}}\right)_T = \mathcal{N} \left(\frac{\partial \mu}{\partial \mathcal{N}}\right)_T. \tag{3.104}$$

Hence

$$
\begin{aligned}
p &= \int_0^{\mathcal{N}} \mathcal{N} \left(\frac{\partial \mu}{\partial \mathcal{N}}\right)_T \, \mathrm{d}\mathcal{N} \\
&= \int_0^{\mathcal{N}} \mathcal{N} \left\{ -\alpha + \frac{k_B T}{\mathcal{N}(1 - \beta \mathcal{N})} \right\} \mathrm{d}\mathcal{N} \\
&= -\tfrac{1}{2}\alpha \mathcal{N}^2 - (k_B T/\beta) \ln(1 - \beta \mathcal{N}).
\end{aligned}
\tag{3.105}
$$

If we make the reasonable assumption (for a gas) that $\beta \mathcal{N} \ll 1$, then $\ln(1 - \beta \mathcal{N})$ can be approximated by the sequence of steps: $\ln(1 - \beta \mathcal{N}) \approx -\beta \mathcal{N} - (\beta \mathcal{N})^2/2 = -\beta \mathcal{N}(1 + \beta \mathcal{N}/2) \approx -\beta \mathcal{N}/(1 - \beta \mathcal{N}/2) = -\beta/(v - \beta/2)$ Hence,

$$p = -\tfrac{1}{2}\alpha/v^2 + k_B T/(v - \tfrac{1}{2}\beta). \tag{3.106}$$

Introducing the Avogadro constant L and putting $\tfrac{1}{2}\alpha = a$ and $\tfrac{1}{2}\beta = b$, we obtain

$$(p + a/V_m^2)(V_m - b) = \mathcal{R}T \tag{3.107}$$

which is the van der Waals' equation of state (3.47) with $n = 1$.

We have shown earlier that b (per molecule $= 2\pi d^3/3$, and here we see that $a = 2\pi C/[(m - 3)d^{m-3}]$, or $-\mathcal{U}_s/2\mathcal{N}$. Thus, we have demonstrated in a simple manner the relationship of the van der Waals' constants a and b to the attractive and repulsive forces between the molecules.

If we have a gas and liquid in equilibrium at activities a_g and a_l, respectively, then their chemical potentials are equal, at a given temperature T:

$$\mu_g + k_B T \ln(a_g) = \mu_l + k_B T \ln(a_l). \tag{3.108}$$

Assuming that $\mu_l \gg \mu_g$, we can approximate (3.108) as

$$-\mu_l = k_B T \ln(a_l/a_g). \tag{3.109}$$

If we replace activities by molar concentrations, which are reciprocal molar volumes, the ratio $V_{m,g}/V_{m,l}$ is approximately 10^3; hence,

$$-\mu_l = k_B T \ln(10^3) = 7 k_B T. \tag{3.110}$$

At the boiling-point temperature T_b, $-L\mu_l = 7 Lk_B T_b$, whence

$$-L\mu_l/T_b = 7\mathcal{R}. \tag{3.111}$$

The energy of vaporization ΔU_v is approximately equal to the free energy of the liquid, that is, $\Delta U_v = -L\mu_l$. Thus, the molar enthalpy of vaporization at the boiling point is

$$\Delta H_v = \Delta U_v + \mathcal{R}T_b = 8\mathcal{R}T_b \tag{3.112}$$

Hence, $\Delta H_v / T_b$ (the molar entropy of vaporization) is approximately 67 J K^{-1} mol^{-1} from this argument, and approximately 85 J K^{-1} mol^{-1} according to Trouton's empirical rule, as an average over many species. For argon, nitrogen and carbon dioxide, the Trouton values are 74.6, 72.2 and 86.8 J K^{-1} mol^{-1}, respectively.

It follows that the boiling-point temperature is related, not suprisingly, to the cohesive energy of the liquid and, from (3.112), we may say that a monatomic gas will condense to a liquid once its cohesive self-energy is in excess of approximately $9.5k_BT$ per molecule. The difference between this value and the calculated value of $8k_BT$ is an average kinetic energy of translation of $\frac{3}{2}k_BT$ which must also be overcome in the condensation process. We recall that for polyatomic species, the energies of rotation and vibration lead to values higher than $\frac{3}{2}k_BT$ for the total 'translational' energy.

3.2.2   Some thermodynamics of equilibrium

Equilibrium properties, such as heat capacity and compressibility, may be deduced if we have expressions for the internal energy and the equation of state. We have the general relation

$$dU = T\, dS - p\, dV. \tag{3.113}$$

Hence, at constant temperature,

$$\left(\frac{\partial U}{\partial V}\right)_T = T\left(\frac{\partial S}{\partial V}\right)_T - p. \tag{3.114}$$

Using the Maxwell equation $(\partial S/\partial V)_T = (\partial p/\partial T)_V$ with (3.114), we obtain

$$\left(\frac{\partial U}{\partial V}\right)_T = T\left(\frac{\partial p}{\partial T}\right)_V - p. \tag{3.115}$$

If we are considering an ideal gas, the right-hand side of (3.115) evaluates to zero, which is a basic property of an ideal gas. We have remarked that the difference between a liquid and a gas in the critical region is indistinct. Thus, it is not unreasonable to explore the properties of a simple liquid first by means of the van der Waals' equation (3.47) in place of the ideal gas equation (1.7).

For $n = 1$, we have

$$p = \frac{\mathscr{R}T}{V_m - b} - \frac{a}{V_m^2}. \tag{3.116}$$

Using (3.115), we find

$$\left(\frac{\partial U}{\partial V}\right)_T = \frac{a}{V^2}. \tag{3.117}$$

Integrating (3.117) leads to

$$U_m = -a/V_m + f(T) \tag{3.118}$$

where $f(T)$ will be the same as that for an ideal gas, which is $\frac{3}{2}\mathscr{R}T$ in the case of a monatomic species. The first term on the right-hand side of (3.118) represents the contribution of the attractive forces to the internal energy of a gas that obeys the van der Waals' equation of state.

The molar heat capacity at constant volume is given by

$$C_{V,m} = \left(\frac{\partial U}{\partial T}\right)_V. \tag{3.119}$$

If we were to assume that the van der Waals' equation were obeyed by a simple liquid such as argon, its molar heat capacity would be $3\mathscr{R}/2$, from (3.119). This value is the same as the ideal gas value, and cannot be correct. In fact, the molar heat capacity for liquid argon is close to $5\mathscr{R}/2$; that for the corresponding solid, being monatomic, is $3\mathscr{R}$. It is evident that a better equation of state is essential for the study of a liquid.

3.2.3   Radial distribution function

The radial distribution function characterizes the average structure of a liquid or, more precisely, the average distribution of its molecules relative to one another. For a completely uniform distribution of molecules, the number $N(r)$ of them lying within a spherical shell of radius r and thickness dr is given by

$$N(r) = 4\pi r^2 \mathscr{N} dr \tag{3.120}$$

where $\mathscr{N}$ is the number density of the species.

We define the radial distribution function $g(r)$ for a simple liquid by

$$N(r) = 4\pi r^2 \mathscr{N} g(r) dr \tag{3.121}$$

and $g(r)$ thus measures the variation, with r, in the probability of observing one species at a given distance r from another.

More explicitly, consider a system of n atoms, and take any one of them as an origin. For a sequence of values of r, count the number $N(r)$ of other atoms whose centres lie within spherical shells of volume $4\pi r^2 dr$. This procedure is repeated with each of the other n atoms as centre, and finally the results for each value of r are averaged. Thus, we obtain a radial distribution function $g(r)$ as a function of the parameter r, in the form $g(r) = N(r)/4\pi r^2 \mathscr{N} dr$.

The molar internal energy U_m of a simple liquid consists of the kinetic energy ($3/2\, k_B T$ per molecule) and the potential energy: we may write U_m in the form

$$U_m = 3/2\mathscr{R}\, T + 4\pi\mathscr{N}\, L/2 \int_0^\infty r^2 V(r) g(r)\, dr \tag{3.122}$$

where $V(r)$ is a pair potential function of the type described in section 3.1.5, such as (3.90), for example; the factor $1/2$ arises because each pair of interactions in $V(r)$ is counted only once. For a more detailed discussion of radial distribution functions, the reader is referred to a more specific treatment of this subject[1]. The determination of $g(r)$ can be achieved by diffraction studies, with X-ray or neutrons, or by computer simulation techniques.

3.2.3.1   *Diffraction studies*

Evidence of a degree of order in liquids is provided by diffraction studies. Argon gas does not give a diffraction pattern because there is no characteristic spacing between

---

[1]See, for example, Y. Marcus (1977) *Introduction to Liquid State Chemistry*, Wiley.

its atoms.[1] In the liquid, however, diffraction peaks are obtained, thereby indicating a degree of order, although not of the extent that exists in solids.

In Fig. 3.12, A and A' represent two argon atoms separated by the distance r. The two typical diffracted rays have a path difference of $r \sin 2\theta$, and where this length is equal to an integral number of wavelengths of the radiation, a maximum in the diffraction pattern occurs at a position that is characteristic of the distance r. For X-ray diffraction, it can be shown[2] that the intensity of the scattered radiation is given by the function $\sin(Sr)/Sr$, where S is $4\pi\lambda^{-1}\sin\theta$. Since $\sin(Sr)$ is sinusoidal, the scattered intensity oscillates, and decays with increasing θ because the atomic X-ray scattering factor decreases[3] with increase in θ. Hence, the diffraction pattern on a flat photographic film consists of concentric rings, somewhat diffuse, decreasing in intensity (blackening) as θ increases (Fig. 3.13). More conveniently, the diffraction pattern can be recorded as a diffractometer trace.

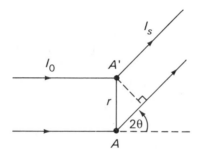

Fig. 3.12.   Scattering of X-radiation of incident intensity I_0 by two argon atoms A and A' separated by a distance r. The angle of scatter is 2θ, and the scattered radiation has an intensity I_S.

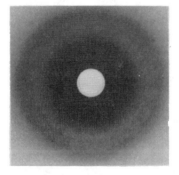

Fig. 3.13.   X-ray scattering pattern from a liquid, showing the characteristic concentric maxima decreasing in intensity as θ increases (from the centre outwards). The white spot at the centre arises from the lead direct-beam trap.

---

[1] Polyatomic gases give diffraction effects arising from the spacings between atoms in the molecule, and not from intermolecular distances themselves.

[2] See, for example, M. M. Woolfson (1970) *An Introduction to X-ray Crystallography*, CUP.

[3] See, for example, M. F. C. Ladd and R. A. Palmer (1993) *Structure Determination by X-ray Crystallography*, 3rd edition, Plenum Publishing Corporation.

The intensity of scattering from the liquid is then given by the product of the scattering function and the radial distribution function, integrated over all space. Since the integrals over θ and ϕ evaluate to 4π, we have

$$I(S) = 4\pi C(S) \int_0^\infty r^2 g(r) \sin(Sr)/Sr \, dr \tag{3.123}$$

where $C(S)$ is related to the X-ray atomic scattering factor. The Fourier inversion of (3.123) is

$$g(r) = \frac{1}{2\pi^2 r} \int_0^\infty [I(S)/C(S)] S \sin(Sr) \, dS \tag{3.124}$$

which represents a formal definition of the radial distribution, and shows how it may be calculated from X-ray scattering data.

3.2.3.2   Solids and fluids

The radial distribution function can be used to characterize a solid. Argon crystallizes with a face-centred cubic unit cell, like that of gold (Fig. 1.5). If the unit-cell side is a, the first three (sharp) maxima will occur at positions S inversely proportional to $a/2$, $a/\sqrt{2}$ and $a/\sqrt{3}$; other maxima will be found at higher values of S (smaller distances). Crystalline solids are not usually studied in this way because the regularity of unit-cell packing imposes a long-range constructive interference on the X-ray scattering pattern, giving rise to an image of the lattice (reciprocal lattice) of sharp, spot maxima.

If we examine a monatomic gas by X-ray diffraction, we obtain a graph of the form shown in Fig. 3.14. In a gas, such as argon, there is no order; no particular value of r

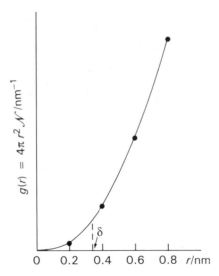

Fig. 3.14.   Radial distribution function for a gas: the parabolic form derives from a dependence of $g(r)$ solely on volume; realistically, the curve drops discontinuously to zero at $r = \delta$, because of strong repulsion for $r < \delta$.

predominates. Hence from our discussion $g(r)$ is unity, and the number of atoms in spherical shells of radius r will be constant, proportional to the square of the given value of r. Thus, we expect the parabolic form shown by Fig. 3.14; the proportionality constant is the number density $\mathcal{N}$, equal to L/V_m for one mole of gas. If the temperature of the gas were increased the parabolic form would remain, but the value of $g(r)$ at any value of r would decrease because the spherical shell would then contain fewer molecules, that is, $\mathcal{N}$ would be smaller in magnitude. There is a certain artificial nature about Fig. 3.14: at a value of r less than the value δ, which is approximately 0.3 for argon, $g(r)$ would in fact tend to zero. Thus, more realistically, the curve should follow the dashed line for $r < \delta$.

If liquid argon is subjected to X-ray diffraction, we obtain a trace for $g(r)$ as a function of r, typified by Fig. 3.15. The radial distribution function for the gas is now modified by the correlation between pairs of atoms arising from the localized order in the liquid. The first peak, at ~ 0.37 nm, represents the nearest neighbours, and the area under the peak is the number of such neighbours:

$$A(r) = 4\pi \mathcal{N} \int_0^{r_{max}} r^2 \, g(r) \, \mathrm{d}r. \qquad (3.125)$$

The upper limit r_{max} is not easy to determine, and an alternative calculation is to take the upper limit in (3.125) as the value of r at the peak, and then double the result of the integration. The coordination number may be taken as the mean of two determinations. The curve of Fig. 3.15 tends to the parabolic form of Fig. 3.14 as r increases, because the distribution in the liquid becomes uniform at large distances.

If we return to solid argon for a moment, we know what values of A to expect for the first three peaks mentioned above. If we refer to the structure shown in Fig. 1.5, the

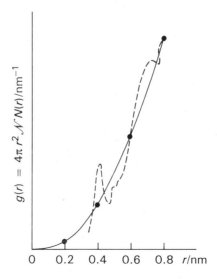

Fig. 3.15. Radial distribution function for liquid argon near the triple-point temperature (83 K). The area under the first peak corresponds to an average coordination number of ~ 10.4 at an average r of 0.37 nm; a second enhanced distribution occurs at ~ 0.70 nm.

areas of the three peaks are, in the order previously given, the integral values 6, 12 and 8 respectively (corresponding to Bragg reflexion from the planes {100}, {110} and {111}), and the value of 12 is the maximum coordination number in a regular cubic close-packed solid of identical spheres.

For the curve shown for the liquid (Fig. 3.15), which is at a temperature near the triple point, the coordination number is approximately 10.4; as the temperature increases, this value decreases continuously to 4.2. Near the triple-point temperature the high coordination number is approaching the maximum for a close-packed solid, but the range and non-integral nature of the coordination number in liquids is one feature that distinguishes them from solids.

In a solid, the coordination number does not change with temperature, unless a phase change occurs which leads to a different structure type, whereupon the coordination number changes discontinuously to another integer value. Thus, for example, at a pressure of about 10 atmosphere, sodium chloride transforms from the structure type shown in Fig. 1.3 (coordination number, 6) to the cesium chloride structure type (coordination number, 8; see Fig. 4.2).

The internal energy of a liquid (3.122) is not easy to evaluate because $g(r)$ is not a simple function. In order to show the quantities involved, we shall adopt a simplified procedure in examining argon as both gas and liquid at the critical point.

Argon gas At the critical point see Table 3.1), $\mathcal{N} = L/V_{m,c}$ (gas) which, from (3.53), is $L/(0.09666 \times 10^{-3}\,\mathrm{m^3\,mol^{-1}})$, or $6.23 \times 10^{27}\,\mathrm{m^{-3}}$. Since there is no correlation in the gas $g(r) = 1$, and we will assume a Lennard-Jones potential energy function (3.90), with $\varepsilon_L = 124k_B$ J and $\delta = 0.34$ nm (see section 3.1.5.1). We now have for the configurational energy term of (3.122)

$$4\pi \, L/2 \, \mathcal{N} \, \varepsilon_L \int_{0.34\,\mathrm{nm}}^{\infty} r^2 \left\{ (\delta/r)^{12} - (\delta/r)^6 \right\} \, \mathrm{d}r$$

and the integral evaluates to

$$\left[-\delta^{12}/9r^9 + \delta^6/3r^3 \right]_{0.3\,\mathrm{nm}}^{\infty} = 5.60 \times 10^{-30}\,\mathrm{m^3}.$$

Introducing the other data from the text leads to a value of -0.36 kJ mol^{-1} for the configurational energy. The kinetic energy, $\frac{3}{2}\mathscr{R}T$ is 1.88 kJ mol^{-1}, so that the total internal energy U_m(gas) = 1.52 kJ mol^{-1}. We remarked earlier that, in the gaseous state, the kinetic energy is necessarily greater than the configurational energy; here, we have given that statement a more quantitative expression.

Argon liquid The differences from the calculation for the gas are first that the value for $\mathcal{N}$ is now (from Table 3.1) $L/(75.2) \times 10^{-6}\,\mathrm{m^3\,mol^{-1}})$, or $8.01 \times 10^{27}\,\mathrm{m^{-3}}$. Second, and more importantly, we need to integrate the complex function in the integrand of (3.122). In practice, it is carried out by simulation procedures. Here, however, we shall use an average value of $g(r)$ over the range $r = 0$ to 1 nm. Introducing these data leads to a value of -2.72 kJ mol^{-1} for the configurational energy in the liquid, and -0.84 kJ mol^{-1} for the total energy. Refined calculations have given a value of -2.48 kJ mol^{-1} for the configurational energy at the critical temperature, compared to an experimental value of -2.47 kJ mol^{-1}, but the results that we have obtained for the gas and liquid states of argon in the critical temperature may be deemed to be satisfactory at this stage.

3.2.4   Equation of state for a fluid

The connection between the intermolecular potential function and the macroscopic p, V, T properties of a fluid can be made in terms of an equation of state. In the case of a gas it may be simplified to a virial equation (3.45). If the interactions in the gas are small, (3.45) may be truncated after the second term. The coefficient $B(T)$ may be related to the pair potential function $V(r)$ through the methods of statistical thermo-dynamics:

$$B(T) = 2\pi L \int_0^\infty [1 - \exp(-V(r)/k_B T)] \, r^2 \, dr. \qquad (3.126)$$

A full treatment of this subject, however, lies outside the remit of the present book, and the reader is referred to specialized texts on the subject.[1]

If the interatomic forces are zero, $B(T)$, and higher virial coefficients, are zero, and (3.45) then reduces to $pV_m = \mathscr{R}T$, which is the equation of state for 1 mol of the ideal gas. The integral in (3.126) can be readily evaluated for the hard-sphere model (section 3.1.5). We have $V(r) = \infty$ for $r \leq d$, but zero otherwise; thus, (3.126) becomes

$$B(T) = 2\pi L \int_0^d r^2 \, dr \qquad (3.127)$$

and, on integrating, we obtain

$$B(T) = 2\pi L d^3/3. \qquad (3.128)$$

If we take the value of d as 0.3 nm (for argon), $B(T)$ evaluates to 34×10^{-6} m^3 mol^{-1}, which is in very good agreement with the value given earlier (see section 3.1.2). It may be noted that the temperature dependence of (3.126) is lost in the simplicity of the hard-sphere model: (3.127) is an equation for the van der Waal's constant b (32×10^{-6} m^3 mol^{-1} for argon; Table 3.1). Thus, we obtain further evidence that b is related to the hard-sphere repulsion part of the intermolecular potential: for, at low density, the van der Waals' equation can be written as $p(V_m - b) \approx \mathscr{R}T$, and we see that $pV_m/\mathscr{R}T \approx 1 + b/V_m$, which is the equivalent to (3.45) to two terms, with $B(T) = b$.

In the case of a liquid the equation of state is more complex, as the radial distribution function must be included explicitly. The statistical mechanics of a simple liquid leads to the equation of state:

$$p = \mathscr{N} k_B T - \frac{\mathscr{N}^2}{6} \int_0^\infty g(r) \left\{ r \frac{dV(r)}{dr} \right\} dr \qquad (3.129)$$

or in energy terms

$$pV = N k_B T - \frac{\mathscr{N}N}{6} \int_0^\infty g(r) \left\{ r \frac{dV(r)}{dr} \right\} dr \qquad (3.130)$$

N being the number of molecules in the volume V; the term $\{r \, dV(r)/dr\}$ is known as the *virial*, Equation (3.128) has been investigated by computer simulation techniques. Two methods that have received most attention are the *Monte Carlo*[2] method, which leads to equilibrium thermodynamic parameters, and the $g(r)$ function, given a form

[1]See, for example, D. Chandler (1987) *Introduction to Statistical Mechanics*, OUP; T. L. Hill (1960) *Introduction to Statistical Thermodynamics*, Addison-Wesley.
[2]See, for example, N. A. Metropolis, A. W. Rosenbluth, M. N. Rosenbluth, A. H. Teller and E. Teller (1953) *J. Chem. Phys.* **21**, 1087; Y. Marcus, (1977) *Introduction to Liquid State Chemistry*, Wiley.

for the potential $V(r)$; and the *molecular dynamics*[1] (time-dependent) method, which provides also measures of transport properties for the liquid. Both methods examine small numbers of particles (*ca* 10^3), being limited by the time length of computation, but include procedures which endeavour to ensure that the results are representative of a bulk sample. The success of the methods may be judged by calculating a parameter that depends upon the potential function, such as the internal energy, U_m. We shall look briefly at these two techniques for studying liquids and liquid structure.

3.2.5   Monte Carlo method

A model is set up in which an elementary cubic unit cell containing an initial configuration of *ca* 10^3 atoms is repeated by side-by-side stacking in three dimensions, so that the macroscopic liquid is generated by lattice-type translations of the cubic cell. Fig. 3.16 is a projection of the model after elapsed simulation time; the cell side a is chosen such that the desired number density, at a temperature T, $\mathcal{N}$ is achieved ($\mathcal{N} = N/V$, where $V = a^3$). The initial configuration $\mathbb{R}$ of N atoms is based on an appropriate Bravais lattice, typically face-centred cubic for simple, monatomic liquids.

The model of Fig. 3.16 contains a translational symmetry that is not present in the liquid. However, it is assumed that provided the range of interaction between atoms is

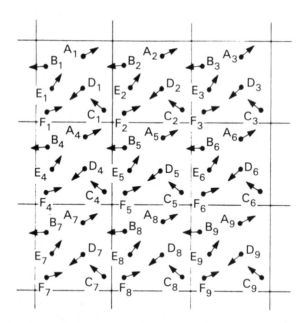

Fig. 3.16. Two-dimensional projection of a configuration of atoms in a Monte Carlo computer simulation of a simple liquid. The positions for the atoms in any one (cubic) cell are random; the other cells are added by equal translations in three dimensions. After sufficient simulation time an equilibrium arrangement is achieved.

[1]See, for example, B. J. Alder and T. W. Wainwright (1957) *J. Chem. Phys.*, **27**, 1208; *idem* (1959) *ibid*, **31**, 459; W. W. Wood and J. D. Jacobson (1957) *ibid.*, **27**, 1207; A. Rahman (1964) *Phys. Rev.* **136**, A405.

less than $a/2$, the potential experienced by any given atom is not affected by the symmetry of the model. It is clear that motion of the atoms will take them out of the unit cell. However, as B_1 moves out of its cell, the image of B_2 moves in to take its place, and so on. In this way, the atom density is conserved in each cell.

For a configuration of atoms, the potential energy ε_i associated with a particular atom i is determined, using a potential function based, for instance, on the hard-sphere or the Lennard-Jones 12–6 model. A given atom is then selected and random increments δx, δy, δz applied to its coordinates, and a new energy ε_i' calculated. If ε_i' is less than or equal to ε_i, then the new configuration is accepted.

If ε_i' is greater than ε_i, a random number ζ is selected, such that $0 < \zeta < 1$. If $\zeta < \exp[-(\varepsilon_i' - \varepsilon_i)/k_B T]$, the new configuration accepted; otherwise, the configuration is rejected. This procedure is repeated many times, typically 10^5 to 10^6. The first 10^4 to 10^5 (ca 10%) configurations are strongly dependent on the initial conditions, and are discarded in order to achieve an equilibrium distribution.

The remaining $\mathcal{M}$ configurations are thus generated with relatively low energies, and are distributed according to the Boltzmann probability $\Pi(\mathbb{R}^N)$, given by

$$\Pi(\mathbb{R}^N) = \frac{\exp[-\varepsilon(\mathbb{R}^N)]/k_B T}{\displaystyle\int_{\mathbb{R}^N} \exp[-\varepsilon(\mathbb{R}^N)]/k_B T} \tag{3.131}$$

Thus the configurational average A of a function of the coordinates $\overline{A}(\mathbb{R}^N)$ is then given by

$$\overline{A} = 1/\mathcal{M} \sum_{i=1}^{\mathcal{M}} A(\mathbb{R}^N) \tag{3.132}$$

Some results for the pressure and internal energy of liquid argon are listed in Table 3.5, and it can be seen that the agreements with the corresponding experimental values are good.

Monte Carlo simulations are subjected to the errors common to this type of procedure; they will be discussed in the next section as molecular dynamics presents similar sources of error.

3.2.6   Molecular dynamics

Molecular dynamics treats the evolution with time of systems of particles that interact through conservative forces under the laws of classical mechanics. In other words, it

Table 3.5 Experimental and calculated values for pressure and internal energy for argon

T/K	$10^6 V_m/m^3\ mol^{-1}$	p/atm		$-U_m/kJ\ mol^{-1}$	
		Calc.	Expt	Calc.	Expt
100.0	29.7	116	115	5.52	5.54
140.0	41.8	18	37	3.81	3.86
150.7	75.2	49	49	2.48	2.47

tracks the motion of molecules in condensed phases by solving Newton's equations of motion:

$$m \frac{d^2\mathbf{R}_i}{dt^2} = \mathbf{F}_i$$

$$\left.\mathbf{F}_i = -\sum_{j \neq i} \frac{\partial}{\partial \mathbf{R}_{ij}} V(R_{ij})\right\}$$

(3.133)

The force F_i on each atom is computed from the derivative of the pair potential with respect to partial coordinate, summed over all neighbours.

Solids are well represented generally by the phonon, or lattice dynamical, theory and gases by the kinetic theory. Consequently, the majority of applications of molecular dynamics have been in the field of liquids, particularly in elucidating transport and equilibrium properties.

In applying molecular dynamics to a simulated liquid system, a set of initial coordinates may be generated from a Bravais lattice set of sites, usually face-centred cubic, at the required density. Initial momenta configurations $\mathbb{Q}$ can be assigned randomly, so that the system is at the desired total energy. Periodic boundary conditions are used as described above for Monte Carlo simulations.

Many molecular dynamics calculations are carried out with the hard-sphere intermolecular potential. It has a useful computational simplicity, but shows also that the structure of simple liquids is almost independent of the chemical nature of the liquid, and well-approximated by the interaction of rigid particles. This idea was present implicitly in the earlier work of Bernal on physical models of liquids (see section 3.2.8). The computational power now available has led to the use of more realistic intermolecular potentials, such as the Lennard-Jones 12–6 model. In turn, it has been shown that real intermolecular forces are long-range, with an R^{-6} attractive potential arising from van der Waals' forces. If the system under consideration is homogeneous, the long-range terms can be treated satisfactorily as a uniform background potential.[1]

Detailed descriptions of the various molecular dynamics algorithms and procedures have been given in the literature.[2]

The initial configuration will evolve into one that is characteristic of a liquid in about 1 picosecond, so that simulation of 100 to 1000 ps suffices to produce averages of both thermodynamic (time-dependent) parameters and time-dependent transport properties, such as the flow of momentum (viscosity; see section 3.2.9) or heat. In this context, 'time' is simulated physical time in the system of N atoms, unrelated to the length of computational time, which may be measured in hours.

As in all numerical processes, errors arise in simulation calculations: they may be number-dependent, potential-dependent and statistical in nature. Checks may be carried out to ensure conservation of energy and momentum, and that molecules do not overlap one another. Intermolecular potentials are subject to truncation errors;

[1] L. Verlet (1967) *Physical Review*, **159**, 98.
[2] See, for example, M. P. Allen and D. J. Tildesley (1989) *Computer Simulation of Liquids*, OUP, and references therein; J. P. Hansen and I. R. McDonald (1986) *Theory of Simple Liquids*, Academic Press; G. Ciccotti and W. G. Hoover (Editors) (1986) *Molecular Dynamics Simulation of Statistical-Mechanical Systems*, North-Holland; G. Ciccotti, D. Frenkel and I. R. McDonald (1987) *Simulation of Liquids and Solids*, North-Holland.

they may be reduced by smoothing the potential at the cut-off distance so that molecules do not experience a discontinuity in potential energy on moving through the cut-off distance.

The most serious errors arise from inadequate statistical sampling. Averages over large blocks should be uncorrelated, with a Gaussian distribution about the mean. The variance σ^2, given by $\overline{X^2} - \overline{X}^2$, leads to a 95% confidence limit expressed as $\overline{X} \pm 2\sigma/\sqrt{n}$, where n is the number of observations. The problem in reducing statistical errors lies in the fact that they are inversely proportional to the square of the run time. Thus, a particular accuracy obtained through a one-day run will need approximately three months running to obtain one more significant figure.

Fig. 1.8a,b (Chapter 1) are the results of molecular dynamics calculations of the crystal liquid interface of argon, using 1500 atom sets. In Fig. 1.8a, the atoms are vibrating about their mean positions in the solid state, the sites of a face-centred cubic lattice. Fig. 1.8b shows the trajectories of the atoms now in a typically liquid phase.

3.2.7   Computer simulation of liquid water

Many attempts have been made to simulate the properties of water by both Monte Carlo and molecular dynamics technqiues. Care is needed in calculating the pair potential between water molecules, because of the relatively long range of the dipolar interactions.[1] However, the major problem is the uncertainty in specifying the interaction between complex molecules like water. Some results for a simple pair potential are listed below.

Table 3.6 lists the configurational energy $\mathscr{E}$, the second term on the right-hand side of (3.130), an equation of state function and the constant-volume heat capacity for the systems studied, and compares the results with those from both molecular dynamics and experiment. The structural properties of water were addressed by computing the radial distribution functions for O—O, O—H and H—H, by sampling the pair distributions after every 250 configurations. Table 3.7 lists results the positions and heights of the maxima for a 256-molecule system, and compares it with both molecular dynamics and experimental results.[2]

The results of this simulation work can be seen to be very satisfactory in their representation of thermodynamic and structural properties of liquid water. Water has been subjected to many subsequent simulation calculations of this nature.[3] and the above general conclusions confirmed and refined.

Table 3.6 Thermodynamic properties of water by computer simulation

System	$\mathscr{E}$/kJ mol^{-1}	pV/ T	C_V/J K^{-1} mol^{-1}
216[a]	-43.1	0.05	100
256[b]	-39.9 ± 0.3	0.6 ± 0.3	70
Expt	-41.4	0.5	75

[1]See, for example, M. P. Allen and D. J. Tildesley (1989) *Computer Simulation of Liquids*, OUP.
[2]F. H. Stillinger and A. Rahman (1974) J. Chem. Phys. **60**, 1545, table reference (a); A. J. C. Ladd (1977) *Molecular Physics*, **33**, 1039, table reference (b).
[3]See, for example, P. Barnes, J. L. Finney, J. D. Nicholas and J. E. Quinn (1979) *Nature* **282**, 459, K. Watanabe and M. L. Klein (1989) *Chemical Physics*, **131**, 157; M. P. Allen and D. J. Tildesly (1989) *Computer Simulation of Liquids*, OUP.

Table 3.7 Structural properties of water by computer simulation: positions r and heights M of maxima in radial distribution functions

	System	r_1/nm	M_1	r_2/nm	M_2
g_{O-O}	216([a])	0.285	3.09	0.470	1.13
	256([b])	0.285	3.11	0.530	1.06
	Expt	0.283	2.31	0.425	1.08
g_{O-H}	216([a])	0.190	1.38	0.340	1.60
	256([b])	0.191	1.24	0.332	1.53
	Expt	0.190	0.80	0.335	1.70
g_{H-H}	216([a])	0.250	1.50	0.390	1.20
	256([b])	0.250	1.15	0.375	1.07
	Expt	0.235	1.04	0.400	1.08

3.2.8　Use of models

The configuration obtained by a Monte Carlo simulation of a simple liquid shows that it is random packed. The hard-sphere model can be simulated by the packing of steel balls into a deformable, irregular-shaped rubber ball. If the centres of the balls in the model are joined by straight lines which are then bisected by perpendicular planes, the resulting polyhedra are Voronoi polyhedra, which fill space completely (seen in projection in Fig. 3.17). The Voronoi polyhedra can be simulated if plasticine balls are used in place of steel; individual balls from the practical simulation then have the shape of Voronoi polyhedra. The theoretical average number of faces and coordination number are 5.1 and 13.6 respectively, and the models show corresponding average values of 5 and 13.

The closest regular packing of equal spheres, as in many metals, has a coordination number of 12, with a volume per sphere of ~ 1.35 (1/0.74). The close packing of irreglar polyhedra leads to a volume per sphere of $\sim 1.35 \times 13.6/12$, or 1.53, an increase of about 13% on regular close packing.

The coordination number can also be obtained from the steel-ball model of the liquid. It is immersed in paint, drained and left to dry. Points of contact in the form of dots, for touching contacts, or polygons—Dirichlet polygons, the two-dimensional equivalent of Voronoi polyhedra—for closer contacts, are found. A model due to Bernal (1959) of the random packed liquid, or *heap*, is shown in Fig. 3.18a; it may be contrasted with the regular array of identical spheres, or *pile* (Fig. 3.18b), for which the coordination number is 12.

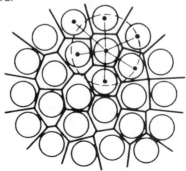

Fig. 3.17. Voronoi polyhedra in projection (Dirichlet polygons) around randomly packed atoms in the model of a sample liquid; the average coordination number here is 5.5.

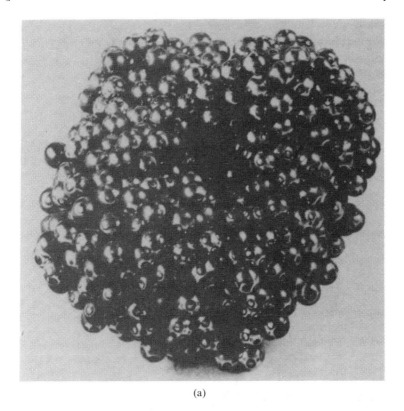

(a)

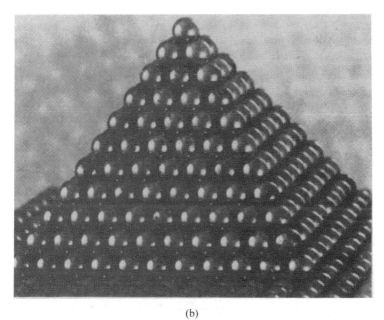

(b)

Fig. 3.18. Stacking of equal spheres: (a) random close-packed *heap*, of average coordination number 8.5; (b) regular close-packed *pile*, of coordination number 12 (after J. D. Bernal (1965) in *Liquids: Structure, Properties and Solid Interactions*, ed. T. J. Hughel, Elsevier, and reproduced with permission).

3.2.9   Viscosity

We take next the property of viscosity, as we did for gases in an earlier section of this chapter. We refer to Fig. 3.11a and, by analogy, the momentum flux $J_{x(y)}$ in the liquid is given by

$$J_{x(y)} = -\eta \frac{dv_y}{dx} \tag{3.134}$$

where η is the coefficient of shear viscosity, and the shearing force per unit area F_y is $-J_{x(y)}$. We saw that the viscosity of a gas was independent of applied pressure, but was proportional to the square root of the temperature; liquids behave rather differently.

Poiseuille's equation for the rate dv/dt of streamline flow of a liquid through a tube of radius r and length d is

$$\frac{dv}{dt} = \frac{\pi r^2 (p_2^2 - p_1^2)}{16 p_0 d\eta} \tag{3.135}$$

where p_1 and p_2 are, respectively, the lower and higher pressures at the ends of the tubes, and p_0 is the ambient pressure, often the same as p_1. Thus, we see that for a liquid there is a linear dependence of η on pressure.

If the viscosity is measured at different temperatures, a linear dependence of $\ln \eta$ on $1/T$ is found, which indicates that

$$\eta = C \exp(E_{m,\eta}/\mathcal{R}T) \tag{3.136}$$

where C is a constant; we alluded to this equation in Chapter 1. It follows, therefore, that the viscosity of a liquid decreases with an increase in temperature. The molar energy of activation for viscous flow $E_{m,\eta}$ is usually considered as a barrier to viscous flow. However, it has been shown[1] that if $1/\eta$ is plotted against $(V_m - V_0)/V_0$, where V_0, in the same units as V_m, is the volume occupied by the molecules themselves, a linear plot is obtained:

$$1/\eta \propto (V_m - V_0)/V_0. \tag{3.137}$$

Thus, the reciprocal of η, or *fluidity*, is proportional to the unoccupied volume in the liquid which, not surprisingly, is sensitive to the thermal expansivity of the liquid. It is possible, then, that the dependence on T expressed by (3.136) may be due to the variation of $(V_m - V_0)$ with temperature.

3.2.9.1   *Liquid viscosity by molecular dynamics*
Computer simulation of viscous flow by molecular dynamics has shown that a difference exists between the structures of a simple liquid at equilibrium and under viscous flow. The liquid at equilibrium has an isotropic structure, which is intuitively expected of a liquid, and the radial distribution function $g(r)$ depends only on the radial parameter r. Under flow, the (spherical) radial surface is distorted to an ellipsoidal shape, so that the radial distribution function is governed both by r and by the shape of the surface.

The flow of simple isotropic liquids may be described in terms of two viscosity coefficients, η and η_v. The first of these coefficients relates to the *shear* viscosity, and is the parameter that we have used in our discussion of viscosity so far. The *bulk* viscosity coefficient η_e describes the irreversible resistance to changes in volume, in

[1]J. H. Hildebrand (1978) *Faraday Discussions, Chemical Society*, **66**, 151.

Table 3.8 Values for η and η_v from molecular dynamics simulations of liquid argon near the triple-point temperature[1]

N	$k_B T/\varepsilon$	$\eta \delta^2/\sqrt{m\varepsilon}$
108	0.728	2.97
256	0.715	2.92
108 − 324	0.722	2.95
108 × 8	0.715	3.0
Exprt	—	2.0

N	$k_B T/\varepsilon$	$\eta_v \delta^2/\sqrt{m\varepsilon}$
108	0.728	1.13
256	0.715	0.89
54	0.715	1.55
Exprt	—	2.0

addition to the reversible resistance governed by the bulk modulus. In fluids of low density, η_v is small and is often omitted in discussions of viscosity. In dense fluids, its importance is increased, but experimental measurement of η_v is difficult. However, simulation calculations by molecular dynamics have permitted η_v to be established.

Table 3.8 indicates some results obtained by molecular dynamics for the shear and bulk viscosity coefficients of liquid argon near the triple point. The Lennard-Jones potential (3.90) has been assumed, and δ (0.34 nm) and $\varepsilon(\varepsilon/k_B = 120$ K) are the parameters of that function.

The results are encouragingly satisfactory: this subject is of both recent and currently active research, and the reader seeking further information is invited to examine more detailed sources of reference.[2]

3.2.10 Quantum liquids

We mentioned quantum liquids at the beginning of section 3.2, as the second of two main classes of liquids. Experimental work first on helium-4 (^{4}He) and then on helium-3 (^{3}He) showed that in certain circumstances they can flow without any viscous effect; such liquids are called *superliquids* (quantum liquids).

Fig. 3.19 shows the phase diagram for helium-4, and it has several unusual properties. It is clear that the solid and gas phases never coexist at any temperature, because of a large zero-point energy of vibration; and only at 25 atmosphere or greater is it possible to solidify helium, whatever the temperature. The liquid phase He_I is a normal liquid, but He_{II} is a superfluid.

The failure of helium to solidify except under pressure cannot be explained by classical mechanics. The attractive energy in helium is the London dispersion energy (3.86), which is very small. However, as $T \rightarrow 0$, the thermal energy likewise becomes vanishingly small. Hence, the reason that a solid does not form at normal pressures rests on the zero-point energy of vibration. If the pressure on the solid is reduced below 25 atm, the zero-point energy effectively 'shakes' the atoms apart.

If we look upon an atom in He_{II} in terms of a particle in a cubic box formed by neighbouring atoms, then from (2.64) the zero-point energy is $3h^2/8ma^2$, where m is

[1]W. G. Hoover *et al*. (1980) *Physical Review A*, **22**, and references therein.
[2]See, for example, M. P. Allen and D. J. Tildesley (1989) *Computer Simulation of Liquids*, OUP; A. J. C. Ladd (1984) *Molecular Physics*, **53**, 459.

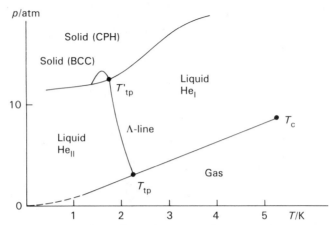

Fig. 3.19. Phase diagram for helium-4 (^{4}He): the solid phase is close-packed hexagonal, with a small region of a body-centred cubic phase; T_c and T_{tp} are the critical point and triple point, respectively; T'_{tp} is an upper triple point; the pressure scale is not linear.

the mass of the atom and a is the length of the box. If the diameter of the atom is d, then in a situation of close contact, a is approximately $2.5d$. Thus, solid helium will not form if

$$3h^2/20md > \tfrac{3}{2}k_B T \qquad (3.138)$$

or

$$0.1\,h^2/md > k_B T. \qquad (3.139)$$

It is clear that the smaller the value of m, the more easily is condition (3.139) realized.

One may reasonably ask how it is that superfluidity does not occur for the even lighter hydrogen. If we examine the London energy term (3.89), we see the dependence on both the square of the volume polarizability α' and the ionization energy I. For helium and hydrogen we have the following data:

	I/eV	$10^{24}\alpha'/cm^3$
He	24.5	2.0
H$_2$	15.5	8.2

The ratio of $\alpha'^2 I$ for H$_2$: He is approximately 10.6, and this is, at low temperatures, the main reason that hydrogen does not form a superfluid; the situation with helium is very rare.

3.2.11   Amorphous solids

We made brief mention of this topic under the heading of 'glasses' in Chapter 1, and we indicated that a glass may be regarded either as a supercooled liquid, or as an amorphous solid, that is, one without long-range order.

There are many solids that can be described as amorphous, such as glass, plastics, resins and toffee. The amorphous condition of matter is a metastable phase. Glass can change extremely slowly from its normal, amorphous condition. In particular, ancient glass may be found in a devitrified condition, that is, the components have crystallized into stable states. The fact that the process is extremely slow implies that it has a very high activation energy. Normal cooling of a liquid produces the crystalline state, which is the thermodynamically stable state.[1] If the liquid is cooled very quickly, the liquid structure may be locked in by the process, so producing an amorphous solid (or supercooled liquid). There is an energy barrier ε_A between the metastable and the stable states, and the probability of the transition to the stable state is proportional to $\exp(-\varepsilon_A/k_B T)$. If $k_B T$ is small with respect to ε_A, the probability factor is very small, and the rate of change may be then so slow as to be experimentally unobservable.

The criterion of crystallinity is the appearance of a spot pattern under X-ray diffraction. Amorphous solids give X-ray patterns very similar to those of liquids (see Fig. 3.13), for which reason they may be considered as supercooled liquids. An amorphous solid may be characterized by a radial distribution function, rather like a liquid, and statistical information on the average coordination number and average nearest neighbour distances may be determined from it.

Glass is, perhaps, the most important amorphous solid. Fig. 1.10 (Chapter 1) illustrates a model for silica glass, and compares it with the structure of α-quartz (see also β-quartz, Fig. 2.48). Evidence for the basic structural unit SiO_4 in silica glass has been obtained through a study of radial distribution functions for this material, obtained by X-ray and neutron diffraction techniques.

Polyethylene ('polyethene') is a long-chain amorphous molecular solid of composition $(-CH_2)_n$. The zig-zag chain is flexible and can take up different conformations. A form of crystallinity, albeit only one-dimensional, can often be induced in polymers by drawing the material into fibres. Fig. 3.20 is an X-ray photograph of an ethene–propadiene polymer with about 70% crystallinity. The photograph shows evidence of a repeat distance, in the fibre direction, superimposed on to the diffuse ring pattern that is characteristic of the amorphous solid.

Crystallinity in a thin section or fibre of polymer can give rise to polarization colours when viewed between crossed polars with a polarizing microscope. It is important to note that these colours can also be produced by strain in the material, so that an inspection of an X-ray diffraction pattern of the polymer is the most sure test of its crystallinity.

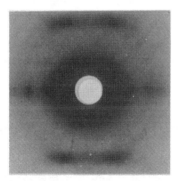

Fig. 3.20.   Fibre X-ray diffraction photograph from an ethene-propadiene polymer at 70% crystallinity; the spot pattern, superimposed on the diffuse ring pattern, is evidence of crystallinity in the vertical (fibre) direction (reproduced by courtesy of Dr. E. J. Wheeler).

---

[1]There may be more than one crystalline state of a solid (polymorphs), but one of them will be more stable than the others at a given temperature.

3.3   Molecular solids

Crystalline solids are characterized first by their three-dimensional regularity. Every crystalline solid has a Bravais lattice as its geometrical basis, and the crystal is built up by the regular placing of the units of structure, atoms or molecules, in a fixed vector relationship at or around each lattice point. The regularity of a crystal structure enables a *unit cell* to be selected, representative of the whole structure. The macroscopic crystal is then obtained (assuming no disorder) by a side-by-side stacking of these unit cells in three-dimensional space.

Often the contents of the unit cell itself are related by symmetry, so that the *asymmetric unit* (of the unit cell) together with the symmetry pattern, or *space group*, represents the whole crystal. To take just two examples, consider first Fig. 1.3, the sodium chloride structure. Its Bravais lattice is designated conventionally by a face-centered (F) cubic unit cell. This unit cell contains four Na^+Cl^- ion pairs. In order to describe the crystal structure, it is sufficient to give the (fractional) coordinates of two species, Cl^- $(0, 0, 0)$ and Na^+ $(0, 0, \frac{1}{2})$, and the space group ($Fm3m$).

In Fig. 3.30c, pentaerythritol molecules are disposed about the origin $(0, 0, 0)$ and the centre $(\frac{1}{2}, \frac{1}{2}, \frac{1}{2})$ of a body-centred tetragonal unit cell. In this example, we need to give the coordinates of one molecule and the space group ($I\bar{4}$) in order to build up the complete crystal structure. A knowledge of the geometry and symmetry of crystals is useful, and the reader is referred to books on that subject, and its application in X-ray crystallography (see also Appendix A1.3.13)[1]

Molecular crystals contain discrete molecules, or in the case of the noble gases just individual atoms, stable in their own right. They are held together by the action of van der Waals' forces, the nature of which we have discussed earlier (see section 3.1.4). Where the molecules are non-polar, the van der Waals' energies are pairwise additive, and the energy of the crystal can be represented by a potential function similar to (3.91), where C is a constant of the attractive potential energy, such as an amalgamation of the species-dependent parameters of (3.88).

The repulsive potential is important in determining the packing in molecular crystals. Both the attractive and repulsive energies in solids include lattice sums that take into account the three-dimensional repeating nature of the unit cell and contents of the crystal structure. Such calculations are complex and will not be considered for molecular crystals. (In the chapter on ionic crystals, the topic will be given a little further discussion). Results for benzene, for example, have given a lattice energy of -53 kJ mol^{-1} at 270 K. This value may be compared with the enthalpy of sublimation for benzene, which is -44 kJ mol^{-1}.

3.3.1   Packing of molecules

In general, chemical species pack together so as to make the most efficient use of the space available—a minimum energy criterion. This principle was formulated first by Barlow in 1895 in respect of simple close-packed structures such as metals or the alkali-metal halides. It was restated and extended by Kitaigorodskii in 1955, for organic crystalline solids. Many organic molecules are irregular in shape, so that in order to achieve the maximum filling of space, the protrusion of a given molecule fits

[1]See, for example, M. F. C. Ladd (1989) *Symmetry in Molecules and Crystals*, Ellis Horwood Ltd; M. F. C. Ladd and R. A. Palmer (1993) *Structure Determination by X-ray Crystallography*, 3rd edition, Plenum Publishing Corporation.

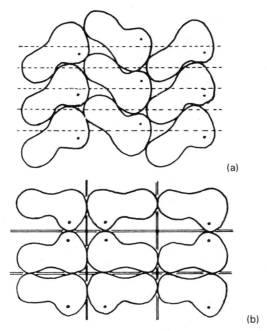

Fig. 3.21. Packing of irregularly shaped organic molecules: (a) economical packing, with the protrusion of any one molecule fitting into a recess in an adjacent molecule; (b) unsatisfactory space-filling, with an increase in the amount of void relative to the packing in (a).

into a recess of an adjacent molecule, a sort of 'lock and key' mechanism. Thus, for a particular molecular shape there exists a preferred mode of packing. Figs. 3.21a, b illustrate, respectively, good space-filling and less economical modes of packing with molecules of a given shape. This aspect of molecular organic crystals has been discussed at length by Kitaigorodskii, and the attention of the reader is directed to accounts of his work.[1]

3.3.2   Classification of molecular solids

Molecular crystalline solids, sometimes called van der Waals' compounds, are of diverse types and extensive in number. Different authors recognize differing numbers of types of molecular compounds; here, we enumerate four so as to provide a basis for classification.

3.3.2.1   Noble gases

Although the noble gases should, strictly, have been considered in Chapter 2, it would not have been convenient to have introduced them there in the detail required before the preparatory work of this chapter had been completed.

The noble gases neon, argon, krypton and xenon crystallize with the face-centred cubic structure exemplified by Fig. 1.5. The coordination number is 12, representative of the closest packing, with a packing efficiency of 74%. Helium is equally close-packed in the solid state but adopts a hexagonal structure, except for a small region of the phase diagram where it is body-centred cubic, with a coordination number of 8.

[1]See, for example, A. I. Kitaigorodskii (1957) *Organic Chemical Crystallography*, Consultants Bureau.

The noble-gas species are non-polar and, hence, the attractive forces are van der Waals' induced dipole–induced dipole (section 3.1.4.6). The weak nature of these forces is clear from the melting-point temperatures of the solids, for example, 83 K in the case of argon. At this temperature, the value of $\frac{3}{2}\mathcal{R}T$ is 1.035 kJ mol^{-1}. If we use ε/k_B (Table 3.1), the energy ε at the equilibrium interatomic distance is -1.032 kJ mol^{-1}. So the kinetic energy and configurational energy are balanced at this temperature, and a solid can exist. The temperature of 83 K is, in fact, the triple point. At a lower temperature, the solid is clearly a stable phase.

3.3.2.2   *Elements*

The elements in periodic groups 15 to 17 form molecular crystals. In the case of the halogens, diatomic molecules X_2 occur in the solid state, linked by London forces Fig. 3.22 is a stereoview of the unit cell of iodine, shared also by the structures of bromine and chlorine; the non-bonded distances range from 0.33 nm to 0.35 nm in these halogens. Fluorine exhibits a close-packed cubic structure, because the molecules are in free rotation and, thus, have time-averaged spherical envelopes of motion.

Among the groups 13–17 structures, an '18 − n' rule[†] tends to hold: *an atom in periodic group n forms 18 − n bonds*. In group 16, this rule is obeyed through the formation of chains or rings; both structures are known for sulphur, with the crown-shaped S_8 molecule[1] (Fig. 3.23) occurring as the thermodynamically stable form. There are, of course, exceptions to this simple 18 − n rule, such as the $[SiF_6]^{2-}$ anion, the nature of which has been discussed in Chapter 2, particularly section 2.14.

Selenium and tellurium form infinite chains, although there is an unstable ring structure for selenium. Fig. 3.24 illustrates the infinite chains of the structures of selenium and tellurium: covalent bonds exist between the atoms in the chains, but the chains are linked by van der Waals' forces.

The group 15 elements arsenic, antimony and bismuth form puckered sheets, each atom being bonded to three others. The bonding between atoms in the sheet takes place through a hybrid of p and s orbitals, so as to give angles greater than 90° (97° in the case of arsenic, and slightly less for antimony and bismuth). The distances between the layers increase down the periodic group from 0.32 nm to 0.35 nm. The atoms have three close neighbours, and three others at a greater distance, forming a distorted octahedral arrangement of nearest neighbours (Fig. 3.25).

Group 14 elements tend to form covalent solids (see section 2.16), with tetrahedral arrangements of covalent bonds, as in the diamond structure of carbon (Fig. 1.2).

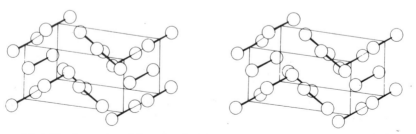

Fig. 3.22.   Stereoview of the unit cell of the structure of iodine (also Br$_2$ and Cl$_2$).

[1]The S_8 molecule contains an S_8 symmetry axis!
[†]This 'rule' was given originally as an '8–n' rule in relation to the earlier numbering of the periodic groups.

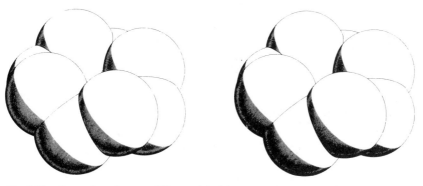

Fig. 3.23.   Stereoview of a space-filling model of the 'crown' structure of the molecule of S_8.

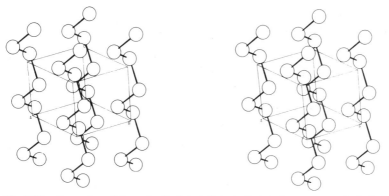

Fig. 3.24.   Stereoview of the unit cell of the infinite helical chain structures of selenium and tellurium

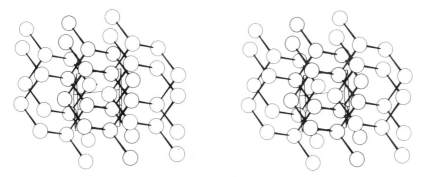

Fig. 3.25.   Stereoview of the puckered sheet structure of arsenic (also antimony and bismuth); the unit cell is shown in thin lines.

There is, however, an increase in metallic character from carbon to tin, and one structure of tin is clearly metallic. Carbon exists in several forms (polymorphs) including graphite (Fig. 3.26). The bonds within the planar layers of the graphite structure arise through sp^2 hybrid orbitals, similar to the bonding in benzene, but the p orbitals normal to the plane of the ring overlap to form π molecular orbitals that are delocalized over entire layers. It is because of this structural effect of conjugated π bonds that graphite is such a good conductor of electricity along directions in the layers; normal to the layers, the electrical conductivity is very small.

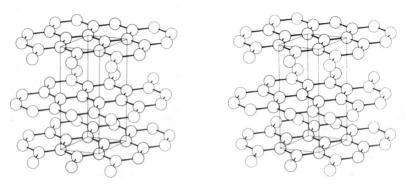

Fig. 3.26.   Stereoview of the planar sheet structure of the graphite allotrope of carbon; the hexagonal unit cell is outlined.

The σ bonds within the ring have a length of 0.142 nm, slightly longer than in benzene, and the distance between the planes is $\sim$ 0.34 nm.

In these structures the tendency of molecules to pack closely can be seen. In the halogens, the molecules are oriented so that one atom of a given molecule is opposite to the centre of the bond of an adjacent molecule, so that the extended planes of molecules, in a zig-zag formation, pack closely over other similar planes. In the helical chain structures, we see that the 'elbow' of one chain fits into a recess in an adjacent helix. In the ring structures, the centres of rings in one layer tend to lie above and below atoms in adjacent layers. All these modes of packing tend to minimize voids, thereby leading to good space-filling in the structures.

3.3.2.3   *Small inorganic molecules*

Although one tends to think of ionic bonding in connection with inorganic compounds, there are many inorganic solids that form discrete molecules which are linked by van der Waals' forces. Examples of these compounds are $HgCl_2$, SnI_4, SF_6, solid CO, N_2O, SO_3, and so on. The structure of $HgCl_2$, for example, is that of linear molecules packed in a manner very similar to that shown by iodine. The coordination may be described as 6:3, reflecting the formula composition, which means that there are 6 nearest Cl neighbours for each Hg, and 3 nearest Hg neighbours for each Cl. Solid carbon monoxide has a close-packed hexagonal structure, like helium. The linear molecules of CO are in free rotation in the solid, and so simulate a close packing of spheres, thus achieving a higher observed symmetry.

There is a gradual transition among groups of related elements and inorganic compounds from ionic types, through layer structures, to the discrete molecules that we have been discussing in this section. We shall consider this topic more fully in the context of ionic compounds, and the reader is also directed to standard works on the crystal chemistry of the elements and of inorganic compounds.[1]

The structures discussed in the last two sections are sometimes treated under the heading of 'covalent solids'. There must always be a certain arbitrariness in the

[1]See, for example, D. M. Adams (1974) *Inorganic Solids*, Wiley; A. R. West (1988) *Basic Solid State Chemistry*, Wiley; H. Krebs (1968) *Fundamentals of Inorganic Crystal Chemistry*, McGraw-Hill; A. F. Wells (1984) *Structural Inorganic Chemistry*, Clarendon Press.

classification of some structures. The author's view is that a classification should rest on the associated structural and physical properties of the compounds. Thus, these structures have been grouped as molecular solids. It is, to some extent, a matter of opinion.

3.3.2.4   Organic compounds

Under this heading, we shall include hydrogen-bonded compounds (although not all examples of hydrogen-bonding occur among organic compounds), clathrate compounds, charge-transfer compounds and π-electron overlap compounds.

Clathrate compounds exist only in the solid state, and may exhibit variable composition. There is often only a very small bonding interaction between the components, the host structure acting as a mechanical trap for the occluded molecule. A well-known example of a host structure is that of 1,4-dihydroxybenzene (quinol), and this compound forms clathrates with small molecules, such as SO_2, CH_3OH and CH_3CN. The quinol–CH_3CN compound has a small dielectric constant, whereas that of the quinol–CH_3OH clathrate is large. We may infer that the occluded CH_3CN molecules are in orientational or dynamic disorder, not strongly linked to the host structure. In the quinol–CH_3OH clathrate, however, the molecules of CH_3OH are locked into position, probably aided by hydrogen-bonding between the —OH groups of methanol and quinol. Ammonia, nickel(II) cyanide and benzene form the clathrate compound illustrated in Fig. 3.27. The Ni^{2+} cations are coordinated octahedrally to two NH_3 molecules, and to four CN^- anions which themselves form a square-planar arrangement. The groups link as shown in the figure to form a cage that retains the benzene molecule.

In contrast, benzene forms *charge-transfer* compounds of the type $C_6H_6X_2$, where X_2 is a molecule of chlorine, bromine or iodine. The typical structure is shown for the chlorine complex in Fig. 3.28. The Cl_2 molecule is oriented normal to the benzene ring plane, and the shortest Cl–ring-centre distance is 0.33 nm. The halogens act as Lewis acids in these compounds; the π electrons of benzene act as donors to σ^* molecular orbitals on the halogens. The movement of electrons is called a charge-transfer transition, and is revealed in the ultraviolet absorption spectra of these compounds.

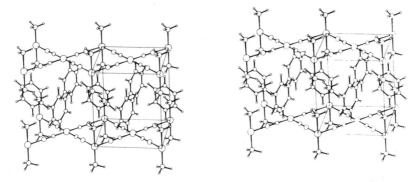

Fig. 3.27.   Stereoview of the nickel(II) cyanide–ammonia–benzene clathrate structure; the unit cell is shown in outline. The circles in decreasing order of size are Ni, N, C and H, and the shortest contact distances are C (benzene) . . . C (CN group), 0.36 nm.

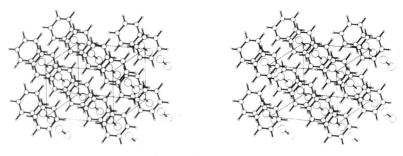

Fig. 3.28. Stereoview of the benzene–chlorine charge-transfer compound; circles in decreasing order of size are Cl, C and H. The shortest distances are Cl . . . benzene ring centre, 0.33 nm.

We shall say a little more about organic charge-transfer compounds after discussing band theory in Chapter 5.

Electron overlap phenomena are important among aromatic hydrocarbons, such as naphthalene and anthracene (see Fig. 3.30f). The molecular planes are oriented such that the delocalized electrons in the π orbitals of adjacent molecules can overlap, leading to enhanced stability. The data in Table 3.9 show an interesting relationship between melting-point temperature T_m and relative molar mass M_r.

Other things being equal, increased molar mass leads to an increased melting-point temperature. In aromatic compounds, however, π electron overlap stabilizes them relative to their alicyclic analogues. Biphenyl $(C_6H_5)_2$ is another interesting compound. In the gaseous (free molecule) state, the planes of the rings are twisted about the central C—C bond by an angle of $\sim 45°$ to each other, so as to achieve a minimum energy conformation. In the crystalline state, the rings are coplanar, and aligned in approximately parallel pairs (Fig. 3.29). The planarity of the biphenyl ring system permits conjugation throughout the whole molecule, and this is revealed by a shortening of the central C—C bond to 0.148 nm. The cohesion in this compound is strongly dependent upon π electron overlap, which is facilitated by the orientation taken up by the planar molecules.

We shall now consider *organic compounds* in general, dividing them for convenience into three main shapes, equant, flat and long. Among these three types, we may consider both polar and non-polar forms of bonding in the solid. In this way, it is

Table 3.9. Melting-point temperatures and relative molecular masses for some aromatic hydrocarbons and their fully hydrogenated counterparts

		T_m/K	M_r
Benzene	C_6H_6	279	78.1
Cyclohexane	C_6H_{12}	280	84.2
Naphthalene	$C_{10}H_8$	353	128.2
Decahydronaphthalene	$C_{10}H_{18}$	230, 241[a]	138.3
Anthracene	$C_{14}H_{10}$	490	178.2
Tetradecahydroanthracene	$C_{14}H_{24}$	335, 366[a]	192.4

[a]Polymorphs.

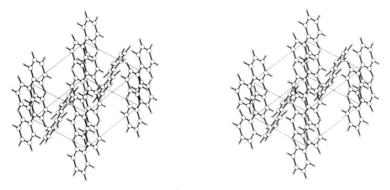

Fig. 3.29. Stereoview of the structure of biphenyl $C_{12}H_{10}$; the shortest intermolecular contact distances are 0.37 nm.

possible to set out a useful classification for a vast range of the organic compounds that are now known. It will be anticipated that this classification is not without anomalies, but it is of considerable assistance in obtaining a general picture of this aspect of solid state chemistry. We shall treat this section only in outline, but with adequate illustrations of typical compounds. The reader may find it of interest to refer to Kitaigorodskii[1] in the context of this particular section.

Among the numerous organic compounds now characterized, we can identify certain structural units that preserve their characteristic shapes, sizes and symmetries within quite small limits; they include bond lengths, bond angles, structural groups and even complete molecular entities. Table 3.10 lists standard values for a range of bond lengths and bond angles. The values depend upon the environment, and the table shows this variation by means of the notation C_4, C_3, N_2, O_1, and so on. The digit indicates the number of atoms directly bonded, the *connectivity*; thus, C_4—C_4 indicates a $C(sp^3)$—$C(sp^3)$ bond, as in ethane, for example.

As an example of a larger structural unit, consider the phenyl moiety: it is usually found to be planar, within experimental error, with C—C bonds of 0.140 nm, C—H bonds of 0.108 nm, and C—C—C and H—C—C angles all close to 120°. According to the nature and position of substituents on the ring, there may be small deviations from these standard values, generally more in the bond lengths than in the bond angles.

It has been found experimentally that intermolecular non-bonded distances between pairs of atoms do not vary greatly. In the absence of hydrogen-bonding, carbon, nitrogen and oxygen atoms exhibit non-bonded, or contact, distances of ~ 0.37 nm among a wide range of compounds. This feature led to the development of van der Waals' radii for atoms, which represent an average minimum distance of approach of atoms in neighbouring molecules. It is, perhaps, not surprising that this value is close to the δ parameter of the Lennard-Jones 12-6 potential function (3.90) for these species.

Table 3.11 lists the van der Waals' radii for a number of common elements. The van der Waals' radius of an atom can be correlated with the size of its outer orbital. Thus,

[1]A. I. Kitaigorodskii, *loc. cit.*

Table 3.10. Standard bond lengths[1] and bond angles

Bond Lengths

Single bond	Length/nm	Single bond	Length/nm
H—H	0.074	C_3—C_2	0.145
C_4—H	0.109	C_3—N_3	0.140
C_3—H	0.108	C_3—N_2	0.140
C_2—H	0.106	C_3—O_2	0.136
N_3—H	0.101	C_2—C_2	0.138
N_2—H	0.099	C_2—N_3	0.133
O_2—H	0.096	C_2—N_2	0.133
C_4—C_4	0.154	C_2—O_2	0.136
C_4—C_3	0.152	N_3—N_3	0.145
C_4—C_2	0.146	N_3—N_2	0.145
C_4—N_3	0.147	N_3—O_2	0.136
C_4—N_2	0.147	N_2—N_2	0.145
C_4—O_2	0.143	N_2—O_2	0.141
C_3—C_3	0.146	O_2—O_2	0.148

Double bond	Length/nm	Double bond	Length/nm
C_3—C_3	0.134	C_2—O_1	0.116
C_3—C_2	0.131	N_3—O_1	0.124
C_3—N_2	0.132	N_2—N_2	0.125
C_3—O_1	0.122	N_2—O_1	0.122
C_2—C_2	0.128	O_1—O_1	0.121
C_2—N_2	0.132		

Triple bond	Length/nm	Aromatic bond	Length/nm
C_2—C_2	0.120	C_3—C_3	0.140
C_2—N_1	0.116	C_2—N_2	0.134
N_1—N_1	0.110	N_2—N_2	0.135

Bond Angles

Apex atom	Geometry	Angle/deg	Example
C_4	Tetrahedral	109.5	CH_4
C_3	Planar	120	C_2H_4
C_2	Bent	109.5	—CHO
	Linear	180	HCN
N_4	Tetrahedral	109.5	NH_4^+
N_3	Pyramidal	107.5	NH_3
	Planar	120	H_2N—CHO
N_2	Bent	109.5	H_2CHN
	Linear	180	HNC
O_3	Pyramidal	190.5	H_3O^+
	Bent	104.4	H_2O

[1]Covalent radii may be defined as one-half of a bond length; thus, $r_{C_4} = 0.77$ nm. However, because of the directionality of the covalent bond, this concept is not of great significance, and is not developed herein.

Table 3.11. Van der Waals' radii of some common species

Atom	Radius/nm	Atom	Radius/nm
H	0.120	C	0.185
—CH$_3$	0.200	Si	0.210
N	0.150	P	0.190
As	0.200	Sb	0.220
O	0.140	S	0.185
Se	0.200	Te	0.220
F	0.135	Cl	0.180
Br	0.195	I	0.215
Half thickness of phenyl ring			0.185

a carbon 2p orbital enclosing 99% of the 2p electron density extends from the nucleus to ~ 0.19 nm, which is very close to the van der Waals' radius for this species.

Two atoms in neighbouring molecules may lie further away than the sum of their van der Waals' radii, because steric effects may inhibit the normal closest approach distances from being realized. In other compounds, non-bonded distances may be significantly less than the sum of the van der Waals' radii. Thus, the distance between two non-bonded oxygen atoms may be as small as 0.24 nm if strong hydrogen-bonding exists between them. We may note that for those atoms that can exist as clearly defined ions, the van der Waals' radius is very close to the corresponding ionic radius. This result is in accord with the fact that repulsive forces increase very rapidly at short distances, because of their $1/r^6$ dependence. For example, the effective size of the bromine species is about the same whether the repulsion energy is balanced against the Coulombic $1/r$ potential function in KBr or against the relatively weaker $1/r^6$ function in C_6H_5Br.

3.3.2.5 *Classification of organic compounds*
We conclude this section with a classification (Table 3.12) of organic compounds, based on the procedure outlined in the previous section. Typical compounds will be illustrated by stereoviews, and the reader is encouraged to extend the study of these classes of compounds through standard reference works.[1]

3.3.3 **Structural and physical characteristics of molecular compounds**
Van der Waals' forces can bond an atom to an indefinite number of neighbours, and they are spatially undirected. In the solid noble gases, van der Waals' interactions are the sole means of cohesion. In other molecular compounds, we find relatively short covalent bonds between atoms, such as Cl—Cl (0.200 nm) in Cl_2, C—C (0.140 nm) and C—H (0.108 nm) in benzene, but with characteristically longer (0.35–0.38 nm) non-bonded, contact distances between adjacent molecular entities.

[1]See, for example, R. W. G. Wyckoff (1963–1971) *Crystal Structures*, Vol. 1–6, Interscience; L. E. Sutton (Editor) (1958, 1965) *Tables of Interatomic Distances and Configuration in Molecules and Ions*, Chemical Society, London; A. I. Kitaigorodskii, *loc. cit.*

Table 3.12. Classification of organic molecular compounds

Molecular shape	Bonding class, and characteristics	
	Non-polar	Polar
EQUANT	Small, isometrically shaped molecules, forming approximately close-packed structures. Examples are methane ethane, hexachloroethane (Fig. 3.30a), adamantane and cubane (Fig. 3.30b). In some structures, such as methane, free rotation is present in the crystal.	Small, isometrically shaped molecules, with dipolar or hydrogen-bonded interactions, rather than close-packing, dominating the structural configuration. Examples are urea, methanol, pentaerythritol (Fig. 3.30c), methylamine and oxalic acid dihydrate (Fig. 3.30d).
FLAT	Molecules lying with their planes nearly parallel. Staggered configurations may be adopted where π electron overlap is possible. Examples are benzene (Fig. 3.30e), hexamethylbenzene, anthracene (Fig. 3.30f) and phthalocyanine.	Molecular packing dominated by dipolar or hydrogen-bonded interactions. Examples are 1,4-dinitrobenzene (Fig. 3.30g) and 4-nitrophenol (Fig. 3.30h).
LONG	Molecules lying in parallel or staggered configurations such as octane (Fig. 3.30i), hexane (Fig. P3.1) and dicyanoethyne (Fig. P3.2). With increasing temperature, some paraffins, such as hexane, rotate in the solid state giving effective cylindrical symmetry to the molecule.	Long, polar molecules, tending to associate in pairs through dipolar or hydrogen-bonded interactions. Alkanammonium salts are ionic, often with the carbon chains in free rotation or static disorder. Example are apidic acid, decanamide (Fig. 3.30j), potassium caprate and propan-1-ammonium chloride.

The dependence of the van der Waals' energy on polarizability is shown clearly by the trends in the melting-point temperatures of the silicon tetrahalides, SiX_4 (Table 3.12). In these compounds, the polarizability of the halogen increases more rapidly from fluorine to iodine than does the corresponding intermolecular distance, with a consequent enhancement of the lattice energy (enthalpy of sublimation). Over a corresponding range of molecular mass among the alkanes, for example, the increase in melting-point is only about 140 K.

Molecular compounds generally form soft, brittle crystals, with low melting-point temperatures and large thermal expansivities. The electrical and optical properties of molecular solids may be said to be the aggregate of those of the component molecules, since the electron systems of the molecules do not interact strongly in the solid state. These properties are, therefore, similar in the solid, melt and solution; in the solid, anisotropy of physical properties is generally marked.

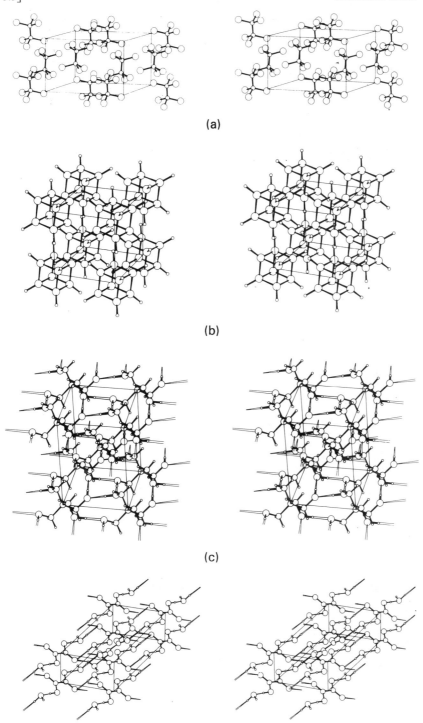

(a)

(b)

(c)

(d)

Fig. 3.30–*continued*

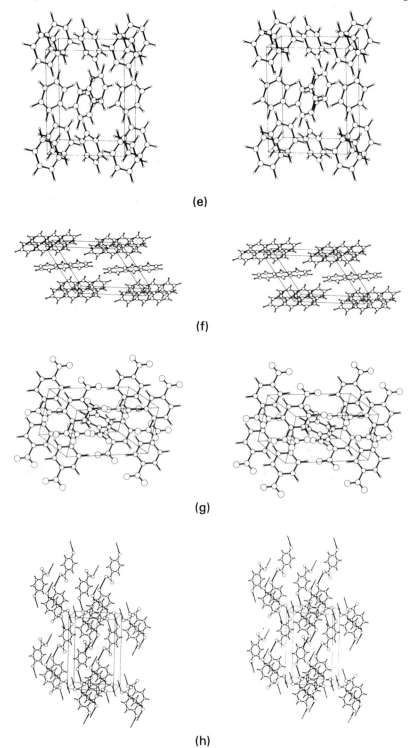

(e)

(f)

(g)

(h)

Fig. 3.30–*continued*

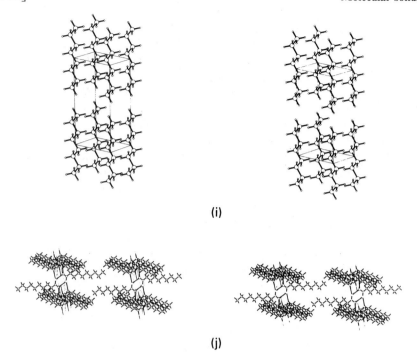

(i)

(j)

Fig. 3.30. Stereoviews of structures to represent the classification of organic structures (Table 3.11). In each example, the unit cell is outlined, and the circles in decreasing order of size represent atoms in the formula in order of decreasing atomic number.

(a) Hexachloroethane, C_2Cl_6: equant non-polar; shortest intermolecular distances 0.37 nm.

(b) Cubane, C_8H_8: equant non-polar; shortest intermolecular distances 0.38 nm.

(c) Pentaerythritol, $C(CH_2OH)_4$: equant, polar; shortest intermolecular $(O—H \cdots O)$ distances 0.27 nm.

(d) Oxalic acid dihydrate $(CO_2H)_2.2H_2O$: equant, polar; shortest intermolecular $O—H \cdots O)$ distances 0.25 nm.

(e) Benzene, C_6H_6: flat, non-polar; shortest intermolecular distances 0.36 nm.

(f) Anthracene, $C_{14}H_{10}$: flat non-polar; shortest intermolecular distances 0.36 nm.

(g) 1,4-Dinitrobenzene, $C_6H_4(NO_2)_2$: flat, polar; shortest intermolecular distances 0.32 nm.

(h) 4-Nitrophenol, $C_6H_4NO_2OH$: flat, polar; shortest intermolecular distances $(O—H \cdots O)$ 0.29 nm.

(i) Octane, C_8H_{18}: long, non-polar; shortest intermolecular distances 0.37 nm.

(j) Decanamide $CH_3(CH_2)_8CONH_2$: long, polar; shortest intermolecular distances $(N—H \cdots O)$ 0.29 nm.

Table 3.12. Melting-point temperatures, polarizabilities and van der Waals' radii sums for the silicon tetrahalides

	SiF_4	$SiCl_4$	$SiBr_4$	SiI_4
Mp/K	183	203	278	394
$10^{40}\alpha$/F m^2	1.0	3.4	4.8	7.3
$\sum r$/nm	0.35	0.39	0.41	0.43

3.3.3.1   *Solubility of molecular compounds*

Molecular compounds are usually soluble in common solvents; the 'rule' like-dissolves-like is well known. Solubility is a chemical reaction between the solute and solvent, and both enthalpic and entropic factors are involved; generally, they are compounded into a free energy for the dissolution process. However, while this parameter describes solubility correctly, it is more informative to consider the interplay of enthalpy and entropy in a study of solubility.

Among ionic solids, enthalpic effects are generally large because strong Coulombic forces are involved. The dissolution of a molecular compound usually involves only a small enthalpy change, and the entropy change may be very significant. We know that naphthalene dissolves in benzene and not in water, and that glucose dissolves in water and not in benzene.

Water is a hydrogen-bonded liquid, and dissolution in water involves *inter alia* the breaking of some hydrogen bonds of the solvent in order to accommodate the solute molecules. In the case of naphthalene and water, there is no mechanism for interaction between the solute and solvent, so that there can be no driving force to break down the solid or to interfere with the hydrogen-bonded structure of the solvent. If a solvent is chosen in which the intermolecular attractions are weak, such as benzene, then although the enthalpic effects are small, the entropy gain of the solution over the (solute + solvent) system is dominant and dissolution occurs.

The reverse situation is met with glucose: with benzene, there are no interactions between this solvent and glucose that will break the (dipole + van der Waals') binding in the solid and, consequently, there is no dissolution. In water, however, the polar —OH groups of glucose interact electrostatically with the water molecules, and the solid is broken down and solvated by the water molecules, leading to dissolution.

In all cases, solubility is governed by the difference between the free energy of the solution and that of the (solute + solvent) system, and a fine balance often occurs. Where sufficient thermodynamic data are available, quantitative results can be obtained, which support the general argument given above (see also section 4.9.4).

3.3.4   Molecular mechanics

Any molecule, as we have seen, is a group of atoms held together by bonding forces. These forces give rise to potential energy functions that govern the conformation of the molecule—bond lengths, bond angles and torsion geometry, for example. More than one type of force is likely to be present in any given molecule. The sum total of them is termed the *force field* for the molecule and, through a configurational energy $\mathscr{E}$, determines the molecular conformation.

Molecular mechanics involves procedures whereby the energy $\mathscr{E}$ of a molecule can be calculated as a function of its conformational parameters and, thence, the preferred minimum energy conformation for the molecules determined. In operating a molecular mechanics procedure for a given system, a trial geometry is postulated in terms of standard bond lengths, bond angles and non-bonded distances (Tables 3.10 and 3.11). Then, by mathematical techniques, such as a simplex search, the geometry is optimized so as to obtain a minimum value for $\mathscr{E}$.

The specification of a force field for a given molecule draws heavily on the experience and data gained through earlier similar work, and a number of computer programs is now available to carry out molecular mechanics calculations.

3.3.4.1 *Some types of procedure*

An elementary form of molecular mechanics was carried out by Ladd (1968)[1] in locating the positions of hydrogen atoms in $BaCl_2 . 2H_2O$ and in $NaBr . 2H_2O$, using an electrostatic force field. The data used specified the water molecule as O—H = 0.099 nm, H—O—H = 104.5°, with $q_O = -0.31$ ($q_H = +0.155$) determined from the dipole moment of 1.84 D, and employed x, y, z atomic coordinates of the heavier atoms determined from X-ray crystallographic studies on the compounds.

The coordinates x', y' and z' of the positive end of the water molecule effective dipole were located by a vector sum of the weighted electrical field strengths around the water oxygen atom. Thus, for the x'-coordinate,

$$x' = \sum_i -q_i x_i / d_i^2 \qquad (3.140)$$

where x_i is the x-coordinate of the ith atomic species of charge (weight) q_i (including sign), and d_i is its distance from the oxygen atom of the water molecule; the sum over i extended to ~ 0.4 nm from the water oxygen atom. The two hydrogen atoms of a water molecule form the ends of a diameter, equal in length to the interproton distance of 0.16 nm, of a circle of centre x', y', z'. The total electrostatic energy was then calculated and minimized with respect to rotation of the H $\cdots$ H diameter about the point x', y', z' from 0° to 180°. The positions of the hydrogen atoms so obtained were in good agreement with those obtained by neutron diffraction, for $BaCl_2 . 2H_2O$, and by proton magnetic resonance, for $NaBr . 2H_2O$.

This method provided a way of locating hydrogen atom positions without recourse to neutron diffraction studies. Hydrogen atoms, in the presence of atoms of high atomic numbers, are not easily revealed by X-ray diffraction techniques, because of the small scattering power of hydrogen for X-rays. In an electrostatic calculation, however, hydrogen atoms are much more significant in relation to the heavier atoms present in a structure.

A large proportion of current work on molecular mechanics is based on the MM1 program (and extensions) of Allinger.[2] In the particular case of organic molecules, the atoms are linked by covalent bonds of given length and directionality, together with non-bonded van der Waals' interactions. Standard values of conformational para-meters (see Tables 3.10, 3.11) enable trial structure models to be constructed.

A force field can be constructed so as to allow for various geometrical effects on the overall molecular conformation. A nonlinear molecule of N atoms can be specified generally in position by $3N$ coordinates. It follows that the molecule has $3N$ degrees of freedom, which may be expressed in another way. The position of the centre of mass of the molecule is governed by three coordinates. Then, three coordinates are needed to express the molecular rotations. Thus, there are $3N - 6$ vibrational degrees of freedom, or $3N - 5$ for a linear molecule because rotation about the molecular axis does not alter the molecular orientation.

The force field is set up with respect to coordinates that are associated with changes in bond lengths and bond angles, and are related to the Cartesian coordinates by linear transformations; they are termed *normal coordinates*. The reader is referred to

[1]M. F. C. Ladd (1968) *Zeitschrift für Kristallographie*, **126**, 141.
[2]N. L. Allinger (1976) *Advances in Physical Organic Chemistry*, **13**, 1.

standard works for more detail of these procedures.[1] Generally, one considers bond stretching, bending, bending–stretching co-operative movements, torsional motion and van der Waals' non-bonded interactions. The total configurational energy $\mathscr{E}$ is then the sum of these effects, in order:

$$\mathscr{E} = \mathscr{E}_s + \mathscr{E}_b + \mathscr{E}_{s-b} + \mathscr{E}_t + \mathscr{E}_v \qquad (3.141)$$

where the minimum value of $\mathscr{E}$ sought corresponds to the energetically preferred conformation. We note, in passing, that the term $\mathscr{E}_v$ is a function such as the Lennard-Jones potential (3.90) or the exp–6 potential (3.91).

Molecular mechanics type calculations may be employed at a lesser level, whereby torsion angles, for example, alone may be varied. The X-ray crystal structure analysis of O,O,O-propan-1,2,3-trithionobenzoate[2] (Fig. 3.31) revealed a short intramolecular distance ($S_2 \ldots C_{38}$) of 0.344 nm. The compound undergoes solid state isomerization by ring closure at the $S_2 \ldots C_{38}$ atoms. Indeed, the significant negative departure from the sum of the van der Waals' radii for carbon and sulphur (0.365 nm) may indicate a latent tendency towards covalency between these atoms that is realized on isomerization.

A potential energy surface was calculated as a function of the torsion angles Φ (C_{28}—O_2—C_{27}—S_2) and Ψ (C_{18}—C_{28}—O_2—C_{27}), and is shown in Fig. 3.32. The shallow minimum occurs at $\Phi = 2.8°$, $\Psi = 155°$, compared to the values 2.80°, 154.9° found by the structure analysis. It shows that there is no significant energy restriction on variations in the values for the torsion angles Φ and Ψ, as might occur in a solid state reaction. Rotations of just 3° in Φ and 25° in Ψ would lead to a minimum $S_2 \ldots C_{38}$ distance of 0.235 nm.

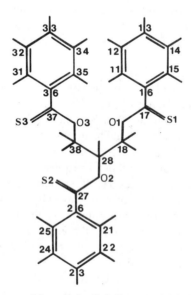

Fig. 3.31.   Chemical structural formula for O,O,O-propan-1,2,3-trithionobenzoate; hydrogen atoms are show as terminal bars.

[1] U. Burkert and N. L. Allinger (1982) *Molecular Mechanics*, A.C.S. Monograph **177**; A. Barlow and M. Diem (1991) *J. Chemical Education* **68**, 35.
[2] J. C. Moore and R. A. Palmer (1991) *Journal of Crystallographic and Spectrographic Research*, **21**, 511.

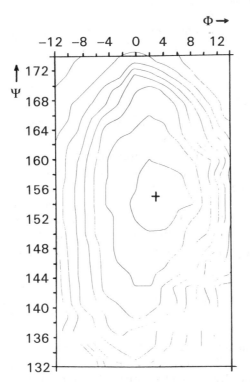

Fig. 3.32. Contour map showing a plot of the non-bonded potential energy function for the O,O,O-propan-1,2,3-trithionobenzoate molecule as a function of the torsional angles Φ (C_{28}—O_2—C_{27}—S_2) and Ψ (C_{18}—C_{28}—O_2—C_{27}). The contours are drawn at intervals of $\sim -3\,kJ\,mol^{-1}$, from $-56.5\,kJ\,mol^{-1}$ to $-77.5\,kJ\,mol^{-1}$; the minimum energy of -79.1 $kJ\,mol^{-1}$ occurs at $\Phi = 2.8°$, $\Psi = 155°$ [after Moore and Palmer, *J. Cryst. Spect. Res.* **21**, 511 (1991)].

3.3.4.2 *Importance of molecular mechanics*

In certain areas of theoretical chemistry, such as the relationship between drug activity and structure, molecular mechanics calculations give the desired results in much less time than with molecular orbital calculations, and they can handle larger structures. The nature of molecular mechanics is more straightforward than that of quantum mechanics, and the conformational input parameters are well recognized.

Certain problems must be recognized as existing within the calculations themselves. An energy surface is a very complex mathematical function, and there are likely to be several positions that correspond to energy minima. Minimization procedures are geared to find the bottom of the nearest potential energy well. Others, maybe even deeper minima, could be missed altogether. A modification of the procedure is to scan the energy surface coarsely to find the approximate locations of wells, and then to optimize on each one, so as to find the deepest minimum.

The reader is directed to reviews of this subject[1] for further information. It is of

[1]D. B. Boyd and K. B. Lipkowitz (1982) *Journal of Chemical Education*, **59**, 269; M. B. Hursthouse *et al*, (1978) *Annual Reports of Progress in Chemistry*, **B75**, 23; S. R. Niketic and K. Rasmussen (1977) *The Consistent Force Field*, Springer-Verlag; N. L. Allinger, *loc. cit.*; O. Ermer (1976) *Structure and Bonding*, **27**, 161, Springer-Verlag.

interest that molecular mechanics now figures in undergraduate degree courses in organic chemistry.

PROBLEMS FOR CHAPTER 3

3.1   Given that, for any gas, $pV = \frac{1}{3}Nm\,\overline{v^2}$, show how (a) Avogadro's, (b) Boyle's, (c) Charles's and (d) Graham's laws follow.

3.2   If the collision cross-section σ for argon at 298 K is 0.34 nm², calculate (a) the pressure at which the mean free path of argon is equal to its collision diameter, (b) the number of collisions per second when the pressure is 5 atm.

3.3   The collision diameter of nitrogen gas is 0.34 nm at 298 K and 1 atm pressure. Assuming that the gas behaves ideally, calculate the mean free path for nitrogen at an altitude where the temperature is 195 K and the pressure 0.045 atm.

3.4   Show that, for a gas that obeys the kinetic theory, the ratio of speeds $v_{max}: \bar{v}: \sqrt{\overline{v^2}}$ is as 1:1.128:1.225.

3.5   The probability that a molecule in a gas has an energy ε is given by the Boltzmann distribution $\exp(-\varepsilon/k_BT)$. By invoking a normalization criterion, show that this equation can lead to the Maxwell–Boltzmann one-dimensional distribution equation.

3.6   Metallic potassium is heated at 900 K in an oven of volume 2 dm³ containing a minute hole in one wall. The atomic collision diameter is 0.45 nm, and at 900 K the vapour pressure of potassium is 115 mmHg. Calculate (a) the mean velocity of the emergent potassium atoms, (b) the mean speed of potassium atoms inside the oven, (c) the frequency of collision made by a single potassium atom, (d) the frequency of collision for all the atoms in the oven, (e) the mean free path of potassium atoms in the oven.

3.7   Convert the van der Waals' equation of state into a virial equation in powers of $1/V$. Hence, obtain a value for the second virial coefficient $B(T)$ in terms of the van der Waals' parameters a and b. Give an expression for the Boyle temperature in terms of a and b, and find its value for carbon dioxide.

3.8   Determine the critical constants for oxygen, given that the van der Waals' constants for this gas are $a = 1.36$ dm⁶ atm mol⁻² and $b = 3.18 \times 10^{-2}$ dm³ mol⁻¹. Hence, evaluate a molecular radius for oxygen.

3.9   (a) The dipole moment for fluorobenzene is 1.70 D. What are the dipole moments for the three difluorobenzenes; which of the answers is the most certain, and why?

   (b) The dipole moment for hydrogen fluoride is 1.83 D, and the H—F bond length is 0.0927 nm. Determine the effective point charges at the hydrogen and fluorine ends of the dipole.

3.10   Calculate the number of air molecules that collide with a plate in one second, assuming that the effective area of the plate is 2.5×10^4 mm². (RAM: O 16, N 14; composition of air: O 20%, N 80%).

3.11   Show that the equilibrium separation of atoms in a gas that is governed by a Lennard-Jones (12–6) potential is given by $r_e = 2^{1/6}\delta$. Assuming that r_e is 0.37 ± 0.02 nm, what is the value of δ?

3.12   The collision cross-section for oxygen is 0.40 nm², and its relative molar mass is 32.00. Calculate the viscosity of oxygen at 0 °C and 400 °C, and the temperature coefficient of viscosity over this temperature range.

3.13   The isothermal compressibility κ is given by $\kappa = -1/V(\partial V/\partial p)_T$. Assuming that liquid argon follows the van der Waals' equation of state, determine an expression for κ in terms of the van der Waals' constants a and b. Find a value for κ at the critical temperature.

3.14   The following data were obtained for the radial distribution function of a simple liquid (the datum at $r = 0.25$ nm is an extrapolation):

$10^{9}\,4\pi r^2\,\mathscr{N}g(r)$nm⁻¹	0	20	90	70	40	45	60	80	110	160	150	140	160	210
r/nm	0.25	0.30	0.35	0.40	0.45	0.50	0.55	0.60	0.65	0.70	0.75	0.80	0.85	0.90

Plot the radial distribution function, and determine the average number of nearest neighbours (coordination number) and the average corresponding non-bonded distance. How may one explain that $g(r) \rightarrow 1$ as r tends to ∞ ?

3.15   Refer to Fig. 1.5, and show that the packing efficiency in this cubic structure is 0.74.

3.16   How would one describe the crystal structure of gold (Fig. 1.5) in concise crystallographic terms?

3.17   Compare the 'long' structures of hexane (Fig. P3.1) and dicyanoethyne (Fig. P3.2); the shortest intermolecular contact distances are 0.36 nm for C . . . C in hexane, and 0.32 nm for C . . . N in dicyanoethyne. Suggest a reason why the molecules of dicyanoethyne do not pack in the simple manner shown by hexane.

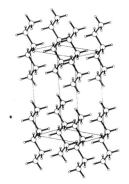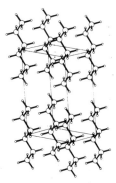

Fig. P3.1.   Stereoview of the structure of hexane C_6H_{14}.

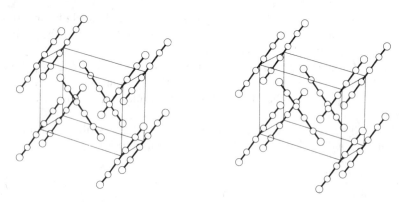

Fig. P3.2.   Stereoview of the structure of dicyanoethyne, NCC≡CCN.

3.18   Measurement of the viscosity coefficient of tetrachloromethane at different temperatures gave the following results:

T/K	278	303	333
η/cP	55.6	46.9	39.6

Use these data to estimate an 'energy of activation' for viscous flow over the given temperature range; 1 cP (centipoise) $= 10^{-3}$ kg m^{-1} s^{-1}.

3.19   If a gas be compressed its temperature rises, and it can be shown in macroscopic terms that the temperature rise ΔT is given by $-p\Delta V/C_V$; ΔV is a negative quantity in compression. How might the same temperature rise be explained in microscopic terms? It might be simplest to think in terms of a monatomic gas.

3.20   (a) Use the particle-in-a-(cubic)-box equation to determine the number $N(\varepsilon)$ of quantum states of energy less than or equal to a given value ε. Show how it can lead to a density of (quantum) states function, that is, a number of states per unit energy range. (b) Why is there a 'hump' in the Maxwell–Boltzmann distribution of speeds? Give your answer in relation to a monatomic gas.

4

Bonding between ions: I

4.1 INTRODUCTION

Compounds that are termed ionic are formed generally between atoms of widely varying electronegativity. One of them, typically a metal, becomes ionized by the loss of one or more electrons, and the other, typically a non-metal, becomes ionized by acquiring one or more electrons. The charged species, or ions, then attract each other by Coulombic forces that are inversely proportional to the square of the distance between them, and vary directly as the product of the charges on the ions.

4.2 ATTRACTIVE ENERGY

If two ions have charges $q_1 e$ and $q_2 e$, considered here as points, separated by a distance $\imath$, then the electrostatic, Coulombic force between them is given by

$$F_E = \frac{q_1 q_2 e^2}{4\pi\varepsilon_0 \imath^2}. \tag{4.1}$$

Earlier we wrote Q_1 and Q_2 for $q_1 e$ and $q_2 e$, respectively. Here, we use the component notation because we shall later need to take cognizance of the individual q values, which may not always be integral.

At infinite distance of separation the interaction between the ions is zero; hence, the Coulombic (potential) energy of the pair of ions at a distance r is

$$U_E(r) = \int_r^\infty \frac{q_1 q_2 e^2}{4\pi\varepsilon_0 \imath^2} \, d\imath$$

$$= \frac{q_1 q_2 e^2}{4\pi\varepsilon_0 r} \tag{4.2}$$

and for ions of opposite sign $U_E(r)$ would be negative, which implies an attractive energy.

In an extended array of ions, as in a crystal, it is necessary to include the effect of all ions present on the result (4.2). In the case of the sodium chloride structure type (Fig.

1.3), for example, we write

$$U_E(r) = -\frac{\mathscr{A} q_1 q_2 e^2}{4\pi\varepsilon_0 r} \qquad (4.3)$$

and $\mathscr{A}$ is a constant for the given structure type, known as the Madelung constant (Madelung, 1918).

4.2.1   Madelung constant

The meaning of a Madelung constant may be considered first in one dimension. Fig. 4.1 represents an infinite row of alternating positive and negative charges of magnitude e, regularly spaced at a distance r. Let any ion (negative in the diagram) be chosen as a reference point, or origin. Its immediate neighbours give rise to an attractive potential energy of $-2e^2/4\pi\varepsilon_0 r$. The next nearest neighbours set up a repulsive energy of $+2e^2/4\pi\varepsilon_0 2r$, and the next nearest an attraction of $-2e^2/4\pi\varepsilon_0 3r$, and so on. The continuation of this process leads to the series for the total electrostatic energy of the row of charges

$$U_E(r) = -\frac{2e^2}{4\pi\varepsilon_0 r}(1 - \tfrac{1}{2} + \tfrac{1}{3} - \tfrac{1}{4} + \cdots). \qquad (4.4)$$

The series of terms in parentheses has a conditional convergence of $\ln 2$. Hence, the effect of this infinite row of charges is to modify (4.2) by the factor $2\ln 2$, or approximately 1.3863, which number may be regarded as the Madelung constant for the one-dimensional structure under consideration.

In three dimensions the corresponding calculation is a little more difficult. Consider again the sodium chloride structure type (Fig. 1.3) and take the central Na^+ ion as a reference origin. The six nearest Cl^- neighbours give rise to an attractive energy of $-6e^2/4\pi\varepsilon_0 r$, where r is the $Na^+ \cdots Cl^-$ distance, or one half of the cell side a for this structure type. The next nearest neighbours, at the centres of the cell edges, set up a repulsive energy of $+12e^2/4\pi\varepsilon_0 r\sqrt{2}$, and the next nearest, at the corners of the unit cell, an attraction of $-8e^2/4\pi\varepsilon_0 r\sqrt{3}$, and so on. Writing these terms as a series, we obtain

$$U_E(r) = -\frac{e^2}{4\pi\varepsilon_0 r}\left(6 - \frac{12}{\sqrt{2}} + \frac{8}{\sqrt{3}} - \cdots\right). \qquad (4.5)$$

This series has a very slow rate of convergence, and is not suitable for computation of the Madelung constant of the structure. Evjen (1932) showed that the convergence can be improved by working with nearly neutral blocks of structure: potential energy falls of more rapidly with distance for a neutral group than with a group of excess charge.

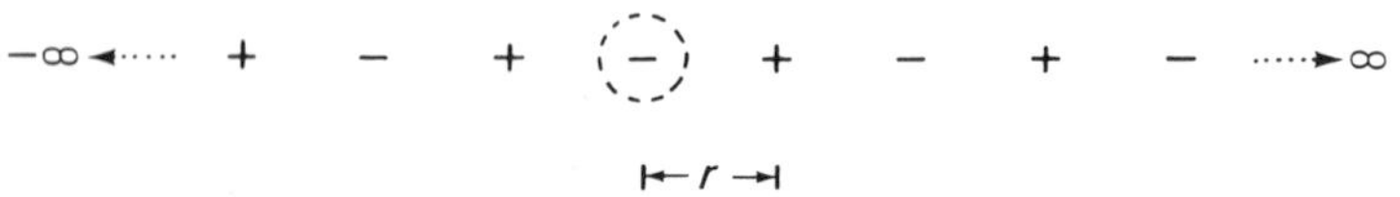

Fig. 4.1. Infinite row of regularly spaced point charges, each of magnitude e, and alternating in sign; the dashed circle indicates an arbitrarily chosen origin for the calculation of the Madelung constant. Since the ions may be considered as being grouped in neutral pairs about the origin, the interactions are effectively ion-dipole (section 3.1.4.3) and the sum is convergent at $\ln 2$.

Consider the unit cell of the sodium chloride structure type (Fig. 1.3) as a nearly neutral block of structure. Evjen's method requires each ion in the given unit cell to have a weighting factor, so as to achieve the neutrality condition. We know that a crystal is built up by stacking unit cells side-by-side in three dimensions. Hence, an ion at the corner of a unit cell is shared by eight adjacent unit cells, so that its contribution, or weight, in one unit cell is one eighth. Similarly, an ion at the centre of a cell edge contributes $\frac{1}{4}$ to the unit cell, and that at the centre of a face $\frac{1}{2}$. If we now restate (4.5), with these weights, we obtain

$$U_E(r) = - \frac{e^2}{4\pi\varepsilon_0 r} \left(\frac{6}{2} - \frac{12}{4\sqrt{2}} + \frac{8}{8\sqrt{3}} \right). \tag{4.6}$$

The sum of the terms in parentheses is 1.46, which is an approximation to the Madelung constant for this structure type. If we take a cube of twice the linear size, the value obtained is 1.75. The Madelung constant calculated by precise methods has the value 1.74756 . . . , so that we are very close by using Evjen's method on a cube of side $2a$. It is interesting to note that these results are a quantitative expression of one of Pauling's empirical rules for crystal structures, which states that *the charge on an ion tends to be neutralized by its immediate neighbours*: a sodium chloride structure of eight unit cells behaves towards a central ion like an infinite structure, to approximately 0.1%.

We can look at the Madelung constant through another model. Imagine an ion of charge e existing in free space. The effect of incorporating it into a crystal structure can be simulated by surrounding it with an earthed, spherical, conducting shell of radius $\imath$; in both situations the electrical field of the ion tends to be neutralized.

The electrostatic potential due to the charge q at a distance $\imath$ from it[1] is $q/4\pi\varepsilon_0\imath$. The work done in decreasing the charge by an amount δq is $q\,\delta q/4\pi\varepsilon_0\imath$. In neutralizing a charge e, the work done per ion w is given by

$$w = \frac{1}{4\pi\varepsilon_0\imath} \int_0^e q\,dq = \tfrac{1}{2}\frac{e^2}{4\pi\varepsilon_0\imath}. \tag{4.7}$$

The volume V of a spherical shell of radius $\imath$ is $4\pi\imath^3/3$; hence (4.7) may be written as

$$w = \tfrac{1}{2}\frac{e^2}{4\pi\varepsilon_0(3V/4\pi)^{1/3}} = 0.8060\,\frac{e^2}{4\pi\varepsilon_0\,V^{1/3}}. \tag{4.8}$$

Let us apply this result to the cesium chloride structure type; a stereoview of its unit cell is shown in Fig. 4.2. If the cell side is a, the unit cell volume is a^3. But the body diagonal of the cube is equal to $2r_e$, where r_e is the equilibrium interionic distance. Thus, the unit cell volume can be equated to $8r_e^3/3\sqrt{3}$ and, since the unit cell contains one pair of Cs^+ and Cl^- ions in identical coordination patterns, the volume per ion is $4r_e^3/3\sqrt{3}$.

Using (4.8) and choosing to identify the radius $\imath$ of the conducting shell with r_e, $V^{1/3}$ becomes $4^{1/3}r_e/\sqrt{3}$, and w is $0.8794e^2/4\pi\varepsilon_0 r_e$, and the work per pair $Cs^+\ Cl^-$ is just $2w$, or $1.759e^2/4\pi\varepsilon_0 r_e$. Thus, a good approximation to the Madelung constant for the cesium chloride structure type is 1.759; the value from a precise calculation is 1.762663 . . . . A selection of Madelung constants is included in Table 4.1.[2]

---

[1]See any good text-book on electrostatics.
[2]The three-dimensional Coulomb sum is also conditionally convergent. By working with neutral blocks of structure (Evjen), or by inhibiting macroscopic polarization (spherical shell model; Ewald sums), convergence can be achieved.

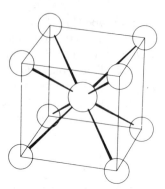 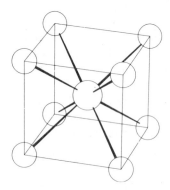

Fig. 4.2. Stereoview of the unit cell of the cesium chloride structure type; the circles represent, in decreasing order of size, Cl^- and Cs^+. It is a primitive (P) cubic unit cell, with one ion of each type per unit volume.

Table 4.1. Madelung constants for some simple structure types

Structure type	q_1	q_2	$\mathscr{A}$	$q_1 q_2 \mathscr{A}$	$-r_e U_E/\text{kJ mol}^{-1}$ nm
CsCl	1	1	1.7627	1.7627	244.90
NaCl	1	1	1.7476	1.7476	242.80
α-ZnS	2	2	1.6407	6.5628	911.80
β-ZnS	2	2	1.6381	6.5524	910.36
CaF_2	2	1	2.5194	5.0388	700.07
TiO_2	4	2	2.3851	19.0808	2650.99
β-SiO_2	4	2	2.2011	17.6088	2446.48

In Table 4.1, the values listed for $\mathscr{A}$ refer to the structure type with unit charges at the ion sites. The corresponding values including the ion charges are given in the next column; in some cases, they are, naturally, the same. Some sources quote the values of $q_1 q_2 \mathscr{A}$ given here as $\mathscr{A}$ itself, so that it is necessary to identify the term given correctly in any reference work. There is no unique value of r_e for TiO_2 (rutile); here, $\mathscr{A}$ has been calculated in terms of the shortest interionic distance in the structure. The sodium chloride structure type is common to many compounds; when magnesium oxide, for example, is being considered, $q_1 q_2 \mathscr{A}$ becomes 6.9904. The final column of Table 4.1 gives an electrostatic energy for each structure type, when divided by the appropriate value of r_e in nm.

4.3 LATTICE ENERGY—ELECTROSTATIC MODEL

Just as we discussed with molecular compounds, so in ionic crystals the attractive energy is balanced by a repulsive energy that is important at small values of r. The Madelung constant incorporates the electrostatic ionic repulsion, but a second term is required to represent the electron–electron repulsion of closed shells, and to achieve an equilibrium state in terms of energy.[1] The earliest repulsion potential function was

[1] The reader may wish to review Earnshaw's theorem in electrostatics.

given in the form of $1/r^n$, where n varied from ~ 9 to 13. As we have already noted, theoretical studies have shown that an exponential form of repulsion potential is more satisfactorily, and the crystal cohesive energy, usually called the lattice energy,[1] may be given as

$$U(r) = -\frac{\mathscr{A} q_1 q_2 e^2}{4\pi\varepsilon_0 r} + B \exp(-r/\rho) \qquad (4.9)$$

where the negative sign implies ions of opposite sign, B is a constant for the structure type, and ρ is a constant of the structure itself. For simplicity of manipulation, we can write

$$U = \frac{\mathscr{A}'}{r} + B \exp(ar) \qquad (4.10)$$

where $\mathscr{A}' = -q_1 q_2 e^2/4\pi\varepsilon_0$ and $a = -1/\rho$. The two terms in (4.9) lead to a curve similar in form to that of Fig. 3.10c, except that in a crystal the actual situation is more complex than that of a diatomic molecule. Nevertheless, we can identify an energy minimum at the equilibrium interionic distance r_e. From (4.10),

$$\frac{dU}{dr} = -\frac{\mathscr{A}'}{r^2} + aB \exp(ar). \qquad (4.11)$$

At $r = r_e$, $dU/dr = 0$, and we obtain

$$B = \frac{\mathscr{A}'}{r_e^2 a \exp(ar_e)}. \qquad (4.12)$$

Hence, from (4.9) and (4.12), inserting the values for $\mathscr{A}'$ and a and multiplying by $10^{-3} L$ so that the energy shall be expressed in the usual units of kJ mol^{-1}, we have

$$U(r_e) = -\frac{10^{-3} L \mathscr{A} q_1 q_2 e^2 (1 - \rho/r_e)}{4\pi\varepsilon_0 r_e} \text{ kJ mol}^{-1}. \qquad (4.13)$$

Although a similar equation could have been obtained with a repulsion potential of the form Br^{-n}, where $1/n$ would replace ρ/r_e, it is worth noting that the quantity ρ/r_e varies very much less over a wide range of compounds than does $1/n$. In fact, ρ/r_e is remarkably close to 0.1 in many 1:1 ionic compounds, although both r_e and ρ themselves show considerable variation; over a similar range, n varies from ~ 9 to 13.

The parameter ρ is related, not surprisingly, to the compressibility of the crystal, because the repulsive energy becomes important as r is reduced below r_e, as would be the case for a crystal under compressive stress. To continue with our analysis, we need an equation of state for a solid. From Appendix 11, we have

$$\left(\frac{\partial U}{\partial V}\right)_T = -p + T\frac{\alpha}{\kappa} \qquad (4.14)$$

where α and κ are, respectively, the coefficients of thermal expansivity and isotropic compressibility. At 0 K, $(\partial U/\partial V)_T = -p$; neglect of the term $T\alpha/\kappa$ leads to an underestimate of the magnitude of the lattice energy by $\sim 1\%$ at 298.15 K. We can

---

[1]This name is strictly a misuse of the term *lattice* but is, nevertheless, traditional usage.

proceed by working at 0 K, subsequently converting the energy result from 0 K to a temperature T, if needed, by the amount ΔU, given by

$$\Delta U = \int_0^T C_V \, dT. \tag{4.15}$$

From (4.14), at $T = 0$, it follows that

$$\left(\frac{\partial p}{\partial V}\right)_T = -\left(\frac{\partial^2 U}{\partial V^2}\right)_T. \tag{4.16}$$

The coefficient κ has been given in Chapter 3 as $-1/V(\partial V/\partial p)_T$; hence, at constant temperature,

$$\frac{1}{\kappa V} = \frac{d^2 U}{dV^2}. \tag{4.17}$$

For isotropic crystal structures, where there is a unique interionic distance r, we can write

$$V = \mathcal{K} r^3 \tag{4.18}$$

where V is here the volume occupied by a pair of oppositely charged ions, and $\mathcal{K}$ is a constant. Differentiation of (4.18) leads to

$$\frac{dV}{dr} = 3\mathcal{K} r^2 = \frac{3V}{r}$$

$$\frac{d^2 V}{dr^2} = 6\mathcal{K} r = \frac{6V}{r^2}. \tag{4.19}$$

By the rules of differentiation, we have

$$\left.\begin{aligned}
\frac{dU}{dV} &= \frac{dU/dr}{dV/dr} \\[2mm]
\frac{d^2 U}{dV^2} &= \frac{1}{(dV/dr)}\frac{d}{dr}\left\{\frac{dU/dr}{dV/dr}\right\} \\[2mm]
&= \frac{(dV/dr)(d^2 U/dr^2) - (dU/dr)(d^2 V/dr^2)}{(dv/dr)^3}.
\end{aligned}\right\} \tag{4.20}$$

From (4.11), we obtain

$$\frac{d^2 U}{dr^2} = \frac{2\mathcal{A}'}{r^3} + a^2 B \exp(ar) \tag{4.21}$$

and from (4.11), (4.17), (4.19), (4.20) and (4.21), we have

$$\frac{1}{\kappa V} = \frac{(3V/r)[2\mathcal{A}'/r^3 + a^2 B \exp(ar)] - (6V/r^2)[-\mathcal{A}'/r + aB \exp(ar)]}{(3V/r)^3}. \tag{4.22}$$

Eliminating B through (4.12) at the equilibrium distance r_e, and rearranging:

$$\frac{9V}{\kappa} - \frac{2\mathcal{A}}{r_e} = a\mathcal{A}' = a\mathcal{A}'\frac{r_e}{r_e}. \tag{4.23}$$

Finally, introducing the values for $\mathscr{A}'$ and a, we obtain

$$\rho/r_e = \frac{Aq_1q_2e^2/4\pi\varepsilon_0 r_e}{9V/\kappa + 2Aq_1q_2e^2/4\pi\varepsilon_0 r_e} \tag{4.24}$$

and this equation for ρ/r_e is used in (4.13) to calculate the lattice energy. We will apply these equations to sodium chloride, given that $r_e = 0.282$ nm and $\kappa = 4.1 \times 10^{-11} \, N^{-1} \, m$, with $V = 2r_e^3$ for this structure type. Executing the calculation gives $\rho/r_e = 0.113$ and $U(NaCl) = -764 \, kJ \, mol^{-1}$. We shall return to the precision of this value and, indeed, of the model that we have used, but it is desirable first to be able to set up a test model with experimentally determined parameters.

4.4   LATTICE ENERGY—THERMODYNAMIC MODEL

We shall consider the stages in the formation of an ionic compound in terms of enthalpy, using the sodium chloride structure as an example. Subsequently, we shall introduce a relationship between the lattice (crystal) enthalpy and the lattice energy.
   The equation

$$Na + \tfrac{1}{2}Cl_2 \to NaCl \tag{4.25}$$

is very familiar. It is both more precise and more descriptive to write this equation as

$$Na(s) + \tfrac{1}{2}Cl_2(g) \to NaCl(s) \tag{4.26}$$

and the corresponding standard[1] enthalpy of formation ΔH_f $(NaCl, s)$ is -412.5 kJ mol^{-1}: but even this equation contains, implicitly, several stages that we need to examine.
   Solid sodium is sublimed:

$$Na(s) \to Na(g) \tag{4.27}$$

and the standard enthalpy of sublimation $\Delta H_s^{\ominus}(Na)$ is $110.2 \, kJ \, mol^{-1}$. The gas is next ionized:

$$Na(g) \to Na^+(g) + e^- \tag{4.28}$$

and the first ionization energy $I_1(Na)$ at 0 K is $495.8 \, kJ \, mol^{-1}$; to refer this quantity to a corresponding enthalpy at 298.15 K, $\tfrac{5}{2}\mathscr{R}T$ is added $(H = U + pV)$ to take account of the increase of one mole of gaseous products over reactants in the reaction, at constant pressure. The total standard enthalpy change for these processes $\Delta H_f^{\ominus}(Na^+, g)$ is $(606 + \tfrac{5}{2}\mathscr{R}T)$, and there is clearly no tendency for (4.27) and (4.28) to occur spontaneously. Considering next chlorine, we have

$$\tfrac{1}{2}Cl_2(g) \to Cl(g) \tag{4.29}$$

and

$$Cl(g) + e^- \to Cl^-(g) \tag{4.30}$$

and the enthalpy changes are, respectively, one half the standard dissociation enthalpy $\tfrac{1}{2}D_0^{\ominus}(Cl_2)$, $121.7 \, kJ \, mol^{-1}$, and the electron affinity $E(Cl)$, $-348.6 \, kJ \, mol^{-1}$ at 0 K;

[1]The symbol $\ominus$ indicates the standard state conditions of 298.15 K and 1 atmosphere.

here, $\frac{5}{2}\mathscr{R}T$ must be subtracted from $E(\text{Cl})$ to give the standard enthalpy value of this parameter. For the anion gas, $\Delta H_f^{\ominus}(\text{Cl}^-, \text{g})$ is $(-226.9 - \frac{5}{2}\mathscr{R}T)$ kJ mol^{-1}, and the total standard enthalpy change for the four processes is 379.1 kJ mol^{-1}, again not indicative of spontaniety. However, the driving force for (4.26) may be said to lie in the stability conferred on the system when one mole of each of the gaseous ions comes together and condenses to form the sodium chloride structure.

The lattice energy may be defined as *that energy liberated when one mole of crystal is formed from its component gaseous ions in the standard state.* The lattice enthalpy is the corresponding enthalpy change, and at 0 K these two quantities are equal. As defined they are negative quantities, and refer to the difference between two thermodynamic states, the crystal and the gaseous ions. Thus, for a substance AX,

$$\Delta U_c(AX) = U_c(AX) - [U(A^+, \text{g}) + U(X^{2-}, \text{g})]. \tag{4.31}$$

We use the subscript c to refer to the crystal in the thermodynamic discussion of lattice energetics.

It is necessary to select a reference state for lattice energy, and the ideal ion-gas at 0 K is defined as the zero for lattice energy. There are no forces of attraction between the ions in the (hypothetical) reference state; hence, for any crystal at 0 K

$$\Delta U_c = U_c. \tag{4.32}$$

Thus, although we are concerned with a change between two states, the lattice energy is usually given by the symbol U_c, referring implicitly to 0 K; at any other temperature the symbol ΔU_c is used for the energy, and ΔH_c for the lattice enthalpy.

The processes that we have just considered, together with the experimental enthalpy of formation, constitute a thermochemical cycle that may be represented diagrammatically as shown by Fig. 4.3; the cycle was put forward independently by Born, Haber and Fajans, and is known by their names (although Fajans is often omitted). Although it is convenient to be able to equate the lattice enthalpy and lattice energy by working at 0 K, it is often more useful to consider the thermodynamic processes at another temperature, such as 298.15 K. At any temperature T, we have

$$\Delta U_c = \Delta H_c - p\Delta V = \Delta H_c - [pV(\text{s}) - pV(\text{g})]. \tag{4.33}$$

At a constant pressure, the volume of a solid is very small in comparison with that of the same mass of gas. Hence, in (4.33), we can neglect $pV(\text{s})$ in comparison with $pV(\text{g})$ and, since the ion-gas is ideal, we replace $pV(\text{g})$ by $n\mathscr{R}T$, where n is the number of moles of gaseous species. From the cycle (Fig. 4.3), we can write generally, for the lattice energy of a compound AX at the standard temperature,

$$\Delta U_c = \Delta H_f^{\ominus}(AX, \text{s}) - \Delta H_s^{\ominus}(A) - I(A) - \frac{1}{2}D_0^{\ominus}(X_2) - E(X) + n\mathscr{R}T + \Delta H_1 \tag{4.34}$$

where $I(A)$ refers to the sum of the ionization energies for the ionic species under consideration; it will be appreciated that the two $\frac{5}{2}\mathscr{R}T$ terms attaching to $I(A)$ and $E(X)$ (Fig. 4.3) cancel.

The quantity ΔH_1 is the enthalpy change for the ion-gas between the reference state at 0 K and another temperature T, according to

$$\Delta H_1 = \int_0^T C_p(\text{g}) \, dT = \frac{5}{2}\mathscr{R}T. \tag{4.35}$$

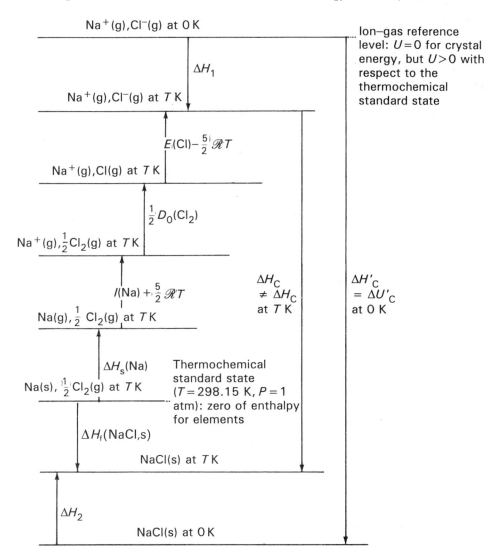

Fig. 4.3. Born–Haber–Fajans thermochemical cycle for the lattice energy (and enthalpy) of an ionic solid, illustrated here for NaCl; be careful to distinguish between the two reference levels, one for the elements in their standard state, and the other for the ion-gas.

For two moles of (ideal) gaseous ions at 298.15 K, $\Delta H_1 = 12.4\ \text{kJ mol}^{-1}$. The term ΔH_2, also shown on the figure, is the enthalpy change for the crystal between 0 K and a temperature T, according to

$$\Delta H_2 = \int_0^T C_p(\text{s})\, \mathrm{d}T. \tag{4.36}$$

In some cases, C_p can be expressed analytically by a function such as

$$C_p = a + b + c/T^2 \tag{4.37}$$

where a, b and c are constants for the substance; in other cases, C_p is plotted against T over the required temperature range, and ΔH_2 evaluated by numerical integration (see Appendix 12).

Completing the evaluation of the lattice energy of sodium chloride from (4.34), we obtain $\Delta U_c = -774$ (± 2) $kJ\,mol^{-1}$. We may note that, in common with other thermodynamic calculations, the evaluation of the lattice energy involves no know-ledge of the particular type of bonding present in the given substance. The infor-mation is locked away in the practical quantity, the enthalpy of formation and, of course, in the lattice energy itself, each with respect to its own reference state.

4.4.1   Precision of the thermodynamic lattice energy
The precision of the thermodynamic lattice energy is governed by the total precision in the quantities set out in (4.34), and we shall examine them briefly here.

4.4.1.1   Enthalpy of sublimation
In the experimental measurement of this parameter, the vapour of a metal A is streamed through a hole that is small compared with the mean free path of the atoms in the vapour. Under these conditions, the rate of effusion is proportional to the pressure p of the vapour over the given temperature range.

The Clausius–Clapeyron equation for the sublimation change of state, at a temperature T, is given by

$$\frac{dp}{dT} = \frac{\Delta H_s}{T[V(g) - V(s)]}.$$

(4.38)

If we make the reasonable assumptions that $V(g) \gg V(s)$, and that the vapour at the low pressures involved behaves ideally, we have for one mole of vapour

$$\frac{dp}{dT} = \frac{\Delta H_s p}{\mathscr{R}T^2}$$

(4.39)

which, by standard manipulation, can be recast in the form

$$\frac{d(\ln p)}{d(1/T)} = -\frac{\Delta H_s}{\mathscr{R}}.$$

(4.40)

The graph of $\ln(p)$ against $1/T$ is fitted by the method of least squares (Appendix 9), and its slope is $-\Delta H_s/\mathscr{R}$. The value at 298.15 K is obtained through (4.41), (4.42) and Fig. 4.4, where ΔH_1 is the required standard enthalpy of sublimation, identified by $\Delta H_s^{\ominus}(A)$ in (4.34):

$$\Delta H_1 = \Delta H_2 + \Delta H_3 - \Delta H_4$$

(4.41)

or

$$\Delta H_s^{\ominus}(A) = \Delta H_{exprt} + \int_{T_1}^{T_2} C_p(s)\, dT - \int_{T_1}^{T_2} C_p(g)\, dT.$$

(4.42)

Results for the alkali metals are listed in Table 4.2.

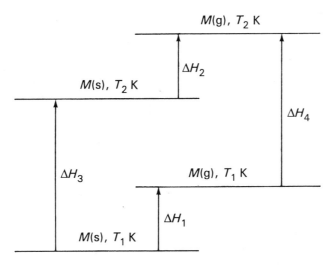

Fig. 4.4. Enthalpy cycle for the extrapolation of the enthalpy of sublimation from the mean experimental temperature T_2 to the standard state temperature T_1.

Table 4.2. Standard enthalpies of sublimation of the alkali metals

A	$\Delta H_s^{\ominus}(A)/\text{kJ mol}^{-1}$
Li	161.5 ± 5.5
Na	110.2 ± 1.5
K	90.0 ± 1.9
Rb	85.8 ± 2.1
Cs	78.7 ± 2.1

4.4.1.2 Ionization energy

We introduced ionization energy in section 2.6.7 in the context of the hydrogen atom; here we consider it more generally, and show how it may be determined for the alkali metals.

The absorption spectrum for sodium (Fig. 4.5) consists of a series of lines known as the principal (P) series. The first two members of this series are the well-known, closely spaced sodium D yellow lines. Subsequent members lie in the ultraviolet region of the spectrum, converging on a series limit $\bar{v}_\infty$ at about 41 500 cm^{-1} which is followed by an energy continuum. At the series limit, also called the convergence limit, the expelled electron has zero kinetic energy at 0 K, which is the reason that ionization energy is defined at the absolute zero of temperature.

The wavenumber of a line in this spectral series follows the equation

$$\bar{v} = \bar{v}_\infty - R_\infty/(n - \delta)^2 \tag{4.43}$$

where R_∞ is the Rydberg constant (109737.3 cm^{-1}), n is the principal quantum number and δ is known as the quantum defect; both $\bar{v}_\infty$ and δ are unknown initially.

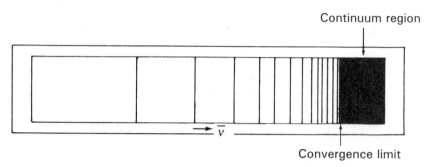

Fig. 4.5. Schematic diagram for the absorption spectrum of atomic sodium; the convergence limit corresponds to $\bar{v}_\infty$, or the energy level as $n \rightarrow \infty$, and leads into an energy continuum.

The wavenumbers for the P spectral series of sodium are listed in Table 4.3. To initiate the determination of the ionization energy, we plot $\bar{v}$ against $1/n^2$, and extrapolate to $1/n^2 = 0$; this process leads to an initial value of 0.81 for δ. It will be evident that extrapolating to ∞ against $1/n^2$ will lead to the same limit as with $1/(n - \delta)^2$. Taking this result, $\bar{v}$ is fitted to $R_\infty/(n - \delta)^2$ by least squares, and better estimates obtained for δ and $\bar{v}_\infty$. This procdedure is repeated until constant values are obtained for δ and $\bar{v}_\infty$; the stages are set out in the table.

The convergence limit of 41449.7 cm^{-1} may be converted into an ionization energy by the relation $I = hc\bar{v}_\infty$. Thus, $I_1(\text{Na}) = 5.1390$ eV atom^{-1}, or 495.84 kJ mol^{-1} Values for the first ionization energies of the alkali metals are listed in Table 4.4; their precision is $\sim \pm 0.001$ kJ mol^{-1}.

4.4.1.3   Dissociation enthalpy
The most precise data on dissociation enthalpies are obtained from the electronic absorption spectra of molecules. When a gas is excited, electronic transactions occur in the molecule and radiation is absorbed. Molecular spectra are complex bands, but in diatomic molecules the banded spectrum gives way to a continuum that signifies the dissociation of the molecule into its component atoms.

Again, a convergence limit must be established, and the frequency at the low-energy end of the continuum corresponds to the required dissociation enthalpy. For chlorine, the appropriate wavenumber is 21189 cm^{-1}, which gives a result of 2.627 eV molecule^{-1}. Since the dissociated atoms are not in the ground state, the excitation energy E_X of 0.109 eV must be subtracted. Hence, $D_0(\text{Cl}_2) = 2.518$ eV molecule^{-1}, or 242.95 kJ mol^{-1}. The results for the halogens are listed in Table 4.5; their precision is better than ± 0.01 kJ mol^{-1}, except for fluorine.

The dissociation energy $D_e(X_2)$ is the value at the minimum of the potential energy curve of the X_2 molecule at 0 K, and it is numerically greater than D_0 by the amount of the zero-point energy of vibration (see also Fig. 2.4).

4.4.1.4   Electron affinity
Single electron affinities have been measured experimentally for the halogens, and for a few other species, such as oxygen; others may be deduced from lattice energies in a manner shortly to be described. Alkali-metal halides, such RbI or CsCl, when heated ultrasonically by shock waves produce a vapour containing I$^-$ or Cl$^-$ species in

Table 4.3. Spectral wavenumbers[a] and the first ionization energy for sodium

n	$\bar{v}/\text{cm}^{-1}$	$[R_\infty/(n-\delta)^2]/\text{cm}^{-1}$ $\delta = 0.81$	$\bar{v}_\infty/\text{cm}^{-1}$	δ
10	40137.2	1299.3		0.867 ⎫
11	40383.2	1056.8		0.871 ⎬ 0.87
12	40566.0	876.4		0.875 ⎭
13	40705.7	738.5		0.880
14	40814.5	630.8		0.888
15	40901.1	545.0		0.896
16	40971.2	475.6		0.905
17	41028.7	418.7		0.914
			41452.7	
		$\delta = 0.87$		
		1316.5		0.853 ⎫
		1069.4		0.852 ⎬ 0.85
		885.9		0.850
		745.8		0.848 ⎭
		636.5		0.847
		549.6		0.845
		479.4		0.842
		421.8		0.838
			41448.4	
		$\delta = 0.85$		
		1310.7		0.857 ⎫
		1065.2		0.858
		882.7		0.859
		743.4		0.859 ⎬ 0.86
		643.6		0.861
		548.1		0.862
		478.1		0.863
		421.8		0.863 ⎭
			41450.1	
		$\delta = 0.856^b$		
		1312.4		0.856
		1066.4		0.857
		883.6		0.856
		744.1		0.855
		635.2		0.857
		548.5		0.857
		478.5		0.857
		421.0		0.856
			41449.71	

[a]$1 \text{ cm}^{-1} = 100 \text{ m}^{-1}$.
[b]Subjective estimate from iterations 2 and 3.

Table 4.4. First ionization energies of the
alkali metals (esd $\pm$ 0.001 kJ mol^{-1})

A	$I(A)/\text{eV atom}^{-1}$	$I(A)/\text{kJ mol}^{-1}$
Li	5.3916	520.211
Na	5.1390	495.835
K	4.3406	418.804
Rb	4.1771	403.029
Cs	3.8938	375.695

Table 4.5. Standard dissociation enthalpies for the
halogens

X_2	$D_0(X_2)/\text{eV molecule}^{-1}$	$D_0(X_2)/\text{kJ mol}^{-1}$
F_2	1.63	157.3 ± 4.0
Cl_2	2.518	242.95 ± 0.01
Br_2	2.319	223.75 ± 0.01
I_2	2.215	213.72 ± 0.01

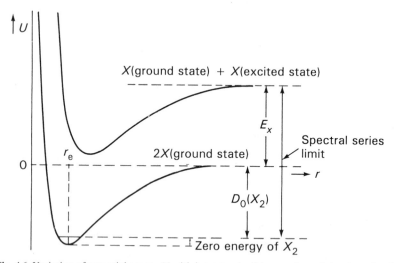

Fig. 4.6. Variation of potential energy U with interatomic distance r for a diatomic molecule X_2; r_e is the equilibrium interatomic distance. For clarity, the value of the excitation energy E_X has been exaggerated. The difference between the theoretical dissociation energy D_e and the experimental value D_0 is the zero-point energy of vibration.

abundance. Their ultraviolet absorption spectra are continua, with sharp low-energy thresholds that correspond to the process of photodetachment of electrons:

$$Cl^-(g) + h\nu = Cl(g) + e^-. \qquad (4.44)$$

The threshold frequencies for the halogens have been ascribed precisely, and the derived electron affinities are listed in Table 4.6. The values for $E(X)$ are accorded a

Table 4.6. Electron affinities for the halogens

X	$E(X)$/eV atom^{-1}	$E(X)$/kJ mol^{-1}
F	-3.399 ± 0.002	-328.0 ± 0.2
Cl	-3.613 ± 0.003	-348.6 ± 0.3
Br	-3.363 ± 0.003	-324.5 ± 0.3
I	-3.063 ± 0.003	-295.5 ± 0.3

negative sign, because the attachment of a single electron (4.30), for which (4.44) is the reverse process, is spontaneous.

4.4.1.5 Enthalpy of formation
The enthalpy of formation of a compound AX refers to the process

$$A(s) + \tfrac{1}{2}X_2(g) = AX(s). \qquad (4.45)$$

At 298.15 K and 1 atmosphere, the enthalpy change for this reaction is the standard enthalpy of formation $\Delta H_f^{\ominus}(AX, s)$. This quantity might be measured for the direct combination of, say, lithium and iodine, but for cesium and chlorine it would be unwise to attempt the direct approach. Instead, a sequence of reactions may be considered:

					$\Delta H^{\ominus}$/kJ mol^{-1}
$H_2(g)$	+	$\tfrac{1}{2}O_2(g)$		$\rightarrow$	-285.9
$Cs(s)$	+	$H_2O(l)$		$\rightarrow$	-191.8
$\tfrac{1}{2}H_2(g)$	+	$\tfrac{1}{2}Cl_2(g)$		$\rightarrow$	-92.3
$HCl(g)$	+	aq		$\rightarrow$	-75.1
$Cs^+(aq)$	+	$OH^-(aq) + H_3O^+(aq) + Cl^-(aq)$		$\rightarrow$	-55.8
$CsCl(s)$	+	aq		$\rightarrow$	$+18.0$
$Cs(s)$	+	$\tfrac{1}{2}Cl_2(g)$		$\rightarrow$	$\Delta H_f^{\ominus}$

The reader is invited to construct a thermochemical cycle to show that $\Delta H_f^{\ominus}(CsCl, s)$ is -433.0 kJ mol^{-1}.

The precision of enthalpies of formation is very variable, even among the alkali-metal halides; Table 4.7 lists the standard values for this group of compounds.

The precision of ± 2 kJ mol^{-1} that we quoted earlier for the lattice energy of sodium chloride derives from the uncertainties in its component quantities. The precision attainable in ΔU_c for sodium chloride and potassium chloride is high, but rarely is it matched with other compounds.

We are now in a position to consider the precision of the lattice energy from the electrostatic model, and that of the model itself, which leads us next to a discussion of polarization in ionic solids.

Table 4.7. Standard enthalpies of formation for the alkali-metal halides; all values are in kJ mol^{-1}, and the precisions are given in parentheses

	Li	Na	K	Rb	Cs
F	−612.1(8.4)	−571.1(2.1)	−562.7(4.2)	−549.4(8.4)	−530.9(16.7)
Cl	−405.4(8.4)	−412.5(0.8)	−436.0(0.8)	−430.5(8.4)	−433.0(8.4)
Br	−348.9(8.4)	−361.7(1.7)	−392.0(2.1)	−389.1(8.4)	−394.6(12.6)
I	−271.1(8.4)	−290.0(2.1)	−327.6(1.3)	−328.4(8.4)	−336.8(10.5)

4.5 POLARIZATION IN IONIC COMPOUNDS—PRECISION OF THE ELECTROSTATIC MODEL FOR LATTICE ENERGY

The lattice energy equation (4.9) was set up on the basis of a point-charge model for ions. The ions in the alkali-metal halides are subject to symmetrical polarization, similar to that which we discussed in section 3.1.4.7. Thus, although ionic compounds have no permanent dipole moment, there are dipolar-type interactions to consider. Calculation shows that dipole–dipole and dipole–quadrupole terms have a significance within the level of precision that is matched by the thermodynamic calculations. We extend (4.9) to the form, applicable at a finite temperature T:

$$U(r) = -\frac{\mathscr{A}q_1q_2e^2}{4\pi\varepsilon_0 r} + B\exp(-r/\rho) - \frac{C}{r^6} - \frac{D}{r^8} + \phi(T\alpha/\kappa) \tag{4.46}$$

where C and D are constants of the dipole–dipole (see section 3.1.4.6) and dipole–quadrupole energies, and the final term arises from the equation of state (4.14) for a solid at a temperature T.

Following an analysis similar to that in section 4.3 leads to the equations for lattice energy:

$$\frac{\rho}{r_e} = \frac{\mathscr{A}q_1q_2e^2/4\pi\varepsilon_0 r_e + 6C/r_e^6 + 8D/r_e^8 - 3VT\alpha/\kappa}{9V\Phi(T,p)/\kappa + 2\mathscr{A}q_1q_2e^2/4\pi\varepsilon_0 r_e + 42C/r_e^6 + 72D/r_e^8} \tag{4.47}$$

$$\Phi(T,p) = 1 + T(\partial\kappa/\partial T)_p/\kappa + \alpha(\partial\kappa/\partial p)_T + 2\alpha/3 \tag{4.48}$$

and

$$U(r_e) = -10^{-3}L\left\{\frac{\mathscr{A}q_1q_2e^2/4\pi\varepsilon_0}{r_e(1-\rho/r_e)} + \frac{C}{r_e^6(1-6\rho/r_e)} + \frac{D}{r_e^8(1-8\rho/r_e)} + \frac{3VT\alpha}{\kappa}\frac{\rho}{r_e}\right\}$$

$$\text{kJ mol}^{-1}. \tag{4.49}$$

Fuller descriptions of these equations have been given elsewhere,[1] and it is not important that they be elaborated here. We may note that their application is limited because there is only a small number of data on the temperature and pressure coefficients of compressibility, $(\partial\kappa/\partial T)_p$ and $(\partial\kappa/\partial p)_T$, respectively. Furthermore, once

---

[1] See, for example, M. F. C. Ladd (1977) *Journal of the Chemical Society Dalton*, 220; M. P. Tosi (1964) in *Solid State Physics*, Vol 16, Academic Press.

we move outside the realm of cubic structures, expansivity and compressibility, for example, become anisotropic.

When the full analysis given by (4.47) to (4.49) is applied to sodium chloride, the value obtained for $U(r_e)$ is -774 kJ mol^{-1}, in excellent agreement with the result from the Born–Haber–Fajans cycle. Not only is this result satisfactory in itself, but it confirms the applicability of the electrostatic model for ionic compounds.

We note also that although our analysis in section 4.3 used the form of (4.14) appropriate to 0 K for simplicity, the values of r_e and κ apply to normal temperatures. These errors are relatively small, but the best precision demands the use of (4.49), and with data appropriate to one and the same temperature.

4.5.1   Approximate calculation of lattice energy

Notwithstanding approximations exist in the lattice energy calculations that we have discussed, they are of value in obtaining information on a range of parameters for which experimental determinations are not possible. However, in view of the paucity of physical data needed for precise calculations of lattice energies, it is pertinent to consider the use of approximate equations.

One equation that is often put forward is that discussed uncritically by Kapustinskii (1956).[1] He gives the equation for the lattice energy as

$$U(r_e) = -(120.2 \text{ kJ mol}^{-1} \text{ nm}) \frac{nq_1 q_2}{r_+ + r_-} \left\{ 1 - \frac{0.0345}{r_+ + r_-} \right\} \qquad (4.50)$$

where n is the number of ions in the formula-weight, q_1 and q_2 are the numerical charges on the ions, excluding their signs, and the sum of the individual ionic radii, given here in nm, replaces r_e. Effectively, this equation reduces all crystals to the sodium chloride type, and inserting the values for the radii of Na^+ and Cl^- leads to the value of -748 kJ mol^{-1} which, not surprisingly, is an acceptable approximation to the best-known value. There is little to commend this equation for more general use.

It is unnecessary to approximate r_e by the sum of the corresponding radii, since so many reliable data are available readily from X-ray structure analyses. Furthermore, the calculation of a Madelung constant is not a problem, given even a small personal computer (see Appendix 13). It has been shown[2] that ρ/r_e is very close to 0.1 over a wide range of 1:1 ionic compounds. Hence, it is possible to approximate (4.13) by

$$U(r_e) = -(125.0 \text{ kJ mol}^{-1} \text{ nm}) \frac{\mathscr{A} q_1 q_2}{r_e} \qquad (4.51)$$

and with q and r_e in the units as before, the lattice energy is given in kJ mol^{-1}. Inserting the data for sodium chloride gives -775 kJ mol^{-1} for $U(r_e)$, which is highly satisfactory. This simple type of calculation can be very useful because, in the absence of sufficient data, it may be the only approach available for the estimation of lattice energies.

[1] A. F. Kapustinskii (1956) *Quarterly Review Chemical Society*, **10**, 283.
[2] M. F. C. Ladd and W. H. Lee (1959) *Journal of Inorganic and Nuclear Chemistry*, **11**, 264.

4.6   USES OF LATTICE ENERGIES

Lattice energies provide certain information about the stabilities of solids with respect to their components. They can be used also to investigate a wide range of other parameters; here, we shall consider three of them.

4.6.1   Electron affinities and thermodynamic parameters

When the electrostatic model for the alkali-metal halides was first set up by Born and Landé in 1918, there were no experimental values for either dissociation energies or electron affinities. From an approximate electrostatic model for $U(r_e)$, together with the other terms in (4.34), the quantities $[\frac{1}{2}D_0(X_2) + E(X)]$ for the halogens among the twenty alkali-metal halides were determined—the crystal structures had been determined by Bragg and Bragg two years earlier. The constancy of these terms for each halogen was regarded as good evidence for the model.

As D_0 became measurable, so E was obtained from the average value of $(\frac{1}{2}D_0 + E)$ for each halogen. For many years, $E(F)$ was given as -393 kJ mol^{-1}. However, when the value for $D(F_2)$ was amended in 1950 from 266 kJ mol^{-1} to 155 kJ mol^{-1}, $E(F)$ was immediately revised to -338 kJ mol^{-1}, which is close to the accepted value.

Single electron affinities have been measured for the halogens and for some other species, such as oxygen, sulphur, and the hydroxyl group OH, but the affinities for oxygen and sulphur for two electrons, applicable to the simple ionic oxides and sulphides, for example, must be deduced by inserting the result of the electrostatic calculation of lattice energies into the appropriate Born-Haber-Fajans cycles. In the case of oxygen, a value of 607 ± 25 kJ mol^{-1} has been found for $E(O \rightarrow O^{2-})$. It should be noted that whereas the addition of a single electron to oxygen (and other species) is a thermodynamically spontaneous process, the addition of a second electron requires an expenditure of energy on the system; the driving force for attaching the second electron lies in the combination of the gaseous ions to form the ionic solid—the lattice energy of magnesium oxide, for example, is[1] -3784 kJ mol^{-1}.

In considering polyatomic ions, such as $[NO_3]^-$ or $[SO_4]^{2-}$, the separate terms D_0 and E do not have a clear meaning. They may be considered collectively as the term $\Delta H_f^{\ominus}(NO_3^-, g)$, for example, representing the process

$$\text{elements in standard state} \rightarrow \text{ion-gas in standard state.}$$

If we can set up an electrostatic calculation of $U(r_e)$ for a suitable nitrate, the enthalpy of formation of the ion-gas can be deduced, as were electron affinities; for the nitrate ion, a value of -351 kJ mol^{-1} for $\Delta H_f^{\ominus}(NO_3^-, g)$ has been calculated.[2] The precision of these derived parameters will generally be noticeably less than that of corresponding quantities obtained by experiment.

4.6.2   Compound stability

We have discussed the sodium chloride structure type in some detail, and we may wish to consider the feasibility of an ionic compound NeCl. If it were to exist, it would be, perhaps, not unreasonable to assume that it would have a simple structure type, such as that of sodium chloride.

[1]M. F. C. Ladd (1974) *Journal of Chemical Physics*, **62**, 4583.
[2]M. F. C. Ladd and W. H. Lee (1960) *Journal of Inorganic and Nuclear Chemistry*, **13**, 218.

From (4.34), we may write

$$\Delta H_f(\text{NeCl, s}) \approx U_c(\text{NeCl}) + I_{\text{Ne}} + \tfrac{1}{2}D_0(\text{Cl}_2) + E(\text{Cl}) - n\mathscr{R}T - \Delta H_1. \quad (4.52)$$

From (4.51), $U(r_e) = -(218 \text{ kJ mol}^{-1} \text{ nm})/r_e$, and inserting the known quantities into (4.52) gives $\Delta H_f(\text{NeCl, s}) = U_c(\text{NeCl}) + 1836 \text{ kJ mol}^{-1}$. If the enthalpy of formation of NeCl is to be negative, which is a reasonable requirement, then $U_c(\text{NeCl})$ must be less than $-1836 \text{ kJ mol}^{-1}$, that is, $-218/(r_e/\text{nm}) < -1836$, or $r_e < 0.12 \text{ nm}$. Since the radius of the Cl^- ion alone is $\sim 0.18 \text{ nm}$ (Table 4.10) the required situation cannot be achieved, and we conclude that NeCl cannot form a stable ionic compound. Because the ionization energy of neon is so large ($\sim 2080 \text{ kJ mol}^{-1}$), the amount of energy that has to be expended in forming the ion-gas cannot be recovered in forming a crystalline ionic solid of the type AX.

There are, however, interesting compounds formed by the noble gases under suitable conditions. In 1962, Bartlett prepared, by accident, a crystalline compound $O_2[\text{PtF}_6]$ by the gas-phase reaction

$$O_2(g) + \text{PtF}_6(g) \rightarrow O_2[\text{PtF}_6](s). \quad (4.53)$$

The compound was found to be crystalline, and X-ray analysis showed that it consisted of the species O_2^+ and $[\text{PtF}_6]^-$; the oxygen molecule had been oxidized. The lattice energy was calculated as approximately -502 kJ mol^{-1} from an equation like (4.51), taking the experimental r_e of 0.435 nm and assuming the Madelung constant of 1.7476 for an NaCl-like structure.

The ionization energies for $O_2(g) \rightarrow O_2^+(g)$ and $\text{Xe}(g) \rightarrow \text{Xe}^+(g)$ were known to be 1182 kJ mol^{-1} and 1175 kJ mol^{-1}, respectively. Thus, it seemed feasible to conduct the process of (4.53) with xenon in place of oxygen. It was found that xenon was oxidized in a similar manner, and the crystalline compound $\text{Xe}[\text{PtF}_6]$ with a lattice energy of -460 kJ mol^{-1} was formed. These experiments heralded the chemistry of the noble gases, and a number of compounds, such as KrF_2, XeF_n ($n = 2, 4, 6$), XeO_3 and $\text{XeNF(SO}_2\text{F})_2$, have been prepared. A recent summary of the chemistry of the noble gases has been given by Holloway.[1]

A literature report of the existence of Cu(I)F was challenged on the grounds that thermodynamic calculation showed that its free energy of formation was positive. This criterion refers to thermodynamic stability with respect to the elements in their standard states, and does not preclude the existence of the compound. Another example is silver cyanide, AgCN. This compound, too, has a positive free energy of formation (164 kJ mol^{-1}), but it can be prepared by precipitation from aqueous solutions of silver nitrate and potassium cyanide. Silver cyanide is thermodynamically unstable with respect to its elements in their standard state, but not with respect to the aqueous ions (Fig. 4.7). We must always exercise caution in referring to 'stability'; the reference level should be precisely defined.

4.6.3 Charge distribution on polyatomic ions
The combination of the electrostatic and thermodynamic models for lattice energy permits an estimation of the distribution of charge on polyatomic ions; we shall illustrate the procedure with respect to the cyanide ion.

[1]J. Holloway (1987) *Chemistry in Britain*, 658.

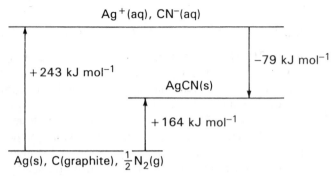

Fig. 4.7. Free-energy level diagram for silver cyanide; although the solid is unstable with respect to the elements by 164 kJ mol^{-1}, it is stable with respect to the aqueous ions by 79 kJ mol^{-1}.

At 298.15 K, KCN and NaCN, but not LiCN, have the sodium chloride structure type (Fig. 1.3), with the cyanide ions in orientational disorder (see also section 1.3.3). At 279 K, NaCN and at 233 K, KCN transform to an orthorhombic structure (Fig. 1.7). The cyanide ions lock into the fixed positions of the orthorhombic structure of the sodium salt at the higher temperature because of the stronger interaction with the Na$^+$ cation.

At the transition temperature, two polymorphs have the same free energy: hence, at constant pressure, the free energy change at the transition temperature is given by

$$\Delta G = \Delta U - T\Delta S + p\Delta V. \tag{4.54}$$

The volume change from one solid polymorph to the other is negligible, and since $\Delta G = 0$ at equilibrium we have

$$\Delta U = T\Delta S. \tag{4.55}$$

In other words, the difference in the lattice energies of the two polymorphs may be determined from the entropy (or enthalpy[1]) change at the transition temperature (Fig. 4.8). The values of ΔU are 2.9 kJ mol^{-1} and 1.3 kJ mol^{-1} for NaCN and KCN,

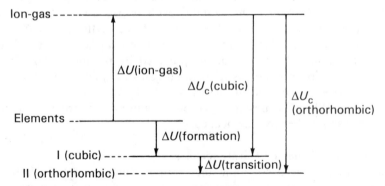

Fig. 4.8. Energy (ΔU) levels for two polymorphs I and II at a transition temperature T; the two forms may be related in energy through the value of ΔU ($T\Delta S$) at the transition temperature.

[1]At the transition temperature, $\Delta H_t / T_t = \Delta S_t$.

respectively. The lattice energies for NaCN and KCN have been calculated in their cubic form from (4.34), at the transition temperatures, and the corrections ΔU from (4.55) applied.

In the orthorhombic structures at the transition temperatures, the CN^- ions are in fixed positions, and the electrostatic energy must be calculated in two parts. The negative charge on the cyanide ion may be considered to be divided as $[C^{q_C}N^{q_N}]$, where $q_C + q_N = -1$. In Fig. 4.9, the components of the electrostatic energy for a cyanide ACN are shown: U_M is the electrostatic energy obtained with a Madelung constant calculated on the lattice array of charged species A^+, C^{q_C} and N^{q_N}; U_E is the electrostatic energy of the array of species A^+ and CN^- actually present in the crystal; the difference between them is U_S, the electrostatic self-energy of the cyanide ion. This term arises from the hypothetical process

$$C^{q_C}(g) + N^{q_N}(g) \rightarrow CN^-(g) \qquad (4.56)$$

which is included in the cycle in Fig. 4.9.

The term U_S may be defined formally by

$$U_S = \sum_{\substack{i,j \\ i \neq j}} \frac{q_i q_j e^2}{4\pi\varepsilon_0 r_{ij}} \qquad (4.57)$$

where r_{ij} is the distance between the ith and jth species of charges q_i and q_j, respectively, in the complex ion. For the cyanide ion it reduces to

$$U_S = \frac{q_C q_N e^2}{4\pi\varepsilon_0 r_{e,\,CN^-}}. \qquad (4.58)$$

Thus, we are able to determine $U(r_e)$ as a function of q_N. Then, by plotting $U(r_e)$ as a function of q_N (Fig. 4.10), we find that the equivalent value of U from the corresponding cubic structure is satisfied by q_N equal to either -0.4 or -0.6. The two values arise because the orthorhombic structure is symmetrical with respect to spatial

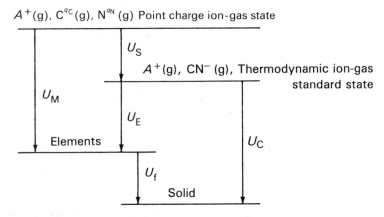

Fig. 4.9. Energy levels showing the relationship between electrostatic energies U_M, U_S and U_E and the lattice energy U_C for NaCN and KCN; U_f refers to the formation of the crystal from its elements, and completes the cycle.

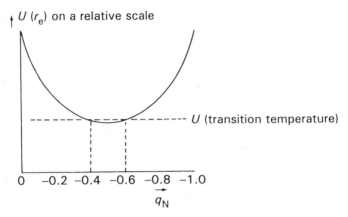

Fig. 4.10. Variation of $U(r_e)$ for KCN, in its orthorhombic structure, as a function of q_N; the symmetry of the structure is reflected in the shape of the curve, and the value of U at the transition temperature intersects the curve of $U(r_e)$ at $q_N = -0.4$ and -0.6. A similarly symmetrical curve was obtained from NaCN but not for LiCN.

interchange of the C and N species for both NaCN and KCN. This situation does not occur with LiCN, and the following results were obtained:

	KCN	NaCN	LiCN
$U(q_N = -0.4)$	-669	-732	-682
$U(q_N = -0.6)$	-669	-732	-791

It is unreasonable that the lattice energy value for LiCN should lie between those for NaCN and KCN, because of the important proportionality to $1/r_e$; thus, the value $q_N = -0.6$ is preferred. Subsequently, this result was supported by quantum mechanical calculations on the CN⁻ ion, which gave $q_N = -0.60$.[1]

Similar calculations have been carried out by the author, both by energetics and quantum mechanics, for a number of species. For example, in the CO_3^{2-} ion, the charge on carbon is $\sim +1$, with a charge of -1 on each oxygen atom. Several detailed reviews of lattice energy calculations and allied topics have been given.[2]

4.7   ASPECTS OF CRYSTAL CHEMISTRY

Crystal chemistry is concerned with structures themselves, and with the relationship between the properties of solids and their internal structure and bonding. It attempts to interpret properties in the light of known structures and, conversely, to associate structural characteristics with measured properties.

A systematic approach to crystal chemistry may be said to have begun in about 1920, after the publication of measurements of the interionic distances r_e for the totality of alkali-metal halides (Table 4.8).

[1]M. F. C. Ladd (1969) *Transactions of the Faraday Society*, **65**, 2712; idem. (1977) *Journal of the Chemical Society, Dalton Transactions*, 220.
[2]M. F. C. Ladd and W. H. Lee (1964) *Progress in Solid State Chemistry*, **1**; (1965) idem. ibid. **2**; idem. (1966) ibid.; **3**, and references therein; T. C. Waddington (1959) *Advances in Inorganic Chemistry and Radiochemistry*, **1**.

Table 4.8. Equilibrium interionic distances/nm for the alkali-metal halides

	Li	Δ	Na	Δ	K	Δ	Rb	Δ	Cs	Δ̄
F	0.201	0.030	0.231	0.036	0.267	0.015	0.282	0.018	0.300	
Δ	0.056		0.050		0.047		0.046		0.056	0.051
Cl	0.257	0.024	0.281	0.033	0.314	0.014	0.328	0.028	0.356	
Δ	0.018		0.017		0.015		0.015		0.015	0.016
Br	0.275	0.023	0.298	0.031	0.329	0.014	0.343	0.028	0.371	
Δ	0.025		0.025		0.024		0.023		0.024	0.024
I	0.300	0.023	0.323	0.030	0.353	0.013	0.366	0.029	0.395	
Δ̄		0.025		0.033		0.014		0.026[a]		

[a] It should be noted that CsCl, CsBr and CsI have the cesium chloride structure type under normal conditions of temperature and pressure; all the others have the sodium chloride structure type. The values underlined are the simple averages for the given row or column.

The values of Δ show that the differences between the r_e values for two halides of a given cation are almost independent of the nature of the cation. Similarly, the differences between the r_e values for two halides of a given anion are almost independent of the nature of the anion. These features may be explained by a model in which the ions are represented by spheres, each of a characteristic radius, and where the sums of the radii are equivalent to the corresponding interionic distances. Thus,

$$r_e(KCl) = r(K^+) + r(Cl^-) \qquad\qquad (4.59)$$

and

$$r_e(NaCl) = r(Na^+) + r(Cl^-) \qquad\qquad (4.60)$$

whence the difference Δ for the two chlorides becomes

$$\Delta(KCl - NaCl) = r(K^+) - r(Na^+) \qquad\qquad (4.61)$$

which is independent of the nature of the halogen considered.

4.7.1   Ionic radii
The interionic distance r_e is an experimental quantity, and it is necessary to consider how it may be divided into its ionic components, a problem that has received much attention. The earliest method is due to Landé (1920). Six compounds that have the sodium chloride structure type are listed in Table 4.9, together with their interionic distances. The constancy of r_e for the selenides and sulphides, irrespective of the change in cation, was taken to indicate that the anions were in close contact, with the smaller cations occupying the interstices, which situation is shown in Fig. 4.11.

In the close-packed arrangement of anions, it is evident that $2r_e\sqrt{2} = 4r_-$. Hence, $r(Se^{2-}) = 0.193$ nm and $r(S^{2-}) = 0.184$ nm. The oxides cannot be treated in this way because their interionic distances indicate a lack of, or uncertainty about, close packing in these compounds. If we accept the additivity of ionic radii, that is, $r_e = r_+ + r_-$, as implied by Table 4.8, then other radii can be deduced; the Goldschmidt ionic radii values (1926) were deduced in this manner.

Pauling (1930) showed that the radius of an atom is governed mainly by the configuration of the outermost electrons in the corresponding atom. For an isoelectronic series of ions, he gave the relationship

$$r_i = c/(Z_i - \sigma) \qquad\qquad (4.62)$$

where c is a constant for an isoelectronic series of ions, Z_i is the atomic number of the ith ion of radius r_i and σ is the screening constant given by Slater's rules (see section

Table   4.9. Interionic distance for some compounds with the sodium chloride structure type

	r_e/nm		r_e/nm
MgO	0.210	MnO	0.222
MgS	0.260	MnS	0.261
MgSe	0.273	MnSe	0.273

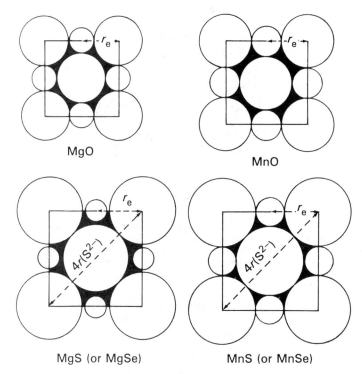

Fig. 4.11. Close-packed arrays of ions, as seen in projection on to a cube face; note the contact of the anions across the face of a cube diagonal for MgS (MgSe) and MnS (MnSe), but their separation in MgO and MnO.

2.6.5.1). Considering NaF, for example, we have from (4.62)

$$r(\text{Na}^+) = c/(11 - 4.15) \qquad (4.63)$$

and

$$r(\text{F}^-) = c/(9 - 4.15) \qquad (4.64)$$

whence

$$r(\text{Na}^+)/r(\text{F}^-) = 0.708. \qquad (4.65)$$

Since $r_e(\text{NaF}) = 0.231$ nm (Table 4.8), and using $r_e = r_+ + r_-$, it follows that $r(\text{Na}^+) = 0.096$ nm and $r(\text{F}^-) = 0.135$ nm.

Ladd (1968)[1] showed that by applying Landé's method to LiI, for which, among the alkali-metal halides, a close-packed array of anions was most likely to exist, the same set of radii was produced as was deduced from direct measurements on electron density contour maps that were obtained by X-ray crystallographic studies[2] on the alkali-metal halides (Fig. 4.12). Table 4.10 lists Ladd's radii and Pauling's radii for several species, together with values for cations from Shannon and Prewitt (1969). It may be seen that some uncertainty still attaches to the precise value of an ionic radius: however, the close correspondence between the results given by Ladd and by Shannon and Prewitt[3], from different evaluations, would tend to indicate a greater reliability of these values.

[1]M. F. C. Ladd (1968) *Theoret. Chim. Acta* **12** 333
[2]H. Witte and Wölfel (1955) *Zeit. physik. Chemie* (Frankfurt) **3**, 296; *idem* (1958) *Rev. Modern Phys.* **30**, 51
[3]R. D. Shannon and C. T. Prewitt (1969; 1970) *Acta Crystallographica*, **B25**, 925; *idem., ibid.* **B26**, 1046

Fig. 4.12. Idealized drawing of the electron density contour map for LiF, as seen in projection on to a cube face; it may be noted that the anions are not in contact in this structure. However, the dashed lines represent the contours of zero electron density, within experimental error, and so define the spatial limits of the ions.

Table 4.10. Radii/nm of some ionic species[a]

	Ladd (1968)	Pauling (1930)	Shannon and Prewitt (1969)
Li^+	0.086	0.060	0.088
Na^+	0.112	0.095	0.116
K^+	0.144	0.133	0.152
Rb^+	0.158	0.148	0.163
Cs^+	0.184	0.169	0.184
NH_4^+	0.166	0.148	–
Ag^+	0.127	0.126	0.129
Tl^+	0.154	0.140	–
Be^{2+}	0.048[b]	0.031[b]	0.041[b]
Mg^{2+}	0.087[b]	0.065[b]	0.086[b]
Ca^{2+}	0.118	0.099	0.114
Sr^{2+}	0.132	0.113	0.130
Ba^{2+}	0.149	0.135	0.150
H^-	0.139	0.208	–
F^-	0.119	0.136	–
Cl^-	0.170	0.181	–
Br^-	0.187	0.195	–
I^-	0.212	0.216	–
O^{2-}	0.125	0.140	–
S^{2-}	0.170	0.184	–
Se^{2-}	0.181	0.198	–
Te^{2-}	0.197	0.221	–

[a]It is interesting to note that Witte and Wölfel gave $r(Na^+) = 0.117$ nm and $r(Cl^-) = 0.164$ nm from measurements on their electron density map for NaCl.
[b]4-coordination; otherwise 6-coordination.

With the exception noted, the radii in Table 4.10 refer to 6-coordination. Measure-ments on polymorphic species show that the radii are dependent on the coordination number. With reference to 6-coordination as the standard, the following changes apply:

4-coordination	standard − 5%
6-coordination	standard
8-coordination	standard + 3%.

4.7.2   Radius ratio and AX structure types

The radius ratio $\mathbb{R}$ is defined as the ratio of the cationic radius to the anionic radius, r_+/r_-. It may be used as a guide to predicting structure, but it is not completely successful in all cases. We shall review the concept first in relation to simple structures of the type AX.

Consider the cesium chloride structure type (Fig. 4.2), in which the coordination pattern is 8:8. Let the ions be of such a size that adjacent anions at the corners of the unit cell are in contact with one another and with the central cation. Then the unit-cell side a is equal to $2r_-$, and the body-diagonal $a\sqrt{3}$ is equal to $2(r_+ + r_-)\sqrt{3}$. Hence, it follows that

$$r_-\sqrt{3} = r_+ + r_- \tag{4.66}$$

and

$$\mathbb{R}_8 = r_+/r_- = 0.732. \tag{4.67}$$

As the cation is made smaller for a constant anion radius, the contact is lost between the cation and the anions. Thus, there is a separation of charged regions, although the distance between the ion centres has not changed, and one may ask if a more stable structure can be attained by an alteration in the coordination pattern. In fact, the sodium chloride structure type, with a 6:6 pattern, offers an energetically more stable configuration at the smaller value of $\mathbb{R}$ now under consideration. Referring to Fig. 4.11 for MnS, we can see that the maximum contact situation corresponds to the new value for $\mathbb{R}$ given by

$$\mathbb{R}_6 = \sqrt{2} - 1 = 0.414. \tag{4.68}$$

Continuing in this manner, we obtain $\mathbb{R}_4 = 0.225$ for the 4:4 coordination in the Würtzite and Blende structures of zinc sulphide (Fig. 4.13).

We can study these results by means of the lattice energy equation. Let (4.51) be written in the form

$$U(r_e) = \alpha/r_e \tag{4.69}$$

where $\alpha = -10^{-3} L \cdot \sqrt{q_1 q_2 e^2 (1-\rho/r_e)/4\pi\varepsilon_0}$ kJ mol^{-1} nm, and let $q_1 = q_2 = 1$ and $\rho/r_e = 0.1$. Assuming that $r_e = r_+ + r_-$, and keeping a fixed value of r_-, we can write

$$U(r_e) = \beta/(\mathbb{R} + 1) \tag{4.70}$$

where $\beta = \alpha/r_-$ and $\mathbb{R}$ is the radius ratio.

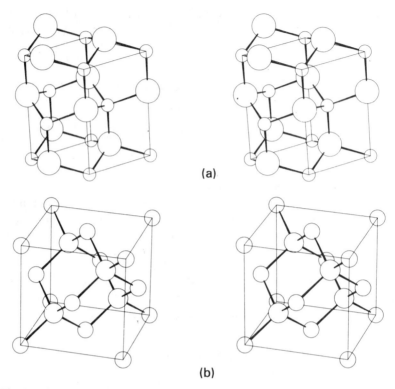

(a)

(b)

Fig. 4.13. Stereoviews of the unit cells of the zinc sulphide structures; circles in decreasing order of size represent S and Zn: (a) α-ZnS—Würtzite, (b) β-ZnS—Blende. Tetrahedral coordination (4:4) is present in both of these structures (compare Fig. 1.2). Blende transforms to Würtzite at 1023°C: both forms occur in nature, so that a return to the equilibrium state at normal temperatures is infinitely slow.

The graph of $U(r_e)$ against $\mathbb{R}$, from (4.70), is shown in Fig. 4.14. Starting at $\mathbb{R} = 1$, the cesium chloride structure type is the most stable. As $\mathbb{R}$ is decreased, keeping r_- constant, $U(r_e)$ decreases as the graph shows. When $\mathbb{R} = 0.732$ the ions are in maximum contact, and cannot become closer packed, even though the central cation may become smaller than its surrounding hole. However mobile we may consider the cation in its central hole, r_e remains constant and the curve for the cesium chloride structure type becomes horizontal at $\mathbb{R} = 0.732$. At this value of $\mathbb{R}$, the energy of the structure may be decreased by its adopting the sodium chloride arrangement of component ions. Similar arguments can be applied to the rest of Fig. 4.14, in respect of $\mathbb{R}$ equal to 0.414 and 0.225, and we can summarize the results as

Structure type	$\mathbb{R}$
CsCl	$\geqslant 0.732$
NaCl	0.414–0.732
α-ZnS/β-ZnS	0.225–0.414

and Table 4.11 lists the values of $\mathbb{R}$ among the alkali-metal halides.

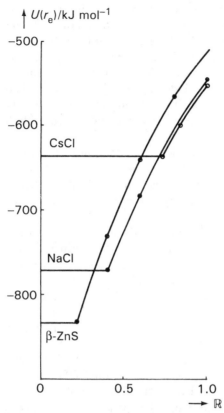

Fig. 4.14. Variation of lattice energy $U(r_e)$ with radius ratio $\mathbb{R}$, at a constant value of r_-; the small difference in electrostatic energy between the CsCl and NaCl structure types is very evident.

Table 4.11. Radius ratios for the alkali-metal halides

	Li	Na	K	Rb	Cs
F	0.72	0.94	0.83[a]	0.75[a]	0.65[a]
Cl	0.51	0.66	0.85	0.93[a]	0.92
Br	0.46	0.60	0.77	0.84	0.98
I	0.41	0.53	0.68	0.75	0.87

For $\mathbb{R} > 1$, $1/\mathbb{R}$ is quoted, which is permissible in $n{:}n$ regular coordination.

Eight values of $\mathbb{R}$, enclosed by the broken boundary in Table 4.11, greater than 0.732 are found for halides that have the sodium chloride structure under normal conditions. The radius ratio is a geometrical concept based on the packing of spheres, and we must not be too disturbed to find that such a simple approach has its limitations. There are three factors to consider.

The energy difference between the cesium chloride and sodium chloride structure types for a given compound is only about 8 kJ mol^{-1}; it depends, in a first analysis, upon the difference in the Madelung constants (Table 4.1). Remembering that this value is based upon the point-charge model (4.13), we may expect that the dipole and quadrupole terms in (4.34) will take on great significance when considering small differences in electrostatic (Madelung) energy. In fact, it has been shown by calculating the lattice energy of potassium chloride, by (4.34), in both the cesium chloride and sodium chloride structure types that the sodium chloride form is more stable by 7–8 kJ mol^{-1}.

There is also a certain covalent contribution to the lattice energy, even in these highly ionic compounds. In the sodium chloride structure, the p orbitals of adjacent ions are directed towards one another, thus facilitating a tendency towards overlap. In the cesium chloride structure type this situation does not obtain, because of the differing coordination pattern.

A change in external conditions that would lead to closer packing, that is, increase in pressure or decrease in temperature, often brings about a transformation. For example, at 83 K, rubidium chloride transforms to the cesium chloride structure type without appreciable change in the radius ratio.

The difference in energy between α-ZnS and β-ZnS is only $\sim 0.2\%$ (Table 4.1). Fig. 4.14 shows that at $\mathbb{R} < 0.414$ the zinc sulphide structure types are more stable than that of sodium chloride: yet they do not occur, even in lithium iodide. Phillips (1970)[1] has pointed out empirically that four-coordinated structures tend to arise where the degree of ionic character is less than about 0.78. Calculations for the alkali-metal halides indicate a range of ionic character from 0.89 in LiI to 0.96 in RbF.

Other attempts to set up demarcation lines between four- and six-coordinated ionic structures based, for example, on different scales of electronegativity, simply restate in other terms that the structural and physical data that are needed to establish good energy calculations are yet to be measured with precision. Where such data are known, the refined electrostatic model (4.34) provides a satisfactory account of the energetics of ionic compounds.

4.7.3   Radius ratio and AX_2 structure types

The most common AX_2 structure types are fluorite (CaF$_2$) and rutile (TiO$_2$); β-cristobalite (SiO$_2$) is less frequently encountered. These structures are illustrated in Figs 4.15–4.17. Since their coordination patterns are 8:4, 6:3 and 4:2, respectively, the radius ratio limits are, in order, the same as for the cesium chloride, sodium chloride and zinc blende structure types. Table 4.12 lists the radius ratios for some structures in this group.

It is noteworthy that, among these compounds, the radius ratio is obeyed without exception. From Table 4.1, we can see that the electrostatic energy differences among these structures are very much greater than among the AX structures. So, although polarization effects are larger among the smaller, more highly charged species, they do not override the packing demands in these structures, because of the large differences between the electrostatic components of their lattice energies.

[1]J. C. Phillips (1970) *Rev Modern Phys* **42**, 317.

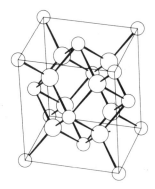 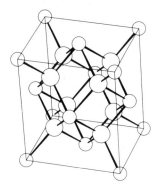

Fig. 4.15. Stereoview of the unit cell of the fluorite (CaF_2) structure type; 8:4 coordination. Circles in decreasing order of size represent F and Ca (compare Fig. 4.2).

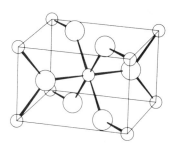 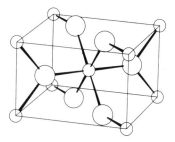

Fig. 4.16. Stereoview of the unit cell of the rutile (TiO_2) structure type; 6:3 coordination. Circles in decreasing order of size represent O and Ti (compare Fig. 1.3).

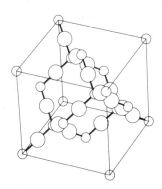 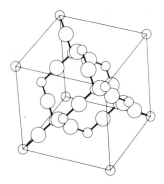

Fig. 4.17. Stereoview of the unit cell of the β-cristobalite (SiO_2) structure type; 4:2 coordination. Circles in decreasing order of size represent O and Si. β-Cristobalite is related to Blende (Fig. 4.13b) as tridymite, another form of SiO_2, is to Würtzite (Fig. 4.13a).

4.7.4 Polarization in AX and AX_2 structures

Polarization, which is expressed in (4.49) through the C and D terms, may be likened to a distortion of the electron density of the ions, with consequent induced dipole–dipole and dipole–quadrupole interactions. Separately, there is also a tendency towards covalent overlap, where the orbitals of the ions have favourable relative

Table 4.12. Radius ratios for some AX_2 structure types

Fluorite		Rutile		β-Cristobalite	
BaF_2	1.25	$CaCl_2$	0.69	$β-SiO_2$	0.30
SrF_2	1.11	$CaBr_2$	0.63	BeF_2	0.23
$BaCl_2$	0.88	MgF_2	0.73		
CaF_2	0.99	MnF_2	0.68		
$SrCl_2$	0.78	ZnF_2	0.62		

Table 4.13. Interionic distances and radii sums in the silver halides

	Structure type	r_e/nm	$\sum r_i/nm$	Δ/nm
AgF	NaCl	0.246	0.246	0.00
AgCl	NaCl	0.277	0.297	0.20
AgBr	NaCl	0.288	0.314	0.26
AgI	β-ZnS	0.281	0.322[a]	0.41

[a]Including -5% for the change of coordination from 6 to 4.

orientations. The presence of these enhancements of the simple electrostatic energy may be indicated by a comparison of ionic radii sums with the corresponding experimental r_e distances; the silver halides provide a good example (Table 4.13).

Among ions of similar radii, it is well known that polarization effects are larger where an outermost d electron configuration exists, probably because of the screening effect of the d electrons. If we compare the screening in Ag^+ with that in the comparable sized Na^+ and K^+, we find the following results:

	Ag^+	Na^+	K^+
Z_{eff}	8	7	10
Z_{eff}/Z	17%	64%	53%

The extent of screening of the nuclear charge in the silver ion results in its electron distribution being much less strongly held than would be expected for an ion of its size, and so more susceptible to distortion. The result of this effect is further manifested in solubility relationships, as we shall discuss presently.

Among AX_2 structures, a decrease in the radius ratio, which may be accompanied by an increase in polarization, may lead to layer structures, of which cadmium iodide is typical (Fig. 4.18). As polarization increases, or ℝ decreases, discrete groups of atoms may segregate into molecules, characteristic of van der Waals' or molecular solids. Mercury(II) chloride is such a case, and we have referred to this compound in section 3.3.2.3. It is an example of an inevitable merging among the arbitrary classification of compounds. The sequence of changes among AX_2 structure types with changes in radius ratio and polarization is shown diagrammatically in Fig. 4.19.

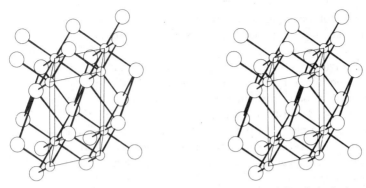

Fig. 4.18. Stereoview of the unit cell of a typical layer structure, CdI_2; circles in decreasing order of size represent I and Cd. The 6:3 coordination consists of a slightly distorted octahedron of iodine around cadmium, and the three cadmium nearest neighbours of any iodine atom lie on one side of it in the composite layer, forming a trigonal pyramid with iodine at the apex.

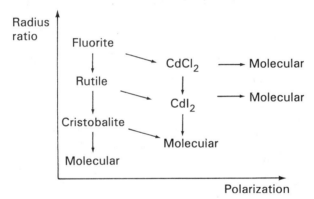

Fig. 4.19. Schematic transformations among AX_2 structure types according to radius ratio and polarization; $CdCl_2$ and CdI_2 are typical layer structures, intermediate in type between the ionic and molecular AX_2 compounds.

4.8 STRUCTURAL AND PHYSICAL CHARACTERISTICS OF IONIC COMPOUNDS

Ionic bonds link an ion to an indefinite number of neighbours. The bond has no particular directionality in space, and ionic structures tend to be governed, to a first approximation, by geometrical considerations, subject to electrical neutrality of the structure as a whole. The Na^+ ion is sufficiently small that twelve of them could be packed round a Cl^- ion. However, the Cl^- ion cannot reciprocate this behaviour and, under electrical neutrality, sixfold coordination obtains. Polarization and partial covalency enhance the value of the lattice energy of ionic compounds, and may modify the predictions of the radius ratio, particularly where the difference in the electrostatic energies of two polymorphs is small.

A refined electrostatic model gives satisfactory lattice energy values. It should be remembered that agreement between the electrostatic equation (4.49) and that based on thermodynamic terms (4.34) need not necessarily imply that the electrostatic model is correct for ionic compounds, but rather that it forms a satisfactory basis for the calculation of lattice energetics.

Molecules do not occur in ionic structures[1], because electrons are almost totally localized in the atomic orbitals of ions and to not overlap with one another sufficiently to form molecules without a change in status of the compound. In solids containing polyatomic ions such as nitrates, carbonates or sulphates, covalent bonding is predominant within the complex ion, with ionic bonding taking place between the polyatomic ions and those species of complementary charge in the structure. Fig. 4.20 illustrates the structure of calcite, which is also shown by sodium nitrate. The planes of the CO_3^{2-} ions are all normal to the vertical direction, the threefold symmetry axis of the rhombohedral unit cell, and the Ca^{2+} ions are each coordinated by six oxygen atoms from different carbonate ions.

Ionic solids form hard crystals of low compressibility and expansivity, but of high melting point. They are electrical insulators in the solid, but when molten, or in solution, they conduct electricity by ion transport: this property distinguishes ionic solids most clearly from covalent and van der Waals' solids.

Trends in physical properties can be related to lattice energies through the dominant $1/r_e$ proportionality; Table 4.14 lists some results on hardness and melting point.

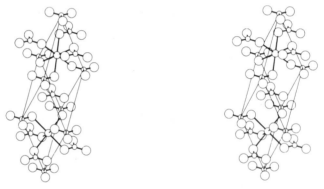

Fig. 4.20. Stereoview of the unit cell of calcite, $CaCO_3$; circles in decreasing order of size represent O, Ca and C; this structure is isomorphous with that of sodium nitrate. The calcium ion is in sixfold coordination with oxygen atoms of different carbonate ions, and each carbonate ion has a trigonal planar structure.

Table 4.14. Hardness and melting point for some ionic solids

	BeO	MgO	CaO	SrO	BaO
r_e/nm	0.165	0.210	0.240	0.257	0.276
Hardness (Moh's scale)[a]	9.0	6.5	5.5	4.1	3.3

	NaF	NaCl	NaBr	NaI
r_e/nm	0.231	0.282	0.298	0.323
T_m/K	1266	1074	1020	934

[a]On Moh's scale, 1–10, of scratch hardness, diamond is 10 and talc 1.

[1]In hydrated ionic compounds, such as $BaCl_2 . 2H_2O$, molecules of H_2O exist, but the main components are ions.

4.9 SOLUBILITY OF IONIC COMPOUNDS

The solubility in water of ionic compounds can be treated quantitatively by thermodynamics, which shows at the same time the relationship between solubility and lattice energy. Solubility is a well-known property. It is often stated that ionic solids are soluble in water, but that covalent solids are not. The latter part of this statement is true, particularly with covalent solids defined as in this book, but the first part of the statement is only partly correct. Again, one can read that solubility decreases with increasing covalent character in compounds; however, consider the following series of compounds:

AgF	AgCl	AgBr	AgI
$\rightarrow$	Decreasing solubility		
$\rightarrow$	Increasing polarization/covalent character		
CaF_2	$CaCl_2$	$CaBr_2$	CaI_2
$\rightarrow$	Increasing solubility		
$\rightarrow$	Increasing polarization/covalent character.		

Evidently, the opening statements about solubility are too simplistic and we shall consider a thermodynamic treatment of this topic, first defining the reference states that will be needed.

4.9.1 Reference states for solubility

We shall restrict this part of the discussion to solids of the type AX, where the species A and X form singly charged ions; the results, however, may be applied generally.
  Consider the equilibrium

$$AX(s) \rightleftharpoons A^+ (aq) + X^-(aq). \tag{4.71}$$

The equilibrium constant K is given in the usual way by

$$K = a(A^+)a(X^-)/a(AX) \tag{4.72}$$

and we recall that it is a constant at a given temperature. The activity a of a pure, crystalline solid in the standard reference state, 298.15 K and 1 atm, is defined as unity; hence,

$$K = a(A^+)a(X^-) \tag{4.73}$$

or

$$K = c^2 f_{\pm}^2 \tag{4.74}$$

where c is the *solubility* (concentration at saturation[1]) in mol dm^{-3} and $f_{\pm}$ is the *mean activity coefficient* for the elecrolyte in its saturated solution; we shall assume a temperature of 298.15 K unless otherwise stated.
  The reference state for solution is the infinitely dilute solution, in which the ratio of the activity of the solute to its molar concentration is unity. The fact that this reference state is hypothetical does not invalidate the arguments that follow. We may regard the reference state as a solution of mean concentration 1 mol dm^{-3} and unit mean activity coefficient, and in which the partial molar heat content of the solute is the same as at infinite dilution (see Appendix 14).

---

[1]In a saturated solution, the dissolved solute is in equilibrium with excess solid solute.

4.9.2 Solubility relationships
Let us simplify (4.71) to

$$A \rightleftharpoons B \tag{4.75}$$

so that

$$K = a(B)/a(A). \tag{4.76}$$

The chemical potentials of the components in the system (4.75) are given by

$$\mu_A = \mu_A^\ominus + \mathscr{R}T \ln a(A) \tag{4.77}$$

and

$$\mu_B = \mu_B^\ominus + \mathscr{R}T \ln a(B). \tag{4.78}$$

At equilibrium, the solid A and the saturated solution B are at the same chemical potential. Hence, $\mu_A = \mu_B$, and the standard free energy change for (4.75) is given by

$$\Delta G^\ominus = \mu_B^\ominus - \mu_A^\ominus = -\mathscr{R}T \ln[a(B)/a(A)] = -\mathscr{R}T \ln K. \tag{4.79}$$

By analogy, for (4.71), we may write

$$\Delta G_d^\ominus = -\mathscr{R}T \ln K = -\mathscr{R}T \ln c^2 f_\pm^2. \tag{4.80}$$

This equation represents the standard free energy change of dissolution for the process

solid in the reference state $\rightleftharpoons$ aqueous ions in the reference state,

and governs solubility.

4.9.3 Two example calculations
We shall consider two calculations based on (4.80) to determine the quantities involved and to see what further analysis might be needed.

4.9.3.1   Silver iodide
The solubility of silver iodide in water at[1] 298 K is 1.02×10^{-8} mol dm^{-3}. At this concentration $f_\pm$ is sensibly unity. Hence, from (4.80), $\Delta G_d^\ominus$ is $+91.2$ kJ mol^{-1}, so that silver iodide is highly insoluble in water.

4.9.3.2   Lithium fluoride
The solubility of lithium fluoride at 298 K is 0.09 mol dm^{-3}. Using the same procedure as with silver iodide, $\Delta G_d^\ominus$ is $+11.9$ kJ mol^{-1}. However, we are not justified here in taking $f_\pm$ as unity, because the saturated solution is not sufficiently dilute. From the Debye limiting equation (see Appendix 15) $f_\pm$ is approximately 0.70, and $\Delta G_d^\ominus$ now becomes $+13.7$ kJ mol^{-1}. If we apply the Debye limiting equation to silver iodide, $f_\pm$ is unity to within 0.01%.

Why do we need any further theory on solubility? We see that these calculations require activity coefficient data for saturated solutions. Where solubilities are less than ~ 0.1 mol dm^{-3}, $f_\pm$ may be calculated with sufficient accuracy (see Appendix 15). In more concentrated solutions the calculation of $f_\pm$ is not reliable. For example,

---

[1]For convenience, we shall write 298 K to mean 298.15 K.

potassium chloride is saturated at 4.8 mol dm^{-3}. At this concentration $f_{\pm}$ would be given as 0.08 from the Debye limiting equation, whereas it is 0.574. There is, unfortunately, a paucity of data on activity coefficients for saturated solutions. Furthermore, we need to analyse $\Delta G_d^{\ominus}$ in order to obtain a clearer understanding of solubility.

4.9.4 Solubility and energy

The important quantity $\Delta G_d^{\ominus}$ may be expanded in the usual way as

$$\Delta G_d^{\ominus} = \Delta H_d^{\ominus} - T \Delta S_d^{\ominus} \tag{4.81}$$

where $\Delta H_d^{\ominus}$ is the standard enthalpy change for the dissolution process, referred to infinite dilution, and $\Delta S_d^{\ominus}$ is the corresponding change in entropy. We may write

$$\Delta S_d^{\ominus} = \sum \bar{S}_i^{\ominus} - S_c^{\ominus} \tag{4.82}$$

where $\sum \bar{S}_i^{\ominus}$ is the sum of the standard relative partial molar entropies of the hydrated ions (see Appendix 14), and $S_c^{\ominus}$ is the standard molar entropy of the crystal; these data are readily available in the chemical literature.[1]

The relationship between solubility and lattice energy is indicated by Fig. 4.21, which is given in terms of enthalpy changes:

$$\Delta H_d^{\ominus} = \Delta H_h^{\ominus} - \Delta H_c^{\ominus} \tag{4.83}$$

where $\Delta H_h^{\ominus}$ is the standard enthalpy change for the hydration of the reference ion-gas to infinite dilution. This equation shows how, from lattice enthalpy and enthalpy of dissolution data, we can obtain hydration enthalpies that are not otherwise readily available.

It is important that the experimental measurements of $\Delta H_d^{\ominus}$ are extrapolated to infinite dilution, because the process of dilution itself can produce significant enthalpy changes. For example, ΔH_d for cadmium sulphate in 200 mole of water is

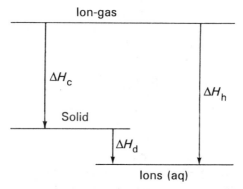

Fig. 4.21. Enthalpy level diagram relating parameters of solubility, lattice enthalpy and hydration enthalpy.

[1]See, for example, F. D. Rossini *et al.* (1952) *Circular No. 500, and later Supplements*, National Bureau of Standards.

$-43.9\ \mathrm{kJ\,mol^{-1}}$, but at infinite dilution $\Delta H_d^{\ominus}$ becomes $-53.6\ \mathrm{kJ\,mol^{-1}}$. The difference between two such values may be commensurate with, or even larger than, $\Delta G_d^{\ominus}$ itself.

Table 4.15 lists thermodynamic data related to solubility for a range of ionic compounds. The interaction between ions and water molecules on the one hand, and between ions in the crystal structure on the other, both increase as the ions become smaller or more highly charged, because the dominant Coulombic energy is proportional to $q_1 q_2 e^2/r_e$. Both ΔH_d and ΔG_d depend on the difference between two quantities, generally of large magnitudes, one concerned with the solid and the other with the hydrated ions.

A decrease in the value of (more negative) ΔG_h tends to stabilize the hydrated state, with respect to the ion-gas, and so promote solubility. However, a decrease in either ΔG_c or ΔS_d tends to decrease solubility, ΔG_c by stabilizing the crystal with respect to the ion-gas, and ΔS_d by making the hydrated state relatively less probable. The term ΔS_d becomes very important with small or highly charged ions, such as in LiF or MgF_2, for example.

On transferring a gaseous ion isothermally into water, two important processes occur. There is a structure-breaking reaction on the water itself because of the interaction between ions and water molecules, including a disruption of some hydrogen bonds in the water. Then there is a structure-making reaction arising from the coordination of ions by water molecules in hydration shells around the ions. The first of these reactions is important with large ions as it tends to increase the $\bar{S}_i^{\ominus}$ values, as may be seen by comparing SrF_2 and BaF_2, for example. The second process is significant with small ions, since it acts so as to decrease $\bar{S}_i^{\ominus}$, as exemplified by a comparison of CaF_2 and MgF_2. Any given case involves an interplay of these factors and, while solubility may be less easy to explain in molecular terms, the thermodynamic analysis is quantitative and the results precise, within the limits of experimental error.

We can now explain the solubility results with which we introduced this section. In the series of calcium halides, ΔG_h increases (becomes more positive) less rapidly than ΔG_c from fluoride to iodide. As the anionic radius increases, the demands of regular packing in the solid determine the closest r_e distance. In solution, however, the same packing demands do not exist, and the total ion–solvent interaction remains strong, despite the increase in anionic radius. Hence, ΔG_d becomes more negative in this direction, despite some large, negative ΔS_d terms.

For the series of silver halides an opposite tendency exists: ΔG_c decreases from fluoride to iodide more rapidly than would be expected for halides containing an ion the size of r_{Ag^+}. A full calculation of the lattice energies of the silver halides using (4.49) gives results that agree with the thermodynamic model (4.34) for silver fluoride, but show increasing discrepancies from chloride to iodide. A partial covalent character in these compounds, in addition to the C and D terms of (4.49), is largely responsible for the greater stability of the crystal with respect to the ion-gas and, hence, for the decrease in solubility along this series.

Substances such as MgO and CaO, for example, are also highly ionic, but effectively insoluble in water. The lattice energies are numerically large because the electrostatic energy includes the product of the ionic charges: these compounds have the sodium chloride structure type, so the Madelung energy is proportional to $4\mathscr{A}(\mathrm{NaCl})$. The

Table 4.15. Thermodynamic data relating to the solubilities of some ionic halides at 298.15 K.

Halide	$\Delta H_d^{\ominus}$ /kJ mol^{-1}	$\sum \bar{S}_i^{\ominus}$ /J mol^{-1}	$S_c^{\ominus}$ /J mol^{-1}	$T\Delta S_d^{\ominus}$ /kJ mol^{-1}	$\Delta G_d^{\ominus}$ /kJ mol^{-1}
LiF	4.6	4.6	36.0	-9.4	14.0
LiCl	-37.2	69.5	55.2	4.3	-41.5
LiBr	-49.0	95.0	69.0	7.8	-56.8
LiI	-63.2	123.4	75.7	14.2	-77.4
NaF	0.4	50.6	58.6	-2.4	2.8
NaCl	3.8	115.5	72.4	12.9	-9.1
NaBr	-0.8	141.0	85.8	16.5	-17.3
NaI	-7.5	169.5	92.5	23.0	-30.5
KF	-17.6	92.9	66.5	7.9	-25.5
KCl	17.2	157.7	82.8	22.3	-5.1
KBr	20.1	183.3	96.7	25.8	-5.7
KI	20.5	211.7	104.2	32.1	-11.6
RbF	-26.4	114.6	72.8	12.5	-38.9
RbCl	16.7	179.5	94.6	25.3	-8.6
RbBr	21.8	205.0	108.4	28.8	-7.0
RbI	25.9	233.5	118.0	34.4	-8.5
CsF	-37.7	123.4	79.9	13.0	-50.7
CsCl	18.0	188.3	97.5	27.0	-9.0
CsBr	25.9	213.8	121.3	27.6	-1.7
CsI	33.1	242.3	129.7	33.6	-0.5
AgF	-20.5	64.4	83.7	-5.8	-14.7
AgCl	66.5	129.3	96.2	9.9	56.6
AgBr	84.1	154.8	107.1	14.2	69.9
AgI	111.7	183.3	114.2	20.6	91.1
TlF	-2.5	111.7	87.9	7.1	-9.6
TlCl	43.5	182.4	108.4	22.1	21.4
TlBr	57.3	207.9	119.7	26.3	31.0
TlI	74.1	236.4	123.0	33.8	40.3
MgF$_2$	-18.4	-137.2	57.3	-58.0	39.6
MgCl$_2$	-155.2	-7.5	89.5	-28.9	-126.3
MgBr$_2$	-186.2	43.5	123.0	-23.7	-162.5
MgI$_2$	-214.2	100.4	145.6	-13.5	-200.7
CaF$_2$	13.4	-74.5	69.0	-42.8	56.2
CaCl$_2$	-82.8	55.2	113.8	-17.5	-65.3
CaBr$_2$	-110.0	106.3	129.7	-7.0	-103.0
CaI$_2$	-120.1	163.2	142.3	6.2	-126.3
SrF$_2$	10.5	-58.6	89.5	-44.2	54.7
SrCl$_2$	-51.9	71.1	117.2	-13.7	-38.2
SrBr$_2$	-71.5	122.2	141.4	-5.7	-65.8
SrBr$_2$	-90.4	179.1	164.0	4.5	-94.9
BaF$_2$	3.8	-6.7	96.7	-30.8	34.6
BaCl$_2$	-13.0	123.0	125.5	-0.7	-12.3
BaBr$_2$	-25.5	174.1	148.5	7.6	-33.1
BaI$_2$	-47.7	230.9	171.1	17.8	-65.5

O^{2-} ion hydrates to a pair of OH^- ions, and $\Delta H_h(OH^-, g)$ is ~ -1000 kJ mol^{-1}. Nevertheless, the hydration energy does not overcome the very large, negative lattice energy (~ 3000–3500 kJ mol^{-1}), and these oxides are not soluble in water.

Where activity coefficient data at saturation are available, the above arguments are confirmed. For example, sodium chloride has a concentration of 6.1 mol dm^{-3} at saturation. At this concentration, the mean activity coefficient, measured by the isopiestic technique, is 1.02. Hence, from (4.80), $\Delta G_d^\ominus = -9.06$ kJ mol^{-1}, which is in excellent agreement with the value reported in Table 4.15.

4.10   VIBRATIONS AND DEFECTS IN IONIC COMPOUNDS

In putting forward our model for ionic crystals, we treated them first as perfectly regular, three-dimensional arrays of point charges. We considered modifications of this picture, by introducing terms to allow for van der Waals' interactions between symmetrically distorted ions, and also admitted the existence of a covalent contribution to the lattice energy. We have, nevertheless, left the impression of a static system of particles, each on its correct lattice site, and the whole system possessing a perfect, lattice geometry. On an atomic scale, real ionic solids possess defects in their structure; they undergo transitions that may result in the appearance of colour; and the ions execute vibrational motion, about their lattice sites, that makes a major contribution to the heat capacity of these compounds.

4.10.1   Absorption spectra

All ionic solids absorb in the ultraviolet region of the spectrum; some compounds absorb also in the visible region. The absorption does not produce opacity, as is the case for metals; often, apparently opaque crystals are simply intensely coloured, like permanganates, and transmission of light can be observed in thin section. Ionic solids are coloured for two main reasons. Some of them may contain ions that give rise to a characteristic colour through transitions among their d electrons, such as salts of $[Ti(H_2O)_6]^{3+}$, $[Fe(H_2O)_6]^{2+}$ or $[CoCl_4]^{2-}$. We have seen in Chapter 2 that the d electrons in a free atom are degenerate. In a polyatomic ion, however, the spherical environment of the atom is lost through combination with ligands, and the d electrons can absorb energy by making transitions among the allowed d energy, t_{2g} and e_g (see section 2.13). The ligand-field-splitting energy parameter Δ is such (~ 2.5 eV) that transitions occur in the visible range of the spectrum, and the compound appears coloured. The colour in some species, such as the $[MnO_4]^-$ ion, arises from charge-transfer transitions of electrons between the ligands and the d orbitals of the central ion of the complex.

In other coloured solids, the ions that impart no colour when in solution undergo polarization in the solid state, and colour arises from transitions among partially delocalized electrons. If the absorption moves from the ultraviolet just into the visible, the wavelengths absorbed would lie at the blue end of the spectrum, and the compound would appear yellow to red. Examples of this situation are found with the compounds AgI, Ag_3PO_4 and PbO.

4.10.2 Heat capacity
From (4.15), it follows that the constant volume heat capacity is defined by

$$C_V = \left(\frac{\partial U}{\partial V}\right)_V. \tag{4.84}$$

As the temperature of a solid is increased its vibrational energy increases. If, as in the case of cesium chloride, for example, there are no translational energy modes, and since rotation of a species about its own axis does not constitute a degree of freedom, all of the energy imparted to the solid in the form of heat enhances its vibrational energy. The vibrational degrees of freedom are the number of independent square terms (see section 1.3.1) needed to specify the total energy of the species in the solid. Cesium chloride contains $2L$ ions per mole of substance, and we may assume that each ion vibrates independently of the others present in the solid. A vibration can be resolved along Cartesian axes into three mutually perpendicular squared terms, so that there are $6L$ degrees of freedom per mole. Each squared term contributes an amount of energy kT (because it involves both kinetic and potential energy components of $\frac{1}{2}k_B T$ each), so that the total vibrational energy is $6Lk_B T$, or $6\mathcal{R}T$ per mole. Thus, from (4.84) C_V is $6\mathcal{R}$, or $\sim 50\ \mathrm{kJ\,mol}^{-1}$. For monatomic species, $C_V = 25\ \mathrm{kJ\,mol}^{-1}$, which result is enshrined in the historic law of Dulong and Petit.

The principal molar heat capacities are related by the equation

$$C_p - C_V = \alpha^2 TV/\kappa \tag{4.85}$$

where the symbols have the meanings as before. For sodium chloride at 298 K, we have $\alpha = 1.1 \times 10^{-4}$, $\kappa = 4.1 \times 10^{-11}\,\mathrm{N}^{-1}\,\mathrm{m}$ and $V = 2.7 \times 10^{-5}\,\mathrm{m}^3\,\mathrm{mol}^{-1}$. Hence, $C_p - C_V = 2.4\ \mathrm{kJ\,mol}^{-1}\,\mathrm{K}^{-1}$. For the ideal gas, $C_p - C_V = \mathcal{R}$; the smaller value for the solid arises from the fact that its heat capacity is determined mainly by the vibrational energy, and this parameter does not vary appreciably between constant pressure and constant volume conditions.

According to the Planck theory, a vibrating atom acquires energy in quanta of $h\nu$, and the probability that a vibrating atom receives this amount of energy is proportional to $\exp(-h\nu/k_B T)$. At room temperature, almost all of the vibrational degrees of freedom are active, and C_V tends to its limiting value (Fig. 4.22). As the temperature is decreased vibrations of increasingly lower frequency cease to be excited. Thus, C_V decreases with decreasing temperature, because the average energy of the solid decreases. In the limit as $T \to 0$, $C_V \to 0$ for a pure crystalline solid, in accordance with the third law of thermodynamics.

The effect of a change in temperature on the atomic vibrations of a solid can be gleaned from the following example. Sodium chloride gives a single absorption band at $164\ \mathrm{cm}^{-1}$, and the ratio of the probabilities that this oscillator will acquire the corresponding energy $hc\bar{\nu}$ at 500 K and 50 K is ~ 70, from the exponential proportionality given above. The energy absorbed is used mainly in increasing the amplitudes of the vibrating systems.

The requirement that a vibration shall be active in the infrared region of the spectrum is that the species shall contain an oscillating dipole, and the rate of change of dipole moment with time determines the ability of a species to absorb infrared radiation. In ionic solids, each pair of vibrating ions of opposite sign is equivalent to

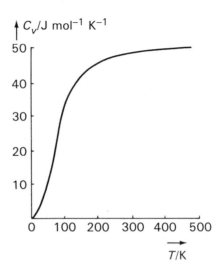

Fig. 4.22. Variation of constant volume heat capacity C_V with temperature T for crystalline sodium chloride.

an oscillating dipole, and its vibrations are excited by infrared radiation; the heavier the species, the smaller the wavenumber (energy) of the vibration.

4.10.3 Defects in crystals
From perfect crystals of vibrating atoms, we consider finally in this chapter the more realistic situation of such crystals containing imperfections. Almost all crystalline substances contain defects. Defects may be classed as *intrinsic*, where they occur in an otherwise perfect, pure substance; or they may be termed *extrinsic*, where they derive from the presence of added impurities. The subject of crystal defects is extensive, and we shall investigate here only that aspect of imperfections known as *point defects*, which can exist in both intrinsic and extrinsic forms. Point defects are limited to random single sites; they may also be linked in one or more dimensions to form *extended* defects, but we shall not consider this class here.

The introduction of defects into a perfect solid requires an expenditure of energy on the system, but the creation of defects brings about an increase in disorder and, hence, in entropy. Provided that ΔA, the change in the Helmholtz free energy for the creation of defects at constant volume, given by the equation

$$\Delta A = \Delta U - T\,\Delta S \qquad (4.86)$$

is negative, it follows that the system with defects is thermodynamically the more stable.

4.10.3.1 Schottky defect
The simplest intrinsic point defect in an otherwise perfect crystal is the Schottky defect. An ion (or atom) is transferred from the interior of the crystal to its surface (Fig. 4.23a). In a crystal at thermal equilibrium with its surroundings, a certain number of vacancies at atom sites throughout the lattice array always exists and contributes to the entropy of the crystal, at any temperature $T > 0$.

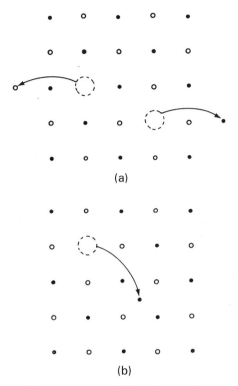

Fig. 4.23. Intrinsic point defects: cations are the filled circles, anions open circles. (a) Pair of Schottky defects in KCl; the ions are removed to the surface. (b) Frenkel defect in AgCl; the cation is under compressive stress in its interstitial position.

Let us consider a constant volume of crystal, and distribute n defects over N lattice sites. The first defect can be distributed in N ways, the second defect in $(N-1)$ ways, and the nth in $[N-(n-1)]$ ways. Thus, the total number of arrangements w of the n defects is given by

$$w = N(N-1)(N-2)\cdots[N-(n-1)] = N!/(N-n)! \qquad (4.87)$$

However, the defects are indistinguishable from one another: we cannot differentiate between defects i and j on site k, for example. There are n indistinguishable ways of obtaining the first defect, $(n-1)$ ways of obtaining the second, $(n-2)$ ways for the third, and so on, and in all $n!$, Then, the total number of arrangements, or probability W, that we require is

$$W = \frac{N!}{(N-n)!\, n!} \qquad (4.88)$$

Let Δu represent the energy change for the formation of a single defect; then

$$\Delta A = n\, \Delta u - T\, \Delta S \qquad (4.89)$$

for the formation of the n defects. The statistical entropy equation, connecting entropy with probability, has been given by Boltzmann, in the form

$$S = k_\mathrm{B} \ln W. \qquad (4.90)$$

The change implicit in (4.89) is from perfect order, $W = 1$, to the given defect state of probability W; hence

$$\Delta A = n\,\Delta u - k_B T \ln\{N!/[(N - n)!\,n!]\}. \tag{4.91}$$

At equilibrium

$$\left(\frac{\partial \Delta A}{\partial n}\right)_T = 0 \tag{4.92}$$

and using Stirling's approximation for factorials

$$\ln X! = X \ln X - X \tag{4.93}$$

we obtain

$$\Delta u - k_B T \ln[(N - n)/n] = 0. \tag{4.94}$$

If we assume that N is of the order of L, then $n \ll N$, and

$$n = N \exp(-\Delta u/k_B T). \tag{4.95}$$

In ionic crystals, it is energetically more favourable to form Schottky defects in pairs, one ion of each sign, so as to maintain the best electrical balance. In this situation, W is squared, and we then obtain

$$n_\pm = N \exp(-\Delta u/2k_B T). \tag{4.96}$$

The average energy of a nearest neighbour bond in a solid is of the order of 1 eV, so this amount of energy has to be expended on the system to create a vacancy. The structure can then relax around the vacancy, and approximately two-thirds of the energy expended is recovered in the relaxation process. The entropy change is positive, and contributes to the driving force for the creation of defects, according to (4.89).

As an example calculation, consider potassium chloride at 300 K. The value of Δu for a pair of K^+, Cl^- defects is ~ 1 eV, or 96 kJ mol^{-1}. Hence, using (4.96),

$$n_\pm/N = \exp(-96\,000\text{ J mol}^{-1}/300\text{ K} \times \mathscr{R}) = 1.9 \times 10^{-17}.$$

If N is set equal to the Avogadro constant, $n_\pm$ becomes 1.2×10^7, or approximately 1 defect pair per 5.2×10^{16} sites per mole. At 600 K, this ratio is increased to 1 in 2.3×10^8.

4.10.3.2   Frenkel defect

Another type of intrinsic point defect, found in silver chloride, for example, is the Frenkel defect. It implies the existence of a vacant cation site, the Ag^+ ion having been transferred to an interstitial position in the structure (Fig. 4.23b).

We now need to consider N lattice sites and N' interstices, as well as the n defects. The probability can be evaluated as for the Schottky defect, and the result for W is now

$$W = \frac{N!\,N'!}{(N - n)!\,n!(N' - n)!\,n!} \tag{4.97}$$

Following the analysis as before leads to the expression

$$n = \sqrt{(NN')} \exp(-\Delta u/2k_B T) \tag{4.98}$$

for the creation of n Frenkel defects consisting of the interstitial ion and its hole. The reason that silver chloride exhibits, preferentially, Frenkel defects is not a size effect; more probably it is connected with the greater tendency towards covalent character than in sodium or potassium chlorides. Moreover, the silver cation is under compressive stress in an interstitial position, and this situation would increase the tendency towards covalent character.

4.10.3.3 Doping and colour centres

Perfection of composition in compounds is effectively unattainable, so that defects of some degree cannot be avoided. In certain applications, such as in semiconductors, impurities are deliberately added. Electron carriers can be introduced if a material is doped with traces, ~ 1 in 10^8, of another species having a greater number of electrons. Thus, arsenic can be substituted into silicon, or calcium chloride into sodium chloride, and very small concentrations produce measurable effects. In the latter example, electrical neutrality is maintained because one Na^+ site remains vacant for each Ca^{2+} ion introduced. We shall say more about semiconduction in the next chapter.

If solid sodium chloride is heated in sodium vapour it acquires a yellow colour. The crystal contains an excess of occupied Na^+ sites, so that some Cl^- vacancies are inevitable, with single electrons remaining delocalized in the holes (Schottky sites). Such an electron constitutes an F centre.[1] The F-centre electron has hydrogen-like character, and $1s \rightarrow 2p$ transitions are responsible for the colour: the electron environment is rather like that of a cubic box of cations, similar to a situation that we considered in section 2.4.2.

4.10.4 Defects and ion mobility

Notwithstanding we have indicated that ionic solids are, in general, electrical insulators, there are some compounds in this class that exhibit electrical conduction. This effect is dependent upon the nature and concentration of defects in the ionic crystal. The defect concentration can be increased by heating the solid to a high temperature and then quenching the system rapidly. The defects associated with the high-temperature condition then become locked into the structure and, because ion mobility is related to defect concentration, an enhanced electrical conductivity results. In fact, in the alkali-metal halides, the electrical conductivity is proportional to the ion mobilities and to an exponential factor, as indicated by (4.95). The mobility of an ion may be defined as its drift speed under unit applied field strength, and it is often quoted in the units $cm^2 \, s^{-1} \, V^{-1}$, which implies the rate of movement of a unit surface under an applied potential difference of 1 V.

If a crystal structure were perfect, it would be difficult to envisage a mechanism for ion transport in the solid. Studies have been carried out with radioactive tracer techniques to determine solid-state conductivity. In a typical experiment, a thin slice of $^{24}NaCl$ was sandwiched between two plates of that substance, and the composite maintained at a constant temperature. The defect concentrations through the sample were determined by cutting thin sections of the sandwich after given times and then

---

[1] German *Farbenzenter*, colour centre.

counting the radioactivity in each section. Assuming that the tracer diffuses by virtue of defects in the substrate, the defect concentration c at a distance d from the surface after a time t was shown to follow the equation

$$\ln c = \alpha - d^2/Dt \qquad (4.99)$$

where α is a constant, and D is the diffusion coefficient for the substance under examination. Further experiments showed that D itself varied with temperature according to a Boltzmann equation:

$$D = D_0 \exp(-E_d/\mathscr{R}T). \qquad (4.100)$$

The activation energy E_d for diffusion in sodium chloride is approximately 173 kJ mol^{-1}, which includes the energy needed both to create the vacancy and to move an ion into the vacant site. Thus, for a single ion, using our datum in section 4.10.3.1,

$$E_d \approx E_\mu + 96 \text{ kJ mol}^{-1} \qquad (4.101)$$

where E_μ, the energy needed to induce migration, is ~ 77 kJ mol^{-1} in sodium chloride; the activation energy is approximately equally divided between the energy needed to create a vacancy and that involved in moving an ion into it.

4.11   MOLTEN SALTS

Pure inorganic salts melt normally at temperatures between ~ 500 K and 1500 K. The melts possess high electrical conductivity because there are free ions in the liquid. On the other hand, inorganic molecular compounds, such as $HgCl_2$, have very small electrical conductivity in the liquid, showing that the molecule remains intact on melting.

Diffraction experiments lead to radial distribution functions for melts, of the kind discussed in section 3.2.3. For molten alkali-metal halides of the NaCl structure type, the coordination number ranges from 4.0 to 5.8 for unlike ions, and from 8.0 to 13.0 for like ions. These values compare with 6 and 12 for the NaCl structure type in the solid state. It may be noted that whereas the coordination number for unlike ions is less than that in the crystalline state, that for like ions can exceed the value of 12, so that very close packing can occur in a melt.

Computer simulation procedures with ionic melts are more complicated than with other liquids, because Coulombic energy, proportional to $1/r$, is of much longer range than is dispersion energy, or even dipolar energy, and many more particles are needed to achieve good simulation.

The alkali-metal halides increase in volume on melting by ~ 10–15%. Computer simulation studies suggest transient voids in the melt structure, which are filled by concerted migration of neighbouring ions. This is both equivalent and preferable to the concept of 'holes', or missing ions, as an explanation of the mechanism of electrical conduction.

There is also an entropy increase on melting of 20–25 J K^{-1} mol^{-1}, indicating the expected decrease in order accompanying fusion. The entropy change on vaporization has a mean value of about 98 J K^{-1} mol^{-1}, which is comparable to the Trouton value of 85 J K^{-1} mol^{-1} for other liquids (section 3.2.1). This result arises because the gaseous states are not very different in the two cases, and the gas phase makes the major contribution to the entropy of the system.

PROBLEMS FOR CHAPTER 4

4.1   The spectral convergence limit for the process

$$Tl(g, \text{ground state}) \rightarrow Tl^+(g) + e^-$$

is 49250 cm^{-1}. Calculate the first ionization energy for thallium in kJ mol^{-1}.

4.2   The following values for the vapour pressure of molten lead were derived from measurements of the rate of effusion of the vapour through a small hole into a vacuum, at different temperatures:

T/K	895.4	922.1	964.5	1009.7	1045.5
$p/\text{N m}^{-2}$	0.0783	0.205	0.539	1.40	3.40

Plot a graph of $\ln p$ against $1/T$, so as to confirm the reliability of each data point, and fit the data by the method of least squares. Determine the enthalpy of evaporation ΔH_e of lead for the experimental temperature range, and the estimated standard deviation in this parameter.

4.3   Refer to problem 4.2. The molar heat capacity of lead may be represented by the following equations:

$$C_p/\text{J mol}^{-1} \text{ K}^{-1} = 23.56 + 0.00975 \text{ K}^{-1} \times T \qquad (298 \text{ K–}600 \text{ K})$$

$$C_p/\text{J mol}^{-1} \text{ K}^{-1} = 32.43 - 0.00310 \text{ K}^{-1} \times T \qquad (600 \text{ K–}1200 \text{ K}).$$

The enthalpy of fusion of lead at the melting-point temperature (600 K) is 4.81 kJ mol^{-1}. If the result already obtained for ΔH_e be taken to apply to the mean temperature of 970 K, extrapolate the result to 298 K to obtain the standard enthalpy of sublimation for lead; it may be assumed that the vapour behaves ideally. It is probably desirable first to construct a thermochemical cycle to show the various processes involved in the calculation.

4.4   From the *standard* data below (I and E refer to 0 K), show how the formation of $MgCl_2(s)$ is preferred to that of $MgCl(s)$:

$I_1(\text{Mg, g})$	6.09 eV	$I_2(\text{Mg, g})$	11.82 eV
$\Delta H_s(\text{Mg, s})$	149.0 kJ mol^{-1}	$D_0(\text{Cl}_2, \text{g})$	243.0 kJ mol^{-1}
$E(\text{Cl, g})$	-348.6 kJ mol^{-1}	$\Delta H_f(\text{MgCl, s})$	-221.8 kJ mol^{-1}
$\Delta H_f(\text{MgCl}_2, \text{s})$	-641.8 kJ mol^{-1}		

4.5   Use the following data to construct a thermochemical cycle for the enthalpy of formation of $NH_4Cl(s)$, and find the standard value of this parameter:

			$\Delta H^{\ominus}/kJ$ mol^{-1}
$\frac{1}{2}N_2(g)$	$+ \frac{3}{2}H_2(g)$	$\rightarrow NH_3(g)$	-46.0
$NH_3(g)$	$+ aq$	$\rightarrow NH_4^+(aq) + OH^-(aq)$	-34.7
$\frac{1}{2}H_2(g)$	$+ \frac{1}{2}Cl_2(g)$	$\rightarrow HCl(g)$	-92.5
$HCl(g)$	$+ aq$	$\rightarrow H^+(aq) + Cl^-(aq)$	-74.9
$NH_4^+(aq) + OH^-(aq) + H^+(aq) + Cl^-(aq)$		$\rightarrow NH_4^+(aq) + Cl^-(aq)$	-52.3
$NH_4Cl(s) + aq$		$\rightarrow NH_4^+(aq) + Cl^-(aq)$	$+15.1$

4.6   Use the spherical conducting shell model to calculate an approximate Madelung constant for the sodium chloride structure; $a = 0.564$ nm.

4.7   From the following structural and standard thermodynamic data (I and E refer to 0 K) on calcium oxide, calculate the affinity of oxygen for two electrons:

Structure type	NaCl
Unit cell side a	0.4811 nm
Madelung constant $\mathscr{A}$	1.7476 (remember the ion charges)
Compressibility κ	0.895×10^{-11} N^{-1} m
$I_1(Ca, g)$	589.5 kJ mol^{-1}
$I_2(Ca, g)$	1145.0 kJ mol^{-1}
$\Delta H_s(Ca, s)$	176.6 kJ mol^{-1}
$D_0(O_2, g)$	489.9 kJ mol^{-1}
$\Delta H_f(CaO, s)$	-635.5 kJ mol^{-1}

4.8   Use the following standard data, together with that on the dissociation energies and electron affinities of the halogens, and the enthalpies of dissolution of the strontium halides, all given in the text, to determine an average value for the standard enthalpy of hydration of the strontium ion (I and E refer to 0 K).

	$\Delta H/kJ$ mol^{-1}
$I_1(Sr, g)$	549.4
$I_2(Sr, g)$	1064.0
$\Delta H_s^{\ominus}(Sr, s)$	163.6
$\Delta H_f^{\ominus}(SrF_2, s)$	-1209.0
$\Delta H_f^{\ominus}(SrCl_2, s)$	-828.0
$\Delta H_f^{\ominus}(SrBr_2, s)$	-715.5
$\Delta H_f^{\ominus}(SrI_2, s)$	-569.4
$\Delta H_h^{\ominus}(F^-, g)$	-513.0
$\Delta H_h^{\ominus}(Cl^-, g)$	-371.1
$\Delta H_h^{\ominus}(Br^-, g)$	-340.6
$\Delta H_h^{\ominus}(I^-, g)$	-301.7

4.9   The sulphate ion may be considered as a regular tetrahedral arrangement of oxygen atoms around sulphur. If the charge on each oxygen is -1 and the S—O distance is 0.130 nm, calculate the electrostatic self-energy of the sulphate ion.

4.10   If $r(Cs^+)$ and $r(I^-)$ are 0.184 and 0.212 nm, respectively, and the departure of r_e from the additivity of radii is -0.001 nm, calculate the density of crystalline CsI.

4.11   Refer to Fig. 4.16 for Rutile: the coordination around Ti is distorted octahedral, and around O that of an isosceles triangle. Use the following X-ray crystallographic data to calculate the Ti—O bond lengths (two different values) and O—Ti—O bond angles (two different values).

Tetragonal crystal system: $a=b=0.4593$ nm; $c = 0.2959$ nm; Two formula-entities per unit cell, at fractional coordinates:

2	Ti	0, 0, 0;	$\frac{1}{2}, \frac{1}{2}, \frac{1}{2}$
4	O	$x, x, 0$;	$\bar{x}, \bar{x}, 0$;
		$\frac{1}{2} + x, \frac{1}{2} - x, \frac{1}{2}$;	$\frac{1}{2} - x, \frac{1}{2} + x, \frac{1}{2}$

with $x = 0.3056$. It may be found helpful first to make sketches of the structure from which the unique distances and angles to be calculated should be clear.

4.12   Show that the radius ratio for the Würtzite structure type with atoms in maximum contact is 0.225.

4.13   In an experiment, precipitated silver iodide was dissolved in aqueous solutions of both potassium iodide and sodium iodide, and the heats evolved were 9.45 kJ mol^{-1} and 7.46 kJ mol^{-1}, respectively. Next, finely divided silver was suspended in similar solutions of the two alkali-metal iodides. On adding iodine, the silver dissolved rapidly to form silver iodide: the heats evolved were 72.0 kJ mol^{-1} (of AgI) in the potassium iodide solution, and 70.3 kJ mol^{-1} (of AgI) in the sodium iodide solution. Set up equations to represent the chemical reactions taking place, and calculate an average value for the standard enthalpy of formation of silver iodide; all processes may be assumed to have been conducted at 298.15 K.

4.14   Determine the standard free energy of dissolution of MgF_2, given that $\Delta H_d^{\ominus} = -18.4$ kJ mol^{-1}, $S_c^{\ominus} = 57.3$ J mol^{-1} K^{-1} and $\sum \bar{S}_i^{\ominus} = -137.2$ J mol^{-1} K^{-1}. If the solubility of MgF_2 is 0.075 g dm^{-3}, calculate the mean activity coefficient for this salt at saturation and 298.15 K.

4.15   Derive an expression for the number n of Frenkel defects in a crystal of silver chloride containing N lattice sites and N' interstices. If the energy needed to set up a single defect in silver chloride is ~ 1 eV, calculate the fraction of Frenkel defects in this substance at 500 K.

4.16   The probability per ion of Schottky defects in KCl may be represented by

$$n/N = \exp(-\Delta u/k_B T).$$

What would be the corresponding expression for $CaCl_2$? Give reasons for your answer.

4.17   Radioactive silver was allowed to diffuse through a silver–indium alloy at 1000 K. The penetration depth x was determined after time intervals t of 6×10^4 s by measuring the radioactivity β_t. Show that the diffusion process follows the equation

$$\beta_t = A \exp(-x^2/Dt)$$

and find the value of the diffusion constant D. The experimental data follow; β_t is dimensionless.

x/mm	β_t	x/mm	β_t
0.000	600	0.329	88
0.084	540	0.376	50
0.132	450	0.425	25
0.183	360	0.470	12
0.230	250	0.520	5
0.279	160	0.568	2

4.18   In further experiments, values of the diffusion coefficient D were obtained as a function of temperature. Given that D is related to T by the Boltzmann equation

$$D = D_0 \exp(-E_d/\mathscr{R}T)$$

determine the value of the activation energy E_d for the diffusion process; the relevant data follow:

T/K	878	1007	1176	1253	1322
D/m^{-2} s	1.6×10^{-18}	4.0×10^{-17}	1.1×10^{-15}	4.0×10^{-15}	1.0×10^{-14}

What is the value of D_0 and what is its significance?

4.19   Calculate the standard molar entropy of nickel from the following data:

T/K	15.05	25.20	47.10	67.13	82.11	133.4	204.1	256.5	283.0
C_p/J mol^{-1} K^{-1}	0.1945	0.5994	3.532	7.639	10.10	17.88	22.72	24.81	26.09

Below 15.05 K the Debye approximation may be used: $C_p \approx C_V = \alpha T^3$, where α is a constant parameter with the units J mol^{-1} K^{-4}. Hence,

$$\int \frac{C_p}{T}\, dT = \int_0^{15.05\ \mathrm{K}} \alpha T^2\, dT = \frac{\alpha T^3}{3} = \frac{C_p}{3},$$

where C_p may be taken as the minimum value recorded, that is, at 15.05 K in this example.

4.20   Consider a face of the cubic unit cell, side a, of the sodium chloride structure type (Fig. 1.3) as a two-dimensional square array of unit charges of alternating sign. Calculate the Madelung constant for this array by Evjen's method. Repeat the calculation for squares of side $2a$, $3a$ and $4a$. By a method of constant second differences, or by a suitable extrapolation, report a probable value of the Madelung constant for this two-dimensional structure.

5

Bonding between ions: II

5.1 INTRODUCTION

Metals are distinguished from other substances by several physical properties, among which their high electrical and thermal conductivities, and their opacity to visible light are well known. About three-quarters of the elements are metals, yet their structure types are few in number and, geometrically, fairly simple. We shall endeavour to show how metallic properties can be understood and, indeed, what is meant by metallic character, and how the metallic bond may be correlated with the molecular orbital theory developed in Chapter 2.

5.2 CLASSICAL FREE-ELECTRON THEORY

A theory of metals was put forward first by Drude (1902); it was elaborated by Lorentz (1916) and subsequently known as the *Drude–Lorentz* (*free-electron*) theory. It considered that electrons in a metal can be of two kinds: those in closed inner shells, the core electrons that belong to a lattice array of positive ions in the metal; and those that are free to move, the valence or *conduction* electrons that permeate the metal as a whole. It is the mobile valence electrons that can be influenced by an applied electrical or thermal field.

Thus, we have the model of a lattice of cations, with their core electrons intact, surrounded by a 'sea' of conduction electrons, or *electron gas*. The cohesive energy of a metal was considered to depend upon the attraction between the positive ions and the electron gas, without considering any interaction involving the electrons of the core.

The theory predated wave mechanics, and provided a satisfactory model in respect of certain properties. The high electrical conductivity was explained by the drifting of valence electrons under an applied potential gradient; thermal conductivity was considered to arise from the redistribution of electrons in a thermal field, carrying thermal energy with them; and opacity was deemed to be brought about through the rapid oscillatory movements of the electrons following the alternations of the electromagnetic field of incident light waves, thus inhibiting the transmission of radiation of visible wavelengths through the metal.

5.2.1 Electrical conductivity

We know from experiment that the resistance R of an electrical conductor of uniform cross-sectional area A and length d is proportional to the length of the conductor and inversely proportional to its area. We write

$$\rho = RA/d \qquad (5.1)$$

where ρ is the electrical *resistivity* of the material of the conductor. The electrical *conductivity* σ is the reciprocal of the resistivity, so that

$$\sigma = d/RA. \qquad (5.2)$$

Since R is equal to V/I, where V is a potential difference applied to the conductor and I is the current through it, we have

$$\sigma = j/\mathscr{E} \qquad (5.3)$$

where j is the current density (I/A) and $\mathscr{E}$ is the electrical field strength[1], that is, the applied potential difference per unit length (V/d). The mobility μ of an electron in a field of strength $\mathscr{E}$ is defined as the drift velocity v_d per unit field strength:

$$\mu = v_d/\mathscr{E}. \qquad (5.4)$$

If the valence electron concentration is $\mathscr{N}$, then

$$j = \mathscr{N}ev_d = \mathscr{N}e\mu\mathscr{E}. \qquad (5.5)$$

Hence, by eliminating $j/\mathscr{E}$ from (5.3) and (5.5)

$$\sigma = \mathscr{N}e\mu. \qquad (5.6)$$

The mean drift velocity $\overline{v_d}$, in the direction of the field, during a time t, before collision with a positive ion is given by

$$\overline{v_d}\tfrac{1}{2}\bar{a}t = e\mathscr{E}t/2m_e \qquad (5.7)$$

where $\bar{a}$ is the mean acceleration of the electron of mass m_e over the same time interval. Hence, the mean mobility is

$$\bar{\mu} = et/2m_e \qquad (5.8)$$

and

$$\sigma = \mathscr{N}e^2t/2m_e. \qquad (5.9)$$

Let τ be a relaxation time during which the electron attains its drift velocity; thus, $\tau = t/2$. Then

$$\sigma = \mathscr{N}e^2\tau/m_e. \qquad (5.10)$$

The mean free path l of an electron is its average distance of travel between collisions with positive ions. Thus,

$$l = \overline{v_d}\tau. \qquad (5.11)$$

---

[1]Strictly, j and $\mathscr{E}$ are vector quantities, but we are concerned here only to derive an expression for σ.

If the electrons in a metal behave like a gas, then the average kinetic energy (Appendix 16) is given by

$$\tfrac{1}{2}m_e\overline{v_d^2} = \tfrac{3}{2}k_B T. \tag{5.12}$$

However, experiment has shown that this equation gives values for $\overline{v_d}$ that are too small; the average kinetic energy is better equated to $\tfrac{3}{5}E_F$, where E_F is the Fermi energy (section 5.3). For copper, E_F is 1.13×10^{-18} J atom^{-1}; hence, $\overline{v_d}$ is $\sim 1.2 \times 10^6$ ms^{-1}. Measurement of the electrical resistivity of copper films has shown that l is ~ 40 nm. Hence, $\tau = 3.3 \times 10^{-14}$ s and, from (5.10), $\sigma = 7.9 \times 10^7$ ohm^{-1} m^{-1}; the experimental value is 6.00×10^7 ohm^{-1} m^{-1}. Furthermore, from (5.10)–(5.12), we see that σ decreases (ρ increases) with increasing temperature, as found by experiment, so that the Drude–Lorentz theory gives a satisfactory account of electrical conductivity.

By analogy with the discussion on gas viscosity (section 3.1.6), it is possible to derive a simple expression for the coefficient of thermal conductivity κ of a gas,[1] in the form:

$$\kappa = \tfrac{1}{3}\mathcal{N} c_V \overline{v} l \tag{5.13}$$

where c_V is the constant volume heat capacity per atom. If we apply this equation to the electron gas in a metal then, using (5.10) and (5.11),

$$\sigma = \mathcal{N} e^2 l / m_e \overline{v_d} \tag{5.14}$$

so that, by identifying $\overline{v}$ with $\overline{v_d}$, we obtain

$$\kappa/\sigma = \tfrac{1}{3} c_V m_e \overline{v^2}/e^2. \tag{5.15}$$

Writing $m_e \overline{v^2} = 3k_B T$, by analogy with (5.12), we have

$$\kappa/\sigma T = \mathcal{K} \tag{5.16}$$

where $\mathcal{K}$ is a constant $(c_V k_B/e^2)$ at a given temperature. This equation expresses the Wiedemann–Franz law, although a more precise analysis is needed to obtain the correct Wiedemann–Franz constant $(\pi k_B/e)^2/3$, or 2.45×10^{-8} W ohm deg^{-2}.

5.2.2 Heat capacity

From Appendix 16, the average classical kinetic energy is given generally by[2]

$$\bar{E}_K = \tfrac{3}{2}k_B T \tag{5.17}$$

and the corresponding constant volume heat capacity is

$$c_V = \left(\frac{\partial E_K}{\partial T}\right)_V = \tfrac{3}{2}k_B \tag{5.18}$$

multiplication by L gives a formulation in molar terms:

$$C_V = \tfrac{3}{2}\mathcal{R}. \tag{5.19}$$

---

[1]See, for example, C. Kittel (1976) *Solid State Physics*, 5th edition, Wiley.
[2]In the context of metals, energy is usually symbolized by E.

5.2.2.1   *Classical particles*

In classical theory, the energies of the electrons in an electron gas are assumed to follow a Maxwell–Boltzmann distribution. Taking the electron energy E_K as $\frac{1}{2}m_e v^2$, it is straightforward to show from (3.4) that the probability of an electron having an energy between E_K and $E_K + dE_K$ is given by

$$\Phi(E_K)\, dE_K = 2\pi/(\pi k_B T)^{3/2}\sqrt{E}\,\exp(-E_K/k_B T)\, dE_K. \tag{5.20}$$

The plot of this equation is similar to that for $\Phi(v)$ in Fig. 3.2: as the temperature increases, the distribution of energies broadens, and the maximum energy moves to higher values of E_K; all values of the energy are possible under this distribution.

The energy of the electron in the gas is related to momentum by

$$E_K = p^2/2m_e. \tag{5.21}$$

Thus, the variation of p with E_K is parabolic (Fig. 5.1). As E_K moves up the parabola, the final energy position is determined by a balance between the thermal energy supplied and that lost in collision with the positive ions.

5.2.2.2   *Classical solids*

The constant volume molar heat capacity of a monatomic solid is $\sim 25\,\text{J mol}^{-1}\,\text{K}^{-1}$, and the form of its variation with temperature has been shown (for sodiim chloride) in Fig. 4.22, where the value of any ordinate must be halved for a monatomic solid. The atoms in a metal crystal may be considered to be oscillating harmonically, each about its mean position, with a common frequency v. Each atom may then be treated as a simple harmonic oscillator, and its total energy is the sum of kinetic and potential energy components, from (2.1) and (1.3). Thus, we can write

$$E = \tfrac{1}{2}mv^2 + 2\pi^2 v^2 mx^2 \tag{5.22}$$

where m is the mass of the vibrating atom, and v and x are, respectively, its speed and linear displacement from the equilibrium position, at any instant. The average thermal energy for a classical oscillator is (Appendix 16)

$$\bar{E} = k_B T. \tag{5.23}$$

Generalizing to N oscillators in three dimensions (there are three mutually perpendicular directions of vibration), we have for the mean total energy

$$\bar{E} = 3Nk_B T. \tag{5.24}$$

If N is set equal to L, the average molar energy becomes

$$\bar{E}_m = 3\mathscr{R}T \tag{5.25}$$

and from (5.18), the constant volume molar heat capacity is

$$C_V = 3\mathscr{R} \tag{5.26}$$

which is $\sim 25\,\text{J mol}^{-1}\,\text{K}^{-1}$, the result given by the law of Dulong and Petit (1819): it is obeyed well at high temperatures but, as Fig. 4.22 shows, it fails dramatically at low temperatures.

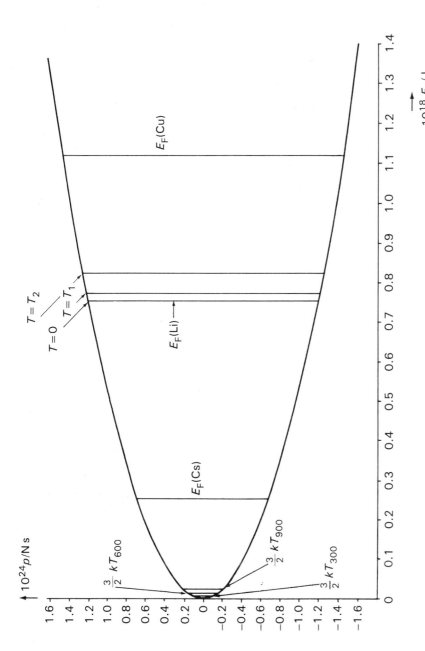

Fig. 5.1. Free-electron theory: parabolic form of the dependence of the electron kinetic energy E_K on its momentum p; some values of the classical average energy $\frac{3}{2} k_B T$ and of the Fermi energy E_F are indicated.

5.2.2.3   *Einstein and Debye solids*

Einstein (1906) treated the vibrations of N atoms as a lattice array of $3N$ independent oscillators, each of frequency v, in one dimension x, but with the vibrational energy quantized according to Planck's equation (2.12). Thermal vibrations are thermally excited *phonons*, and the phonon is the quantized unit of lattice vibration, with energy hv (or $h\omega/2\pi$) and momentum $kh/2\pi$, where k is given by (5.41). The distribution of phonon energies is no longer continuous, and the average energy is, following (3.35) but replacing the integrals with summations,

$$\bar{E} = \frac{3N \sum_{n=0}^{\infty} nhv \exp(-nhv/k_B T)}{\sum_{n=0}^{\infty} \exp(-nhv/k_B T)}. \tag{5.27}$$

Putting $-hv/k_B T = x$, we see that (5.27) can be recast in the form

$$\bar{E} = 3Nhv \frac{d}{dx} \{\ln(1 + e^x + e^{2x} + \cdots)\} = 3Nhv/\{e^{-x} - 1\} \tag{5.28}$$

or

$$\bar{E} = \frac{3Nhv}{\exp(hv/k_B T) - 1}. \tag{5.29}$$

At high temperatures we can neglect second- and higher-order terms of the exponential expansion, whereupon we obtain $\bar{E} = 3Nk_B T$, which is the classical result (5.24). At low temperatures $\exp(hv/k_B T) \gg 1$, whence

$$\bar{E} = 3Nhv \exp(-hv/k_B T) \tag{5.30}$$

and

$$c_V = 3Nk_B(hv/k_B T)^2 \exp(-hv/k_B T). \tag{5.31}$$

From (5.31), c_V tends exponentially to zero as T tends to zero, but experiment shows that $c_V \to 0$ as a T^3 variation.

A satisfactory correction was made by Debye (1912). He proposed that not all oscillators had the same frequency. At wavelengths that are long (low frequencies) relative to the interatomic spacings, large regions of crystal volume may be coupled in vibrational motion. At low temperatures some vibrations correspond to the condition $hv \ll k_B T$ and, for them, the long wavelength vibrations are very important. They make a classical contribution to the energy, thereby modifying the exponential dependence on temperature shown by (5.29); only for the low-temperature region is modification needed.

Debye's treatment is complex, and we state, without proof here, his formulation for the average energy and constant volume molar heat capacity:

$$\bar{E}_m = \frac{9\mathscr{R}T}{(\Theta/T)^3} \int_0^{x_m} \frac{x^3}{e^x - 1} dx$$

$$C_V = \frac{9\mathscr{R}}{(\Theta/T)^3} \int_0^{x_m} \frac{e^x x^4}{(e^x - 1)^2} dx \tag{5.32}$$

where $x = hv/k_B T$ and $x_m = \Theta/T$. The Debye temperature Θ is equal to hv_{max}/k_B, and the higher the value of Θ the lower the heat capacity, at a given temperature. The

Table 5.1. Debye temperature Θ for some solids

Element	Θ/K
C (diamond)	2230
C (graphite)	420
Na	158
K	91
Al	428
Cu	343
Pb	105
NaCl	321
KCl	235

wavelength λ_{min} corresponding to v_{max} is given by $\lambda_{min} = h v_s/k_B\Theta$, where v_s is the velocity of sound in the material. For copper $v_s = 4000 \text{ m s}^{-1}$, and taking $\Theta(\text{Cu}) = 343 \text{ K}$, $\lambda_{min} = 0.56 \text{ nm}$. Values for λ_{min} may be approximated by twice the metallic diameter, which for copper is 0.52 nm. The currently accepted values of Θ for a selection of solids are listed in Table 5.1.

At low temperatures the Θ/T integration limit in (5.32) can, for convenience and without sensible error, be set at infinity. Then, by integration, we have

$$\bar{E}_m = 3\pi^4\mathscr{R}T^4/5\Theta^3 \qquad (5.33)$$

and

$$C_V = 1944(T/\Theta)^3 \text{ J K}^{-1} \text{ mol}^{-1}. \qquad (5.34)$$

Equation (5.34) establishes the T^3 law that we used in problem 4.19; if C_V follows a T^3 law, then C_p does also, from (3.85).

5.2.2.4   *Heat capacity paradox*

If a mole of conduction electrons in a metal behaved as free particles, its contribution to C_V would be, from (5.19), $1.5\mathscr{R}$; the total molar heat capacity of the metal would then become with (5.26), $4.5\mathscr{R}$. Now, the molar heat capacities of monatomic solids do not deviate by more than about 10% from the Dulong and Petit limit of $3\mathscr{R}$, for Θ/T greater than about 0.7. Since there can be no doubt about the nature of atomic vibrations and *their* contribution to the heat capacity, it follows that the contribution from the electrons is negligibly small, a result that is clearly at variance with the classical theory.

5.3   WAVE-MECHANICAL FREE-ELECTRON THEORY

In the Drude–Lorentz theory, the free electrons interacted with the lattice of positive ions only insofar as to use them to arrest the electron motion. In the wave-mechanical free-electron theory, electrons are still treated as being free, but they are bound collectively to the positive ions and interact with them uniformly over the whole crystal.

According to the Pauli exclusion principle (section 2.6.4), the energy states of conduction electrons must, like all other electrons, be specified by sets of quantum numbers. We shall consider first that a conduction electron in a metal behaves like a particle in a three-dimensional box (section 2.4.2). There, we assumed that the potential was zero inside the box and infinite everywhere outside it. In the present application, the potential is required to be periodic, so as to conform with lattice geometry, and the box will be assumed to be cubic, with side a. These two situations may be summarized as follows:

$$\text{Electron-in-a-box} \qquad \psi(x, y, z) = 0 \quad \text{for} \begin{cases} x = 0, a \\ y = 0, a \\ z = 0, a \end{cases} \tag{5.35}$$

$$\text{Electron-in-a-metal} \qquad \psi(x, y, z) = \begin{cases} \psi(x + a, y, z) \\ \psi(x, y + a, z) \\ \psi(x, y, z + a). \end{cases} \tag{5.36}$$

While the electron was confined to the cubic box of zero potential, the wave function was given by the standing wave

$$\psi(x, y, z) = (2/a)^{3/2} \sin(\pi n_x x/a)\sin(\pi n_y y/a)\sin(\pi n_z z/a) \tag{5.37}$$

where n_x, n_y and n_z are positive integers, as before. We need now a wave function that will satisfy the periodic boundary conditions implied by (5.36), that is,

$$\psi(x + a, y, z) = \psi(x, y, z) \tag{5.38}$$

and similarly with the y- and z-directions. A suitable solution of the wave equation is the plane travelling wave

$$\psi_{\mathbf{k}}(\mathbf{r}) = \exp(i\,\mathbf{k}.\mathbf{r}) \tag{5.39}$$

where

$$\mathbf{r} = \mathbf{i}.(x + y + z) \tag{5.40}$$

and the *wave vector* $\mathbf{k}$ has components

$$\mathbf{k} = \mathbf{i}.(k_x + k_y + k_z) \tag{5.41}$$

such that

$$k_x\,(k_y,\,k_z) = 0,\; \pm 2n_x\,(n_y,\,n_z)\,\pi/a \tag{5.42}$$

The components of $\mathbf{k}$, given by (5.42), satisfy (5.36); for example,

$$\exp\{ik_x(x + a)\} = \exp(ik_x x)\exp(ik_x a) = \exp(ik_x x)\exp(i2n_x \pi)$$

$$= \exp(ik_x x) \tag{5.43}$$

and similarly for the y- and z-directions.

The components of $\mathbf{k}$ are the quantum numbers for the electron under the periodic potential and, together with the spin quantum number m_s, make up the four quantum numbers required to specify an electron (analogous to n, l, m_l and m_s for a *bound* electron in an atom). The magnitude of the wave vector is related to wavelength by

$$k = 2\pi/\lambda. \tag{5.44}$$

By the usual manipulation (see section 2.4),

$$E_k = \frac{h^2 k^2}{8\pi^2 m_e} = \frac{h^2}{8\pi^2 m_e}(k_x^2 + k_y^2 + k_z^2) \qquad (5.45)$$

The electrons are still free, and so the curve of E against p remains parabolic. Indeed, by introducing the de Broglie equation (2.28), we find

$$E_k = p^2/2m_e. \qquad (5.46)$$

Since E_k is kinetic, we have

$$\tfrac{1}{2}mv^2 = \frac{h^2 k^2}{8\pi^2 m_e} \qquad (5.47)$$

and

$$mv = kh/2\pi \qquad (5.48)$$

hence, k is a measure of quasi-momentum for a free electron.

The energy levels given by (5.45) are very closely spaced in a solid, and may be considered to approximate to a *band*. At 0 K, the electrons occupy energy levels up to a finite value of energy, the Fermi energy E_F. If we take three-fifths of the Fermi energy as the average kinetic energy, as in section 5.2.1, then the equivalent temperature, the Fermi temperature T_F is in the region of 50 000 K! This temperature is not that of the electron gas. Its interpretation is that, in practice, the heat supplied to a crystal will lead to a temperature so very much less than T_F that it will have a negligible effect on the energy distribution of the electrons. Consequently, the contribution of the electrons to the heat capacity of the solid is vanishingly small, as we predicted in the previous discussion.

The wave vector $\mathbf{k}$ exists in $\mathbf{k}$-space, which is similar to reciprocal space in crystallography; k has the dimensions of reciprocal length, by (5.44). From (5.42) and (5.45), we can write for a cubic crystal

$$\mathbf{k} = -\mathbf{i}(2\pi/a)(n_x + n_y + n_z). \qquad (5.49)$$

Let us now introduce N electrons into our metal crystal, in accordance with the Pauli exclusion principle. For each energy state dictated by the combination of n_x, n_y and n_z, one electron with a spin of $\pm\tfrac{1}{2}$ can be accommodated. As k increases, so do the energy and momentum, represented by the lattice point n_x, n_y, n_z. We may note that although quantum numbers are positive integers, the quadratic form for E in (5.45) can be satisfied by $n_x, n_y, n_z = 0, \pm n$, as implied by (5.42).

In the ground state of a system of N free electrons, the occupied orbitals may be represented by the points of a lattice in $\mathbf{k}$-space lying within a sphere of radius k_F (Fig. 5.2). The energy at the (Fermi) surface of this sphere is the Fermi energy, given by

$$E_F = h^2 k_F^2/8\pi^2 m_e \qquad (5.50)$$

and values of E_F for some metals have been indicated on Fig. 5.1.

The quantity $(2\pi/a)^3$, where a is the side of the cubic unit cell in real space, defines a primitive (reciprocal) unit cell volume in $\mathbf{k}$-space that is identified with a single energy state. One lattice point is associated with the volume of any primitive unit cell; hence,

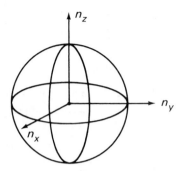

Fig. 5.2. Sphere of radius k_F in **k**-space, enclosing a portion of the (reciprocal) lattice on n_x, n_y and n_z directions as mutually perpendicular axes; the triplet n_x, n_y, n_z may be used to designate a lattice point in **k**-space.

in a sphere of radius k_F, the total number N of electron states is given by

$$N = \frac{2(\frac{4}{3}\pi k_F^3)}{(2\pi/a)^3}. \tag{5.51}$$

The factor 2 arises in (5.51) because each lattice point can have associated with it two electrons, corresponding to spins of $\pm\frac{1}{2}$. The electron concentration N/V in real space, where $V = a^3$, is then

$$N/V = k_F^3/3\pi^2. \tag{5.52}$$

Alternatively, we can write

$$N/V = Zn_e/a^3 \tag{5.53}$$

where Z is the number of atoms per cubic unit cell and n_e is the number of valence electrons per atom. For copper, $a = 0.361$ nm, $Z = 4$ (face-centred unit cell) and $n_e = 1$, whence $N/V = 8.50 \times 10^{28}$ m^{-3}. From (5.45) and (5.53), we obtain

$$E_F = \frac{h^2}{8\pi^2 m_e}(3\pi^2 N/V)^{2/3} \tag{5.54}$$

and the value of E_F for copper is 1.13×10^{-18} J atom^{-1}, or 7.06 eV.

We mention in passing that a *primitive cubic* unit cell for a metal is energetically unfavourable; a face-centred or a body-centred cubic unit cell is obtained in practice. Nevertheless, it is always possible to define a primitive unit cell in any lattice, although its volume would not be simply the cube of the cell side, and our discussion in terms of primitive unit cells has complete reality.

5.3.1   Density of states

The density of states function $g(E)$ is defined as the rate of change with energy of the number of energy states of an atom; thus, $g(E)$ represents the number of states with energy in the range E to $E + dE$. Rewriting (5.54) for N as a function of E and differentiating with respect to E, we find

$$g(E) = \frac{dN}{dE} = 8\sqrt{2}\pi V(m^*/h^2)^{3/2}\sqrt{E}. \tag{5.55}$$

In equation (5.55), m^* is not equal to the rest mass m_e of the electron for all values of the energy E. An electron in a periodic potential field is accelerated as though its mass were effectively

$$m^* = \frac{h^2/4\pi^2}{d^2E/dk^2} \tag{5.56}$$

and band theory (section 5.4) shows that m^* can vary between $\pm\infty$.

Fig. 5.3 is a plot of (5.55); it is parabolic, like Fig. 5.2. At 0 K the energy levels are filled to the sharp cut-off at the Fermi energy $E_F(0)$, and higher-energy states are unoccupied by electrons. As the temperature is increased, thermal agitation moves electrons from some states below $E_F(0)$ to states above it, as shown by the rounded curve for $T > 0$ K, but for which $k_B T \ll E_F$. Only those electrons with an energy range of $\sim k_B T$ below the Fermi energy are excited in this way; more strongly held electrons have insufficient thermal energy to make the transition. The operation of the Pauli exclusion principle ensures that nearly all electrons are unavailable for thermal excitation at low temperatures, because of the stability of paired electrons in filled orbitals.

The constant volume molar heat capacity of a metal at very low temperatures, less than ~ 15 K, may be written as the sum of lattice (α) and electronic (γ) contributions:

$$C_V = \alpha T^3 + \gamma T. \tag{5.57}$$

In practice, it has been shown that C_V/T varies linearly with T^2 at low temperatures, in accord with (5.34), and the intercept at $T^2 = 0$ is γ. Experiments with copper gave a value for γ of $\sim 7 \times 10^{-4}$ J mol^{-1} K^{-2}. Fig. 5.4 is a plot of the lattice and electronic contributions to the heat capacity between 0 K and 14 K. It is clear that the electron

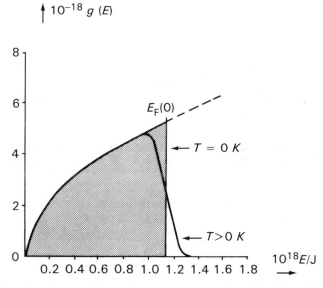

Fig. 5.3. Density of states function $g(E)$ for copper: the sharp cut-off occurs at $T = 0$ K; at higher values of T the curve becomes rounded as some electrons are promoted above the value of the Fermi energy E_F.

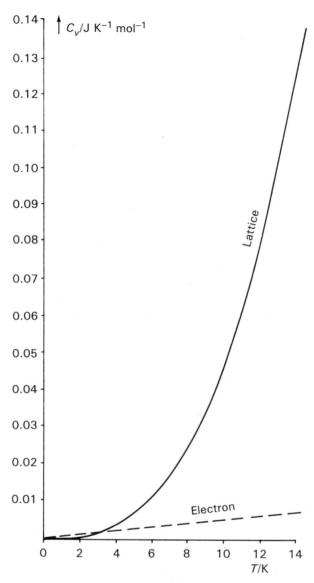

Fig. 5.4. Contributions to the constant volume molar heat capacity of copper from lattice vibrations and electrons, between 0 K and 14 K; the electron contribution becomes significant only in the very low temperature region.

contribution to the heat capacity is less than 10% of the total at temperatures greater than about 9 K. At 298 K, its contribution is about 0.06% of the Dulong and Petit limiting value. We can be satisfied that the apparent heat capacity paradox is explained by the theory developed so far.

5.3.2   Fermi–Dirac distribution
The concentration of valence electrons in a metal is about 10^4 times greater than that of molecules in a unit volume of a gas at stp, and the average spacing between these

electrons is commensurate with their de Broglie wavelength. In these circumstances, the Maxwell–Boltzmann distribution equation is inadequate: any number of electrons can have identical energy and momentum; the high degeneracy is inherent in equation (5.45).

Fermi–Dirac statistics, which applies to particles having half-integral spin,[1] treats all electrons as indistinguishable, and requires that each energy state is either empty or fully occupied by a single electron which is specified by the quantum numbers n_x, n_y, n_z and m_s. A derivation of the Fermi–Dirac distribution is given in Appendix 17. There, we obtain the distribution equation

$$f(E) = \exp\{(E - E_F)/k_B T + 1\}^{-1}. \tag{5.58}$$

The number of states of energies lying between E and $E + dE$ that are occupied by electrons is given by

$$N(E)\, dE = g(E) f(E)\, dE \tag{5.59}$$

where $g(E)$ follows (5.55). From (5.58), $f(E)$ is unity at $T = 0$ up to $E = E_F$ but zero for $E > 0$; hence, $N(E) = 0$ for $E > E_F$ at 0 K. Since $g(E)$ is a continuous function in E, the highest occupied state has the energy E_F. Thus, at 0 K, E_F is also E_{max}.

The form of $f(E)$ is illustrated in Fig. 5.5. At 0 K it is unity up to $E = E_F$, where it falls discontinuously to zero. For $T > 0$, the curve near E_F is smoothed off. It follows from (5.59) that $N(E)$ at 0 K is the function $g(E)$ multiplied by a function that is unity up to $E = E_F$; hence, $N(E)$ has the form of the shaded portion of Fig. 5.3. For values of T greater than zero, $f(E)$ falls below unity for $E < E_F$, and is greater than zero for $E > E_F$. The high-energy 'Maxwellian tail' to the $f(E)$ distribution for $T > 0$

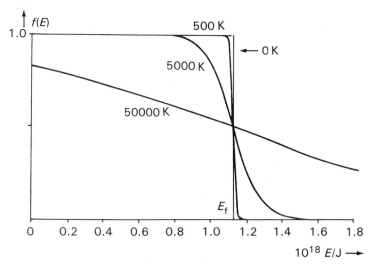

Fig. 5.5. Fermi–Dirac distribution function $f(E)$ of energy states, showing E_F for copper; the curves for different temperatures greater than zero intersect at $f(E) = \frac{1}{2}$, where $E = E_F$.

[1]Particles with integral spin, including zero, such as photons or phonons, are governed by the Bose–Einstein distribution:

$$f(E) = \{\exp(E - \alpha)/k_B T - 1\}^{-1}$$

where α is a constant determined by the condition $\int_E f(E)\, dE = 1$.

corresponds to $(E - E_F) \gg k_B T$, or $\exp(E - E_F)/k_B T \gg 1$. Then

$$f(E) \approx \exp(E_F - E)/k_B T \propto \exp(-E/k_B T) \qquad (5.60)$$

which identifies with the classical distribution.

5.3.2.1 Mean energy of electrons
The mean energy $\bar{E}_{el}$ of the free electrons is given, following (3.35), by

$$\bar{E}_{el} = \frac{\int_0^\infty E \, g(E) f(E) \, \mathrm{d}E}{\int_0^\infty g(E) f(E) \, \mathrm{d}E} \qquad (5.61)$$

and we shall confine our calculation to $T = 0\,\mathrm{K}$, for which $f(E)$ is unity up to E_F and zero beyond it. From (5.55), we can write $g(E) = \alpha \sqrt{E}$, where α is a constant. Then, (5.61) becomes

$$\bar{E}_{el} = \frac{\alpha \int_0^{E_F} E^{3/2} \, \mathrm{d}E}{\alpha \int_0^{E_F} E^{1/2} \, \mathrm{d}E} \qquad (5.62)$$

which is readily shown to be equal to $\tfrac{3}{5} E_F$, the value that we used in section 5.2.1. More detailed calculations show that $\bar{E}_{el}$ is not very sensitive to temperature (cf. Fig. 5.4), and is always close to $\tfrac{3}{5} E_F$.

5.3.2.2 Electronic heat capacity
We have discussed some experimental evidence for the very small electronic contribution to the heat capacity relative to the lattice contribution, except at very low temperatures. We can now look at that problem in terms of the theory that we have just developed.

Only those electrons of energy E_K within $\sim k_B T$ of the Fermi energy can change states as T is raised above $0\,\mathrm{K}$. The number of such electrons is the density of states per unit energy multiplied by the energy of the electrons involved, that is, $g(E_K)k_B T$. These electrons interact with lattice phonons and increase their energy by $\sim k_B T$. Hence, the total increase in energy is the number of electrons multiplied by the energy gain, that is, $g(E_K)k_B^2 T^2$. The electronic contribution to the heat capacity is then given by

$$c_{V,el} = \frac{\mathrm{d}}{\mathrm{d}T} \{g(E_K)k_B^2 T^2\} = 2g(E_K)k_B^2 T. \qquad (5.63)$$

Thus, γ in (5.57) can be identified with $2g(E_K)k_B^2$. From (5.55), the total number N of free electrons is given by

$$N = \int_0^{E_F} \alpha \sqrt{E} \, \mathrm{d}E \qquad (5.64)$$

where $\alpha = 8\sqrt{2\pi} V(m^*/h^2)^{3/2}$. Hence, $N = \tfrac{2}{3}\alpha E_F^{3/2}$, and

$$g(E_F) = \tfrac{3}{2} N/E_F \qquad (5.65)$$

so that γ becomes

$$\gamma = 3N k_B^2/E_F. \qquad (5.66)$$

For one mole of metallic copper, $N = L$; hence, $\gamma = 3 \times 10^{-4} \, \mathrm{J \, mol^{-1} \, K^{-2}}$. A more exact calculation replaces the constant 3 in (5.66) by $\pi^2/2$, whence γ (Cu) becomes $5 \times 10^{-4} \, \mathrm{J \, mol^{-1} \, K^{-2}}$, which is in good agreement with the value of $7 \times 10^{-4} \, \mathrm{J \, mol^{-1} \, K^{-2}}$ obtained from experiment. It follows that the value of $C_{V,\mathrm{el}}$ at 298 K is $\sim 0.2 \, \mathrm{J \, mol^{-1} \, K^{-1}}$, which is approximately 0.8% of the molar heat capacity at that temperature (25 $\mathrm{J \, mol^{-1} \, K^{-1}}$). At 9 K, $C_{V,\mathrm{el}}$ is $\sim 5 \times 10^{-3} \, \mathrm{J \, mol^{-1} \, K^{-1}}$, whereas the lattice contribution is, from (5.34), $3.5 \times 10^{-2} \, \mathrm{J \, mol^{-1} \, K^{-1}}$, so that again we see that the electronic contribution is very significant at low temperatures.

5.4   BAND THEORY

The theory in the preceding sections has been shown to be capable of explaining many properties of metals. However, we cannot yet see how the extreme variations in electrical resistivity among the elements arise. We need to determine the circumstances in which a certain fraction of electrons behave as though they were free. We shall find that the electron energy levels in solids are arranged in bands, which are really very closely spaced energy states approximating to continua within the bands themselves. Between the bands there are forbidden regions, or gaps, within which no energy states are allowed. The occupancy of the bands, together with the width of the gaps between them, serves to determine more fully the electrical properties of differing solids.

Fig. 5.6 shows a schematic arrangement of energy bands for the three main classes of electrical conductors, metals (conductors), semiconductors and insulators. If a band is partially filled the solid behaves as a metal. If all bands are filled except for one or two bands which are either nearly filled or nearly empty, the solid is a semiconductor; and if all occupied bands are filled the solid is an insulator. If all occupied bands are filled but the energy gap between the uppermost filled band (*valence band*) and the band of next highest energy (*conduction band*) is small, then the solid can again show semiconduction. Clearly, there will be gradations throughout these classes, as

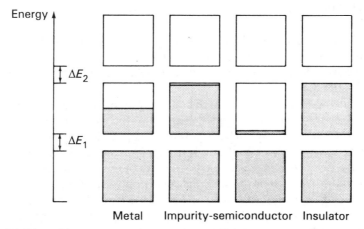

Fig. 5.6. Schematic arrangement of energy bands in solids: the shading indicates electron occupancy; ΔE_1 and ΔE_2 are forbidden energy ranges, or gaps.

measurements of electrical resistivity show. We shall consider next how the gaps
between energy bands arise.

5.4.1 Energy bands and Brillouin zones

On the free-electron model of a metal, the permitted energy states are distributed
quasi-continuously according to (5.45). The free-electron wave functions are, in one
dimension x, of the form $\exp(ikx)$, and represent travelling waves of momentum $kh/2\pi$.
The band theory of solids begins with the Schrödinger equation, and incorporates the
periodic potential field that arises from a lattice of atoms. In this form, it is often called
the Bloch equation:

$$-\frac{h^2}{8\pi^2 m_e}\frac{d^2\psi}{d^2x} + V(x) = E\psi. \tag{5.67}$$

The potential function is periodic, so that $V(x) = V(x + na)$ where a is the repeat
distance along the x-direction in the lattice and n is an integer. The solution of (5.67)
has been given by Bloch as

$$\psi_k = E_k(x)\exp(ikx) \tag{5.68}$$

where $E_k(x)$ is an energy function with the periodicity a of the lattice. The energy
dependence on k is still quadratic, as in (5.45), but for $k = \pm n\pi/a$ discontinuities
appear in the energy, thus giving rise to a band structure (Fig. 5.7).

For most values of k and, hence, the electron wavelength, the electrons behave very
much like free electrons. At values of k equal to $\pm n\pi/a$, the condition for Bragg
reflexion of electron waves in one dimension is realized: $k = \pm n\pi/a$ is equivalent to
the Bragg equation $2a \sin\theta = n\lambda$, where $k = 2\pi/\lambda$ and $\sin\theta = 1$. The first-order
reflexion at $k = \pm\pi/a$ arises because waves reflected from adjacent atoms interfere
constructively, the phase difference being just 2π. The region in **k**-space lying between
$\pm\pi/a$ is called the first *Brillouin zone* (Fig. 5.8). The energy is quasi-continuous within
a zone, according to (5.45), but discontinuous at the zone boundaries. As k increases
towards $n\pi/a$, the eigenfunctions contain increasing amounts of Bragg-reflected wave.
At $k = \pi/a$, for example, the wave $\exp(i\pi x/a)$ reflects as $\exp(-i\pi x/a)$, and the resulting
combinations are standing waves ψ_1 and ψ_2 of the forms $\cos(\pi x/a)$ and $\sin(\pi x/a)$,
respectively.

The probability densities of the two standing waves are $|\psi_1|^2$ and $|\psi_2|^2$, whereas
that for the travelling wave is $(e^{ikx} e^{-ikx})$, which has a constant value. Fig. 5.9
illustrates a one-dimensional periodic potential field and the wave probability functions just described.

The travelling wave distributes charge uniformly along the x-axis; ψ_1 has its peaks
at na and ψ_2 at $(n + \frac{1}{2})a$. The potential energies of the two distributions follow the
order $|\psi_1|^2 < (e^{ikx} e^{-ikx}) < |\psi_2|^2$. Hence, an energy gap ΔE arises, and waves ψ_1 and ψ_2
correspond to the points such as A and B in Fig. 5.7. The combination of the waves
$\exp(\pm ikx)$ at $k = \pm n\pi/a$, that is, at the boundaries of the Brillouin zones, leads to an
energy gap of $2V_k$, where V_k is the potential energy function at the position in **k**-space
corresponding to k. This result may be compared with the bonding/antibonding
situation in MO theory, which also makes use of a core potential energy. Band theory
modifies the density of states function $g(E)$ in Fig, 5,3, and a more correct plot is
illustrated by Fig. 5.10.

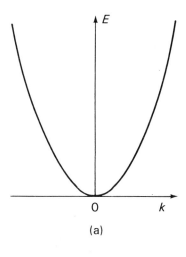

(a)

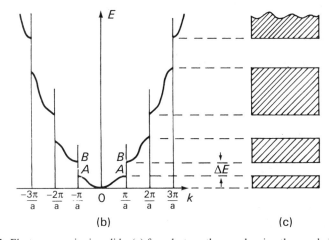

(b)   (c)

Fig. 5.7. Electron energies in solids: (a) free-electron theory, showing the parabolic dependence of E on k ($kh/2\pi$ is a quasi-momentum – cp Fig. 5.1); (b) energy bands, showing gaps, such as ΔE. The positions A and B correspond to waves ψ_1 (cos $\pi x/a$) and ψ_2 (sin $\pi x/a$), respectively; the values of k ($\pm n\pi/a$) delineate the boundaries of the one-dimensional Brillouin zones.

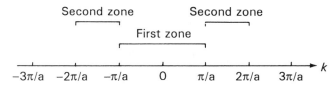

Fig. 5.8. The first two Brillouin zones for a one-dimensional lattice of periodicity a.

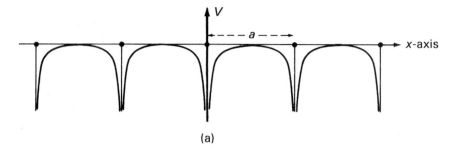

(a)

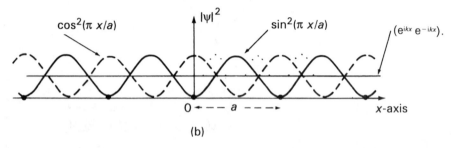

(b)

Fig. 5.9. One-dimensional lattice of periodicity a: (a) periodic potential field $V(x)$; (b) probability densities $\cos^2(\pi x/a)$, $\sin^2(\pi x/a)$ and $(e^{ikx} e^{-ikx})$.

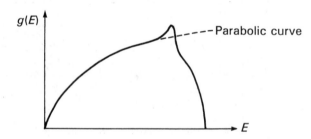

Fig. 5.10. Density of states function $g(E)$, modified in the light of band theory.

Brillouin zones can be extended to two and three dimensions. The zone boundaries are determined by the regions in **k**-space where the Bragg equation is satisfied. Thus, Brillouin zones are governed by crystal structure rather than by chemical composition (see also section 5.8.2.1).

The Bragg equation can be written in the form

$$2(2\pi/\lambda) \cos \phi = d^* \qquad (5.69)$$

where $\phi = 90 - \theta$ and d^* is the reciprocal of an interplanar spacing, d. Multiplying both sides of (5.69) by d^*, we obtain

$$2(2\pi/\lambda)d^* \cos \phi = d^{*2} \qquad (5.70)$$

which may be rewritten as

$$\mathbf{k} \cdot \mathbf{d} = d^{*2}/2. \qquad (5.71)$$

Thus, the Bragg equation is satisfied if **k** terminates on the plane (line in two dimensions) normal to **d*** at the midpoint of d*, and these terminations determine the boundaries of the Brillouin zones.

5.4.2   Energy bands and molecular orbital theory

The Bloch theory is essentially a molecular orbital model of the metallic bond. The important feature of the MO theory as applied to a chemical molecule is that each electron moves in the potential field created by all other atoms in the molecule, the core field. It uses one-electron Hamiltonians, and the solutions of the corresponding wave functions are obtained from an LCAO approximation. We saw that in a compound such as benzene, strong overlap of the π orbitals of the atoms leads to electron delocalization over the whole molecule. In the case of a metal, the number of atoms is infinite or, at least, very large, and we can envisage an extreme case of electron delocalization now over the whole solid, often called the 'tight binding' approximation, because the outer electrons are assumed initially to be associated with the atoms. The molecular orbitals are now the conduction orbitals, and metallic properties depend upon their degree of overlap. In lithium, for example, the overlap integral $\int \psi_1(2s) \int \psi_2(2s) \, d\tau$ for adjacent atoms is approximately 0.5. An extension of this overlap to encompass all atoms in a metal crystal leads to the complete delocalization that establishes metallic character.

In order to highlight the important difference between the MOs in a metal and those that we discussed in Chapter 2, we will consider building up a crystal of lithium; this element has the ground state electronic configuration $(1s)^2(2s)^1$. When we discussed the one-electron species H_2^+ (section 2.8.1), we saw that when the two hydrogen nuclei were brought together, two diatomic molecular orbitals $\Psi_\pm$ were obtained: from the *single* energy level of each atom, *two* energy levels arose in the molecule. If we add a third atom, it overlaps its nearest neighbour strongly (and only slightly its next nearest), and three MOs are formed. The addition of a fourth atom leads to four MOs, and so on. The general effect of adding more atoms is to spread out the range of energies spanned by the MOs while filling in the range of energies with more and more MOs. When a large number N of atoms have been added there are N MOs within a band of finite width. The Hückel determinant for this configuration would then be of the form

$$\begin{vmatrix} \alpha - E & \beta & 0 & 0 & 0 & \cdots & 0 \\ \beta & \alpha - E & \beta & 0 & 0 & \cdots & 0 \\ 0 & \beta & \alpha - E & \beta & 0 & \cdots & 0 \\ 0 & 0 & \beta & \alpha - E & \beta & \cdots & 0 \\ \vdots & \vdots & \vdots & \vdots & \vdots & & \\ 0 & 0 & 0 & 0 & 0 & \cdots & \alpha - E \end{vmatrix} = 0 \qquad (5.72)$$

and the solution of this determinant may be given as

$$E_n = \alpha + 2\beta \cos[n\pi/(N + 1)] \qquad (5.73)$$

where n is an integer ranging from 1 to N, and α and β have meanings as before. As N tends to infinity, the difference between adjacent energy levels becomes vanishingly

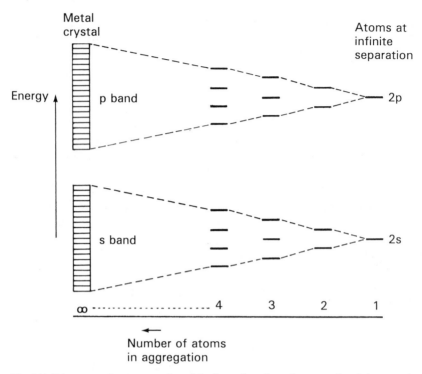

Fig. 5.11. Diagrammatic representation of the formation of s and p energy bands in a metal. The overlap of s AOs gives rise to the s band, and the overlap of p AOs gives the p band; the distance between the top of the s band and the bottom of the p band is the s–p energy band gap. The lowest and highest level of each band correspond to wholly bonding and wholly antibonding situations, respectively.

small, so that an effective band is obtained. The solutions of (5.73) for $n = 1$ and $n = N$, as $N \to \infty$, are $\alpha \pm 2\beta$, so that the width of the band is 4β. Thus, the band consists of N molecular orbitals: if it is formed from s orbitals, it is referred to as an s band; p orbitals lead to a p band. Fig. 5.11 illustrates some aspects of this discussion on the MO model of a metal. In the case of lithium, a cubic crystal of side 1 mm contains approximately 5×10^{19} atoms, and there are $\sim 9.3 \times 10^{19}$ 2s energy states present in this crystal of lithium, forming a quasi-continuous series of energy levels; a continuum is the limiting situation for infinite degeneracy.

5.4.2.1   *Occupation of orbitals*

Consider a metal such as lithium, in which each atom contributes one valence electron. At $T = 0\,\mathrm{K}$, the lowest $N/2$ MOs are filled, and the Fermi level is the highest occupied molecular orbital (HOMO). The important difference from the molecules that we discussed in Chapter 2, even with delocalization as great as in benzene, is that in metals there are unoccupied orbitals very close to the Fermi level, so that very little energy is needed to excite electrons into higher states.

   At $T > 0$, the distinction between occupied and unoccupied MOs in a band is no longer clear-cut, and the thermal energy of the electrons near the Fermi level is sufficient to cause electron excitation. The energy states in a band occur in pairs (n_x,

n_y, n_z, $\pm m_s$), and could represent waves travelling in opposite directions ($\pm m_s$) in the crystal. In the absence of an external electric field, the net momentum of the electrons is zero. However, if a potential difference is applied to the metal, there will be a net resultant electron flow, and some electrons, those lying $\sim k_B T$ below the Fermi energy, may be raised from the half-filled band to an empty, higher-level band in the process. Thus, for a metal with one valence electron, the band theory and the free-electron theory give similar results.

If the metal has two valence electrons per atom, like beryllium or magnesium for example, then the $2N$ valence electrons would fill the N orbitals of the s band. For $T = 0$, the Fermi level now lies at the top of the band. Since there is normally a gap before the next highest band begins, these elements might be expected to be insulators. In order to explain the fact that they are good electrical conductors, it was postulated that the bands actually overlap for these metals. This apparent 'theory to fit the facts' has, in fact, been adequately justified by further studies.

The electrical conductivity of a metal decreases with increasing temperature, even though the increased thermal energy of the electrons causes more electron excitation. However, that same thermal energy brings about an increase in the lattice vibrations of the positive ions, so that the electron scattering, or current-limiting, process increases and the steady electron current decreases in magnitude. At $T = 0$, the lattice vibrations cease, so that the electrical resistivity ρ_1 (reciprocal of conductivity) should tail off to zero. However, the presence of minute amounts of impurity atoms or of point defects (section 4.10.3) causes some electron scattering, so that there remains a small residual resistivity ρ_0 even at absolute zero, as implied by (5.74), where ρ_t is the total electrical resistivity:

$$\rho_t = \rho_1 + \rho_0. \tag{5.74}$$

The purer the sample of the metal the smaller will be the value of the residual ρ_0; the contribution to ρ_0 from point defects is not affected by a change in temperature.

5.4.2.2   *Semiconductors and insulators*
At 0 K, semiconductors have a completely filled valence band separated by a small energy gap, ~ 1 eV or less, from an empty conduction band. There is no electrical conduction because the electrons are unable to cross the gap to the conduction band. At temperatures greater than zero, some electrons may be excited to the conduction band. If an electrical field is then applied, it can act upon the energy state in both bands, and a current will flow. The number of electrons that are promoted across the gap increases with increasing temperature. Hence, the electrical conductivity of a semiconductor increases with increasing temperature, in contradistinction to a metal.

If the gap is large, then only very few electrons may be promoted to the higher energy band. The effect again increases with increasing temperature, but it is so small that the solid is an insulator. Thus, the essential difference between a semiconductor and an insulator resides in the size of the energy gap between the valence and conduction bands, and does not have the absolute distinction possessed by a metal.

The most important semiconducting elements are silicon and germanium; the atoms of these elements have four outer electrons, $(1s)^2(2s)^2(2p)^2$. Where the energies of the AOs are close, the bands formed from them may merge into a sort of hybrid band, rather than remain discrete. This situation arises for silicon and germanium, and the s and p levels are hybridized to form a completely filled hybrid valence band. The valence band is separated from the conduction band by a gap of only ~ 1 eV or less for silicon and germanium, and such pure materials are termed *intrinsic* semiconductors.

Since the active electrons lie in the Maxwellian tail of the Fermi–Dirac distribution, the dependence on temperature of the electrical conductivity can be expected to follow a Boltzmann distribution, since the conductivity will depend on the population of charge carriers. If we let E_n be the energy of the uppermost level of the valence band, then we can write

$$E_F - E_n = \Delta E/2. \tag{5.75}$$

Then from (5.60),

$$f(E) \approx \exp(-\Delta E/2k_B T) \tag{5.76}$$

and the conductivity may then be expressed as

$$\sigma = A \exp(-\Delta E/2k_B T) \tag{5.77}$$

which is like an Arrhenius-type equation where the energy of activation for electrical conductivity is one-half the band gap. The following example illustrates the application of (5.77) to germanium.

The resistance R of a sample of pure germanium has been found to vary with temperature as follows:

T/K	300	350	400
R/ohm	20.4	2.67	0.581

From (5.77) and (5.2) we have $R = A' \exp(\Delta E/2k_B T)$, where A' is a constant. Hence, the plot of $\ln R$ against $1/T$ will be a straight line of slope $\Delta E/2k_B$. By least squares, the slope is 4271.3 K, so that the band gap $\Delta E = 5.90 \times 10^{-20}$ J, or 0.74 eV.

The number of charge carriers in silicon or germanium can be increased by 'doping', that is, by implanting impurity atoms in minute traces, ~ 1 in 10^9, into an otherwise highly pure material, thus forming an *extrinsic* semiconductor (see also section 4.10.3.3). If the dopant atoms have fewer valence electrons than do the atoms in the bulk of the solid, such as gallium in silicon, it can extract electrons from the conduction band leaving 'holes' that allow movements of other electrons. Such materials are termed p-type semiconductors, because the hole is positive with respect to electrons. Alternatively, a dopant may introduce more valence electrons than those of the parent material, such as phosphorus in germanium. The 'extra' electrons can occupy otherwise empty bands, and give rise to n-type (negative) semiconductors.

Fig. 5.12 shows the density of states function for a metal, a semiconductor and an insulator. Where the highest-energy occupied states lies near to the centre of a band, the energy distribution is very similar to that given by the free-electron model. If the energy states are filled up to or very close to a zone boundary, then free-electron theory is unsatisfactory and band theory must be used to account for the electrical properties of the solid.

$g(E)$

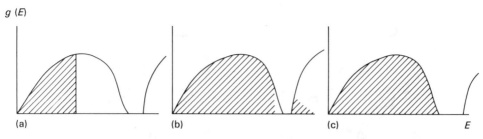

Fig. 5.12. Density of states function $g(E)$ as a function of energy E: (a) metal, (b) semiconductor, (c) insulator. The condition of filled or nearly filled energy states, (b) and (c), requires band theory in order to explain electrical conduction.

The range of electrical resistivity covered by metals, semiconductors and insulators is vast, as can be seen from the following few data on resistivity at 298 K:

	Cu	GaAs	GaSb	InSb	C (diamond)
ρ/ohm m	1.7×10^{-8}	0.40	1.00	1×10^{4}	5×10^{12}

Ultrahigh purity materials are prepared by a fractional crystallization process known as *zone refining*, and a modification of this technique allows controlled doping to be carried out. Many materials other than silicon and germanium form the basis of semiconductors, such as GaAs, GeIn, and InP, for example, where the second symbol refers to the dopant. Evidently, these materials are n-, p- and n-type, respectively.

Later theories of metals have refined and built on to those discussed here, and for such further theoretical studies the reader is referred to more specialized discussions.[1]

5.5   STRUCTURES OF METALS

The metallic bond does not possess directional character, and the structures adopted by metals are determined to a large extent by space-filling criteria. The metals of periodic groups 1, 2, 11, together with the transition-type metals, and certain others, form relatively simple structures: the closed-packed (face-centred) cubic (A1), the body-centred cubic (A2) and the close-packed hexagonal (A3) arrangements; they are illustrated by Figs 1.5, 5.13 (including A1 again, for convenience), and 5.14.

The close-packed types A1 and A3 represent the two modes of regular, closest packing of identical spheres (Fig. 5.13). In both A1 and A3, a first layer is obtained by placing spheres in contact such that their centres form the apices of equilateral triangles (Fig. 5.14a). A second similar layer is added such that the spheres of that layer rest in the depressions of the first layer (Fig. 5.14b). A third layer can then be added in one of two ways. If it is arranged such that the spheres in the third layer lie above voids in *both* the first and second layers, the close-packed cubic (CPC) structure

[1]See, for example, H. M. Rosenberg (1978) *The Solid State*, 2nd edition, Clarendon Press; A. H. Cottrell (1988) *Introduction to the Modern Theory of Metals*, Institute of Metals; J. M. Ziman (1964) *Principles of the Theory of Solids*, Cambridge University Press; P. A. Cox (1987) *The Electronic Structure and Chemistry of Solids*, Oxford University Press.

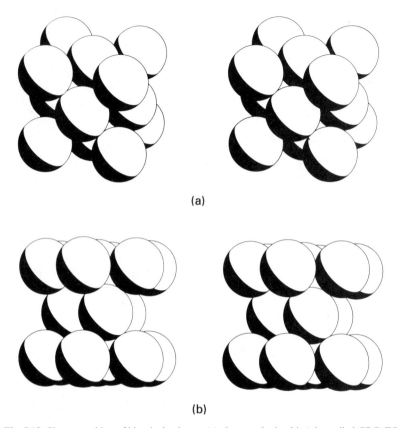

Fig. 5.13. Closest packing of identical spheres: (a) close-packed cubic (also called CPC, FCC or A1), (b) close-packed hexagonal (CPH or A3). The coordination number is 12 and the packing efficiency 0.74, for each structure type.

A1 is obtained (Fig. 5.14c). On the other hand, if the spheres of the third layer lie directly above those of the first layer, the close-packed hexagonal (CPH) structure A3 is obtained (Fig. 5.14d). The sequence of layers in the two arrangements is:

$$\text{A1:} \quad 1 \ 2 \ 3 \ 1 \ 2 \ 3 \ 1 \ 2 \ \dots$$

$$\text{A3:} \quad 1 \ 2 \ 1 \ 2 \ 1 \ 2 \ 1 \ 2 \ \dots$$

In these two structure types, each sphere is in contact with twelve other spheres, the maximum coordination for identical spheres in a regular packing mode in crystals. In the CPC structure, the volume of the unit cell is a^3, where a is the length of the side of the unit cell. The face diagonal is $a\sqrt{2}$ and we have

$$a\sqrt{2} = 4r \tag{5.78}$$

where r is the radius of the sphere, because the spheres are in contact along that diagonal (Fig. 5.13a). Hence the unit cell volume can be written as $(4r/\sqrt{2})^3$, and the volume *occupied* per sphere is $(4r/\sqrt{2})^3/4$, or $5.66r^3$. The actual volume of a sphere is $\frac{4}{3}\pi r^3$; hence, the fraction of space occupied, or the packing efficiency, is 0.74. The same value obtains for the CPH structure type.

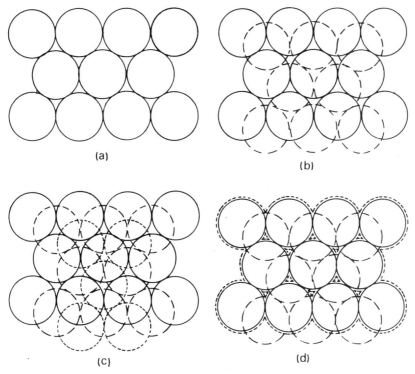

(a)

(b)

(c)

(d)

Fig. 5.14. Close packing of identical spheres: (a) A first layer A, full lines; the centres of each sphere form a succession of equilateral triangles. (b) A second layer B, long-dashed lines; there is only one way in which the second layer can be closed-packed on to the first. (c) CPC, with the layer sequence $ABCABCA\ldots$; the layer C is shown by short-dashed lines; (d) CPH, with the layer sequence $ABABA\ldots$; again the, layer C is shown by short-dashed lines and the C layer circles are shown slightly larger, for convenience.

In the body-centred cubic (BCC) structure type A2, the coordination number is 8; it is a less closely packed structure type. Reference to Fig. 5.15 shows that the spheres are in contact along a body diagonal. Hence,

$$a\sqrt{3} = 4r \tag{5.79}$$

from which the packing efficiency is 0.68.

5.5.1 Metallic radii
From equations (5.78) and (5.79), we can develop the idea of a metallic radius, or atomic radius for metals (cf. section 4.7.1), since a is an experimentally measurable parameter. Table 5.2 lists the radii for a selection of metals.

Comparison with the table of ionic radii (Table 4.10) emphasizes the dependence of the values of radii on bonding and environment. A number of metals exist in polymorphic modifications. From a study of their structures, the following empirical relationship between metallic radius and coordination number has been evolved:

Coordination number	12	8	6	4
Relative radius	1.00	0.97	0.96	0.88

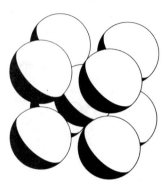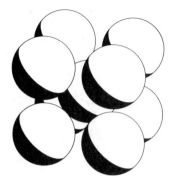

Fig. 5.15. Body-centred cubic packing (BCC or A2) of identical spheres; the coordination number is eight and the packing efficiency is 0.68.

Table 5.2. Metallic radii/nm

Li	0.152	Be	0.112	Cu	0.128
Na	0.186	Mg	0.160	Ag	0.144
K	0.227	Ca	0.197	Au	0.144
Rb	0.248	Sr	0.215	Fe	0.124
Cs	0.265	Ba	0.222	Co	0.125
Fr	0.293	Ra	0.229	Ni	0.125

5.5.2 Interstitial sites

Each sphere in either a CPC or a CPH structure type has tetrahedral sites as nearest neighbours and octahedral sites as next nearest neighbours. The size of the site is significantly less than that of the surrounding spheres, and the radius ratio limits of 0.225 and 0.414 (section 4.7.2) apply here to tetrahedral and octahedral[1] sites, respectively. The arrangement of these sites in the CPC structure is different from that in the CPH structure. In the latter, interstitial sites lie directly above one another forming rows parallel to c (Fig. 5.14d). In addition, there is a third type of interstitial site in the CPH structure type, namely, at the centres of trigonal bipyramids. It is well worth the reader making layers of close-packed spheres (Fig. 5.41a), using polystyrene or table-tennis balls, so as to build the A1 and A3 structure types and identify the interstitial sites in each of them.

5.6 STRUCTURAL AND PHYSICAL CHARACTERISTICS OF METALLIC COMPOUNDS

The metallic bond is spatially undirected, and metal structures have high coordination numbers and high densities. Metals are opaque, and possess high reflecting power. Electrons near the Fermi level in metals can absorb energy, according to (2.12), and

---

[1] It should be remembered that while 'tetrahedral' refers to four neighbours, which is reasonable, 'octahedral' refers, by tradition, to *six* (the apices of an octahedron).

are raised in energy state. If there is only slight scattering interaction between these electrons and the lattice of positive ions, the energy gained is radiated away without change of phase, and the crystal is transparent to that radiation. Metals interact in this way with radiations of wavelengths less than those of the ultraviolet region of the spectrum.

Band shape and structure have been investigated by the technique of electron-spectroscopy for chemical analysis (ESCA). A metal specimen in vacuo is exposed to monochromatic ultraviolet or X-ray radiation of known energy E_v, sufficient to induce electron emission from the metal, and the kinetic energy E_K of the electrons emitted (~ 1 eV) is measured accurately in a β-ray spectrometer. The energy E needed to remove an electron is given, by analogy with (2.23), as

$$E = E_v - E_K. \tag{5.80}$$

Fig. 5.16 shows the overlapping of the 5d and 6s energy bands as determined by ESCA experiments on a single crystal of gold, using Al $K\alpha$ radiation (see also Fig. 2.19). The results apply strictly to the outer layers of the metal, since electron emission can take place through only ~ 2 nm; the corresponding energies deep within the metal can be expected to be different in value.

Metals have variable strength. Deformation by gliding is common in metals, and it takes place most easily along close-packed planes of atoms in the crystal. In the CPC structure type there are four such planes, with Miller indices (111), ($\bar{1}$11), (1$\bar{1}$1) and ($\bar{1}\bar{1}$1), whereas there is only one such plane, (0001), in the CPH structure type, and none of a similar degree of close packing in the BCC structure type. Consequently, we find that the more malleable and ductile metal, such as copper, silver, nickel and γ-iron, crystalline in the CPC structure type, whereas the harder and more brittle metals, such as beryllium, chromium, tungsten and α-iron, have the CPH or BCC structure types. The metallurgical importance of iron is related to its ability to adopt either the A1 or the A2 structure type according to heat treatment. Metals have sharp melting-

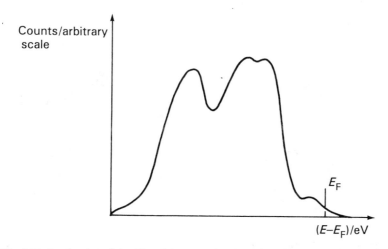

Fig. 5.16. Overlapping of the 5d and 6s energy bands in gold, from an ESCA study (after Shirley, and reproduced by courtesy of the American Institute of Physics).

point temperatures, but they vary widely (Hg, 234 K; W, 3683 K) and the liquid interval is long (Ga, 2370 K; Hf, 3400 K).

A study of the elements of groups 12–17 in the periodic table reveals continuous change in bond type towards increasing metallic character as the atomic number down the group increases. Thus, solid iodine (group 17) is a semiconductor; it has the lustre associated with a metal, and under pressure become metallic. Again, selenium and tellurium in group 16 and, more especially, antimony and bismuth in group 15 show varying degrees of metallic character. Group 14 elements in the solid state show a transition from covalent character in carbon (diamond) to metallic character in tin and lead. Lead has the CPC structure, whereas tin is dimorphic, grey tin (α-Sn) having the tetrahedrally coordinated structure type of Fig. 5.17a (cf. diamond, Fig. 1.2), and white tin, the metallic form, β-Sn (Fig. 5.17b). In β-Sn, a flattened tetrahedral coordination can be seen with the equatorial bonds, and two slightly longer axial bonds complete a distorted octahedral coordination. Group 13 shows a transition in character similar to that in group 14. In group 12 all its members are metallic. Mercury has a unique rhombohedral structure, but zinc and cadmium are close to the CPH structure type. Their axial ratios c/a are $\sim$1.9, whereas the ideal CPH ratio is $\sqrt{(8/3)}$, or 1.633.

Perhaps the most distinctive properties of metals are their high electrical and thermal conductivities, to which we have referred earlier in this chapter.

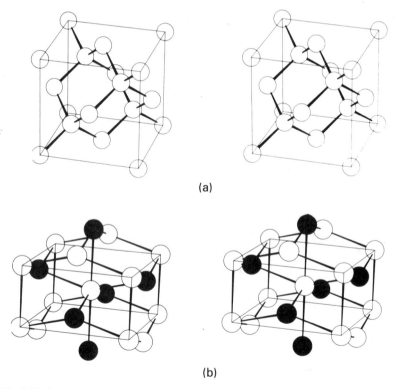

(a)

(b)

Fig. 5.17. Stereoviews of the unit cell of the dimorphic structures of tin: (a) grey tin (α-Sn), semiconducting, with a regular tetrahedral coordination (compare diamond, Fig. 1.2); (b) white tin (β-Sn), metallic, with a distorted octahedral coordination. The nearest neighbour atoms have been 'filled in' so as to highlight the coordination pattern.

5.7 SUPERCONDUCTIVITY

The electrical resistivity of many highly pure metals falls dramatically to zero as the specimen is cooled below a very low (critical) temperature τ_c. Zero resistivity applies under d.c. conditions; for a.c., a superconductor continues to exhibit impedance, although it is very small at low frequencies. Superconductivity was shown to exist first in mercury at $\sim 4\,\text{K}$ by Kamerlingh Onnes (1908), who also succeeded in liquefying helium in that same year. Since then, many elemental metals and numerous alloys have been shown to be superconducting.

A current initiated in a superconducting metal ring by magnetic induction has been shown to persist for several years. An experimental upper limit of resistivity has thereby been deduced as approximately 10^{-25} ohm m, which may be compared with that of copper, $\sim 10^{-8}$ ohm m. The proof that it was zero would require an experiment of infinite duration.

Meissner showed (1933) that the magnetic flux density in a metal in the supercon-ducting state was zero. Thus, a superconductor can act as a magnetic barrier, but the application of a strong external magnetic field, greater than a value B_c, will restore the resistivity. Hence, there exists a critical applied magnetic field strength B_c for superconduction, above which normal resistivity is restored; it has been found to vary with the material and, for any given material, also with the temperature.

5.7.1 Simple theory of superconductivity

In section 5.4.2.1, we discussed the residual resistivity at absolute zero arising from electron scattering by impurity atoms. It is evident, therefore, that superconduction must involve a different mechanism. In fact, electrons form pairs that can be scattered only if the pair can be separated into single electrons. In general, at very low temperatures, this energy cannot be acquired, and the movement of electron pairs is unhampered by the presence of impurity atoms.

The normal Coulombic interaction between electrons is repulsive, and the attrac-tive forces in an electron pair arise in an indirect manner. Let an electron with a wave vector (quasi-momentum) $\mathbf{k}_1$ interact with a positive ion. There will be a transient attraction which will result in the loss of momentum $\delta\mathbf{k}$ as a phonon, and the electron is scattered with a reduced momentum $\mathbf{k}_1'$ (Fig. 5.18). Thus, we have

$$\mathbf{k}_1' = \mathbf{k}_1 - \delta\mathbf{k}. \tag{5.81}$$

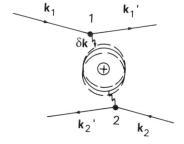

Fig. 5.18. Schematic view of electron pairing: the interaction of the proximal electron 1 with the positive ion perturbs its vibration; it interacts in turn with another, proximal electron 2 which leads to an overall attraction. The pairing may be viewed as the emission of a phonon of momentum $\delta\mathbf{k}$ by electron 1 ($\mathbf{k} \rightarrow_1 \mathbf{k}_1'$) and its subsequent capture by electron 2 ($\mathbf{k}_2 \rightarrow \mathbf{k}_2'$).

If the phonon is captured by another proximal electron of wave vector $\mathbf{k}_2$, it, too, will experience an attraction and pass on with enhanced momentum $\mathbf{k}_2'$, given by

$$\mathbf{k}_2' = \mathbf{k}_2 + \delta\mathbf{k}. \tag{5.82}$$

Hence,

$$\mathbf{k}_1' + \mathbf{k}_2' = \mathbf{k}_1 + \mathbf{k}_2. \tag{5.83}$$

The overall effect is an attraction between the electrons that would not have taken place in the absence of the positive ion, while the total electron momentum is conserved.

Cooper (1956)[1] showed that in an attractive interaction between the two electrons, the lowest energy state at absolute zero was that in which the electron spins were paired; in this context the two electrons are known as a Cooper pair. A simplified form of his argument can begin with a simple metal at 0 K, with all its energy states up to E_F fully occupied, that is, the Fermi sphere is full. If two electrons are then added to the metal, the Pauli exclusion principle requires that they be in states of energy greater than E_F by a positive amount E_B, which is the difference in energy per electron between the uppermost filled band and the lowest unoccupied band. The kinetic energy E_K of the two electrons is clearly greater than $2E_F$ ($|\mathbf{k}| > k_F$), but if there is an attraction between them their potential energy E_P will be negative. Thus, we may write the total energy E_C as

$$E_C = 2E_F - E_B = E_K + E_P \tag{5.84}$$

so that

$$E_K = 2E_F - E_B - E_P. \tag{5.85}$$

Since $E_P < 0$, E_K can exceed $2E_F$, which is reasonable since the added electrons are external to the Fermi sphere. At infinite separation of the two electrons, $E_P = 0$, and E_K is then less than $2E_F$ by the amount E_B, which may be considered as a binding energy per electron for the Cooper pair. Thus, the two electrons would be within the Fermi surface, but since this situation is disallowed by the starting hypothesis, it follows that the two electrons remain bound, and can be separated only if supplied with energy greater than E_B. Provided that the two electrons have the same energy, momenta $\pm\mathbf{k}$ and spins $\pm\frac{1}{2}$, they can form a Cooper pair.

An elaboration of Cooper's theory was given by Bardeen, Cooper and Schrieffer (1957)[2], and is generally known as the BCS theory of superconductivity. They expanded Cooper's concept of two 'added' electrons to the normal electron concentration in a metal ($\sim 10^{29}$ m^{-3}). Free-electron theory had shown that, at 0 K, all states with $|\mathbf{k}| \leq k_F$ were full, and those with $|\mathbf{k}| > k_F$ empty. However, BCS theory permits occupancy of states with $|\mathbf{k}| > k_F$ at 0 k. The Cooper pairs[3] form a collective ground state assembly that is separated from the first excited state by a forbidden energy gap. of $2E_B$, above which the pairs are split. Superconduction arises because the Cooper pair is not split below the critical temperature τ_c and, hence, the electrons are not scattered by positive ions. In terms of Brillouin zones, the whole of the Fermi gas is moved by an applied electrical field, from the region $|\mathbf{k}| = 0$ to a new position in the zone. This displacement of electrons produces a supercurrent that is not attenuated by

[1] L. N. Cooper (1956) *Physical Review*, **104**, 1189
[2] J. Bardeen, L. N. Cooper and J. R. Schrieffer (1957) *Physical Review*, **106**, 162; *idem. ibid.*, **108**, 1175.
[3] Cooper pairs are controlled by Bose–Einstein statistics, because an electron pair has zero spin.

the normal process of thermal electron scattering by the positive ions. In an insulator, an energy gap still exists, but whereas in the superconductor the gap is linked to the mobile Fermi gas, in an insulator it is tied to the positive ion lattice and therefore immutable.

5.7.2 Thermal properties below the critical temperature
From equations (5.57) and (5.66), and the discussion, we obtain

$$C_V = \alpha T^3 + \pi^2 N k_B^2 T / 2E_F \qquad (5.86)$$

and the corresponding equation for temperature less than τ_c, when superconduction can come into play, is

$$C_V = \alpha T^3 + \beta \exp(-E_B/k_B T) \qquad (5.87)$$

where β is a constant and E_B is the energy needed to excite a single electron, although both electrons must be excited simultaneously if they are to remain as a Cooper pair. At $0\,K$, the BCS theory gives E_B as $\frac{7}{2}k_B T$, and this value is confirmed by experimental measurements of heat capacity at temperatures below τ_c.

The normal conduction of heat takes place almost entirely by electron transport. However, Cooper pairs cannot interact with phonons, so that thermal conductivity should decrease in the superconducting state, as is indeed observed. At very low temperatures, heat conduction can take place through phonons. Since the electron–phonon interaction is attenuated by the pairing of the electrons, the mean free path of the phonon is increased significantly so that heat can then be transported by the phonons.

The reader is directed to more detailed works for an enhanced description of the theory and practice of superflow and superconductivity.[1]

5.7.3 Superconduction in organic compounds
Although, strictly within the classification of compounds used herein, these materials should fall within the section on organic molecular compounds, their unusual conductivity in the solid state places them more appropriately in this chapter.

During the 1960s it was discovered that salts of tetracyano-1,4-quinodimethane (TCNQ: I) behaved as semiconductors. About ten years later, a molecular charge-transfer complex was prepared by a 1:1 addition of TCNQ and tetrathiafulvalene (TTF: II). It was found to exhibit electrical conductivity comparable to that of a metal. At ambient temperatures, $\sigma = 5 \times 10^4$ ohm^{-1} m^{-1}, but at $\sim 60\,K$ it is 10^8 ohm^{-1} m^{-1}, comparable with the conductivity of metallic copper.

(I)   (II)

[1]See, for example, J. Bardeen, L. N. Cooper and J. R. Schrieffer *loc. cit.*; A. C. Rose–Innes and E. H. Rhoderick (1978) *Introduction to Superconductivity*, 2nd edition, Pergamon Press; D. R. Tilley and J. Tilley (1974) *Superfluidity and Superconductivity*, Van Nostrand.

Subsequently, many other electrically conducting charge-transfer organic compounds have been prepared. Below a critical temperature, these compounds exhibit
superconductivity. One of the most striking is the 1:2 complex formed between
bis(ethylenedithiolo)tetrathiafulvalene (BEDT-TTF: III) and bis(dicyanamido)cop-
per(I) bromide, $Cu[N(CN)_2]Br$, with a critical temperature of $\sim 12\,K$.

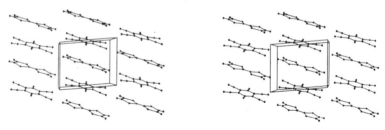

X-ray crystallographic studies have shown that these charge-transfer compounds
are composed of stacks or sheets of planar, or nearly planar, molecules. Each stack
comprises electron donors, such as TTF or BEDT-TTF, alternating with electron
acceptors, such as TCNQ or 2,5-dimethyl-N,N'-dicyanoquinoneimine (DMDCNQI,
IV), and their structures produce strong anisotropy in their electrical conductivity. A
donor such as TTF is a radical cation $TTF^{\cdot +}$, and an acceptor such as TCNQ
behaves as a radical anion $TCNQ^{\cdot -}$, and both compounds are open-shell molecules[1].
Fig. 5.19 illustrates a typical stacking mode in the charge-transfer complex of TTF
(donor) and 1,2,4,5-tetracyanobenzene (TCNB, acceptor).

Supermolecular orbital structures, as in metals, provide a mechanism for electron
delocalization, and the width of the conduction band is governed by the interaction of
molecular orbitals on adjacent molecules. In addition the HOMO-LUMO energy gap
must be small enough for intrinsic semiconducting behaviour. The temperature
dependence of electrical conductivity in these compounds at low temperature is
determined by electron–phonon interactions. The lower the temperature the fewer the
lattice vibrations, and the greater the overlap of adjacent MOs, with a corresponding
increase in conductivity. Superconducting behaviour sets in below a critical temperature for each complex, because of highly coordinated movement of Cooper pairs, and
the BCS (*loc. cit.*) theory has, to some extent, been able to explain superconduction in
organic compounds, as it did for inorganic materials. Recent reviews of this subject
provide interesting additional reading, especially with respect to superconductivity at
higher temperatures.[2].

Fig. 5.19. Stereoview of the crystal structure of the TTF–TCNB charge-transfer compound,
as seen along *b*; for clarity, the hydrogen atoms are not shown (after Bandoli, Lunardi and
Clemente, and reproduced with permission).

[1]In open-shell species, not all of the electrons occur in spin-pairs, whereas in closed-shell species all
occupied orbitals are paired.
[2]See, for example, M. R. Bryce (1991) *Chemical Society Reviews*, **20**, 355; P. M. Chaikin and R. L. Greene
(1986) *Physics Today*, **39**, 24; D. Jerome and H. J. Schultz (1982) *Advances in Physics*, **31**, 299.

5.8 ALLOYS

Metals readily form alloys, and their study has played an important part in the understanding of the metallic state. Alloys are numerous in type, and exhibit variable, non-stoichiometric compositions. They are generally prepared by simply melting the constituents together, a feature that is consistent with the model of the metallic bond.

The band theory of metals assumes that the crystal is periodic with the lattice translations, but that in the presence of impurities or of a second major component, the translational symmetry is perturbed. The effect of this interference is, however, quite small, particularly if the two components are in the same periodic group, and thereby have the same number of valence electrons. We shall consider next some binary systems, in order to develop some rules that govern these types of alloys.

5.8.1 Copper–gold system

Copper and gold both have the close-packed, face-centred cubic structure type (FCC, or CPC) shown by Fig. 5.13a, and also similar radii; they form a complete range of substitutional solid solutions which, like liquid solutions, are homogeneous within the given range. It has been shown that provided the radii of the two component metals do not differ by more than about 15%, complete solid solution may be expected. Where the difference is greater than this value, the range of solid solution will be restricted, or its formation may be completely inhibited. In the case of copper and gold, the difference in radii (Table 5.2) is 12.5%.

If copper is added progressively to gold and the molten alloy quenched rapidly to room temperature, a face-centred cubic structure is obtained (Fig. 5.20) in which there is a random replacement of gold by copper. The cubic unit cell dimension a decreases in proportion to the concentration of copper added:

$$a = \mathcal{K} c_{Cu} \qquad\qquad (5.88)$$

where $\mathcal{K}$ is a constant; this relationship is known as Vegard's law. The distribution of copper and gold atoms on the FCC sites is completely random, so that each site is populated by atoms that may be regarded as a certain fraction x of copper and $1 - x$ of gold. Any given unit cell would not, in general, have cubic symmetry, but the

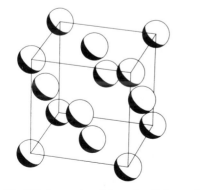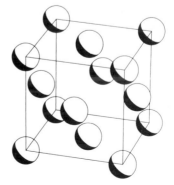

Fig. 5.20. Stereoview of the unit cell of a random solid solution of copper in gold: each sphere represents a fraction x of copper and $1 - x$ of gold; all atoms are statistically identical, so that the high cubic symmetry obtains over the crystal as a whole.

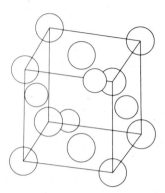 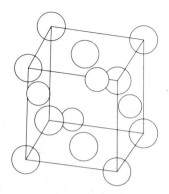

Fig. 5.21. Stereoview of the ordered, pseudo-face-centred tetragonal structure of CuAu; the circles in decreasing order of size represent Au and Cu.

statistical distribution of the atoms over numerous unit cells simulates an overall FCC arrangement.

A different situation arises if the specimens are annealed. Random replacement occurs until the composition reaches CuAu. Then, the atoms segregate into layers in a tetragonal structure (Fig. 5.21), with $c/a = 0.93$. Random replacement of gold by copper, now in the tetragonal structure, continues to the composition Cu_3Au, which crystallizes in a pseudo-face-centred cubic structure (Fig. 5.22); these ordered phases are termed superlattice structures. X-ray powder photographs show extra lines, *superlattice lines*, over and above those to be expected from the underlying cubic and tetragonal crystal types.[1]

The rearrangement of the atoms into an ordered phase obviously lowers the free energy of the system with respect to the random solid solution of the same composition, although the entropy is decreased in the process. If the metallic radii are

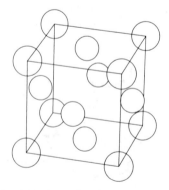 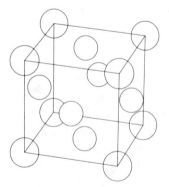

Fig. 5.22. Stereoview of the ordered, psuedo-face-centred cubic structure of Cu_3Au; the circles in decreasing order of size represent Au and Cu. This structure and that of CuAu are superlattice structures. The X-ray powder photographs show lines (superlattice lines) that would be absent in a true face-centred unit cell.

[1]See, for example, M. F. C. Ladd and R. A. Palmer (1993) *Structure Determination by X-ray Crystallo-graphy*, 3rd edition, Plenum Publishing Corporation.

very similar and the atoms have the same numbers of valence electrons, the strain introduced by substitution will be negligible and no superlattice structure will form. Thus, silver and gold, of equal radii (owing to the lanthanide contraction) form a continuous series of solid solutions, but no superstructures. If there is no strain for a possible composition, such as AgAu, then the free energy change for a reaction such as

$$\text{AgAu (solid solution)} \rightarrow \text{AgAu (superstructure)} \tag{5.89}$$

will be governed by the entropy change, whereupon the random solid solution would be expected.

5.8.2 Silver–cadmium system

The silver–cadmium system (Fig. 5.23) is more complex than that just discussed. The α-phase is pure silver, and it can dissolve cadmium in solid solution up to about 42% Cd; then a β-phase appears, corresponding to AgCd at 50% Cd, Fig. 5.24 shows the structures of most of the phases in this system. The β-phase is statistically body-centred cubic; it is not a cesium chloride structure type (Fig. 4.2).

The γ-phase is a complex cubic structure corresponding to the composition Ag_5Cd_8, and is hard and brittle. The ε-phase is approximately close-packed hexagonal, the lattice sites being occupied by silver and cadmium atoms in a random manner. Pure cadmium is represented by the η-phase which, with a c/a ratio of ~ 1.9, is closer to the ideal CPH structure than is the ε-phase. It can take up only about 4% of silver into solid solution.

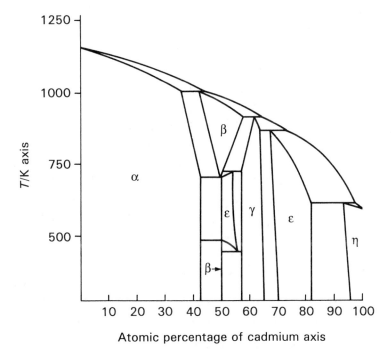

Fig. 5.23. Equilibrium phase diagram for the silver cadmium binary alloy system, showing the ranges for the α- (pure Ag), β-, γ-, ε- and η- (pure Cd) phases (after Westgren (1932) *Angewandte Chemie*, **45**, 33).

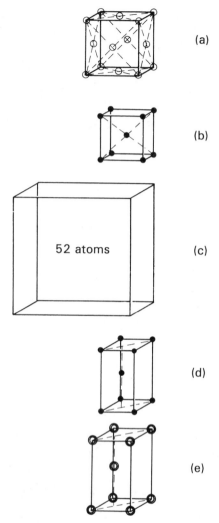

Fig. 5.24. Unit cells of the crystal structures of the phases in the silver–cadmium system: α-phase, pure silver; (b) β-phase, represented ideally by the composition AgCd; (c) γ-phase, a complex cubic structure represented ideally by the composition Ag_5Cd_8; (d) ε-phase, represented ideally by the composition $AgCd_3$; (e) η-phase, pure cadmium. The open circles represent silver, the filled circles a random mixture of silver and cadmium, and the double circles cadmium (after Westgren (1932) *Angewandte Chemie*, **45**, 33).

Fig. 5.25 shows X-ray powder photographs of various alloys of silver and cadmium. The diffraction lines that are characteristic of the phases can be identified quite readily, and these data were used to derive the phase diagram in Fig. 5.22. The regions in which two phases coexist can be seen and compared with the phase diagram. No superlattice structures are formed in this system.

5.8.2.1   *Hume-Rothery rules*
The size factor governing substitutional solid solution that we discussed in section 5.8.1 was put forward first by Hume-Rothery in 1926. In addition, he gave empirical

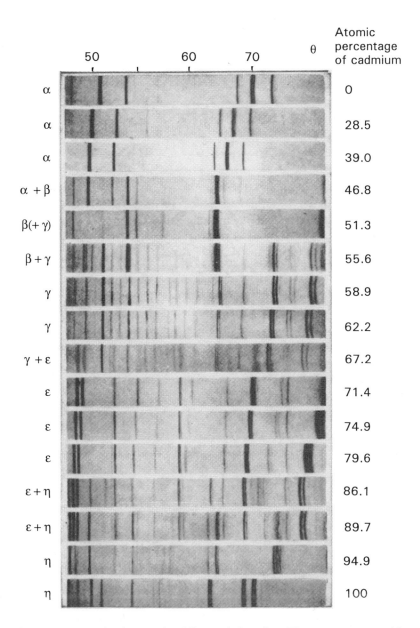

Fig. 5.25. X-ray powder photographs of silver–cadmium alloys. The percentage compositions were determined by measuring the intensities of selected diffraction lines and comparing them against similar lines from alloys of known compositions. The lines that are characteristic of each phase can be seen clearly, and their change in position for a given phase is a manifestation of Vegard's law for solid solutions. The disappearance of a given phase and the introduction of a new phase are clearly marked in the X-ray patterns (after Westgren and Phragén (1928) *Metallwirtshaft*, **7**, 700).

Table 5.3. Electron/atom ratios in some binary alloys

Phase	Composition	Valence electrons	Atoms	Electron/atom ratio
β	AgCd	$1 + 2$	2	3/2 (1.50)
β	Cu_3Al	$3 + 3$	4	3/2
γ	Ag_5Cd_8	$5 + 16$	13	21/13 (1.62)
γ	Cu_9Al_4	$9 + 12$	13	21/13
ε	$AgCd_3$	$1 + 6$	4	7/4 (1.75)
ε	Ag_5Al_3	$5 + 9$	8	7/4

rules that govern the appearance of different phases in the alloys between a true metal and a metal from groups 12–15; they were subsequently shown to have a theoretical basis in terms of the band theory of the metallic bond. The frequent occurrence of the β-, γ- and ε-phases among chemically dissimilar binary systems at widely differing chemical compositions is determined by the ratio of the total number of valence electrons to the number of atoms in the ideal formula (Table 5.3).

In terms of band theory, we find that the limit of the α-phase occurs with an electron/atom ratio of 1.36, in which circumstances the Fermi surface touches the first FCC Brillouin zone boundary. For the β-phase it is 1.48, and for the γ- and ε-phases the calculated limits are 1.54 and 1.69, respectively. Once the filled states reach a zone boundary, it is energetically difficult to add further electrons, but a change in crystal structure, which affects the Brillouin zone, can lead to an energetically more favourable state as the electron/atom ratio increases. The results in Table 5.3 are neatly explained by this theory. For further study of alloys, the reader is recommended to specific works on this topic.[1]

PROBLEMS FOR CHAPTER 5

5.1   The electrical resistivity for lithium (BCC) at 273 K is 8.55×10^{-8} ohm m, the relative atomic mass is 6.94, the density is 537.5 kg m^{-3} and the mean electron drift velocity in the metal is 1.30×10^6 m s^{-1}. Calculate the mean free path and mean mobility of the valence electrons in lithium.

5.2   The electrical resistivity of copper at 20 °C is 1.70×10^{-8} ohm m. Calculate the thermal conductivity of the material at the given temperature.

5.3   Calculate the electronic and lattice contributions to the molar heat capacity of aluminium at 10 K $[\Theta(Al) = 428$ K; $E_F(Al) = 11.6$ eV]. At what temperature are the two contributions equal?

5.4   Calculate, from free-electron theory, the Fermi energy and the Fermi temperature for lithium; the unit cell constant a for the body-centred cubic structure of lithium is 0.350 nm. Determine the density of states function $g(E_F)$ for lithium.

5.5   Draw a two-dimensional square lattice in **k**-space, and mark on it the boundaries of the first three Brillouin zones. What is the first Brillouin zone for a lattice with a primitive unit cell of side a?

[1]See, for example, A. H. Cottrell (1975) *An Introduction to Metallurgy*, Arnold; W. Hume-Rothery, R. E. Smallman and C. W. Haworth (1969) *The Structure of Metals and Alloys*, Institute of Metals.

5.6   If the spherical Fermi surface just touches the first Brillouin zone in a primitive cubic unit cell, side a, of a metal, how many conduction electrons per atom are there present in the metal?

5.7   Calculate the packing efficiency for the closest packing of identical spheres in the CPH (A3) structure type.

5.8   The CPC structure type contains both tetrahedral and octahedral holes. How many of each kind are unique to one FCC unit cell, and what are the fractional coordinates of the centres of the holes?

5.9   CdS is a photoconducting material with a band gap of 2.42 eV. What is the largest wavelength of (visible) radiation that can excite a valence electron to the conduction band? What is the ratio of the resistivities of this material at 300 K and 400 K?

5.10   The anharmonic vibrations of atoms about their mean positions in a metal crystal may be represented approximately by the potential energy function $V(r) = ar^2 - br^3$, where a and b are constants. By determining an expression for the mean displacement $\bar{r}$, show that $\bar{r}$ is directly proportional to temperature, that is, it is consistent with the solid expanding on heating. Note that since r is small, $\exp(br^3) \approx (1 + br^3)$.

5.11   Use the appropriate Debye equation to calculate the molar heat capacity at constant volume of copper at 298.15 K.

Appendix 1 Problem solving and computers

A1.1 INTRODUCTION

An important part of the study of any of the physical sciences involves problem solving. This work may range from a fairly straightforward evaluation of a parameter from a given equation to a full interpretation of a set of experimentally determined data. The ready availability of computers means that one can handle much more data than would be reasonable with a simple hand calculator, and thereby study a wide range of topics in significant numerical detail.

At the same time, it is well worth noting that an involvement with numerical work brings with it an understanding of, or 'feel' for, the magnitudes of various physical quantities. Calculators and computers invariably produce correct answers—provided that they are supplied with the appropriate data. Thus, an appreciation of the approximate size of the parameter to be expected from a given calculation is a valuable facility to acquire.

A1.2 SOLVING NUMERICAL PROBLEMS

Numerical problems give practice in relating experimental observations to theoretical models. The insertion of magnitudes into a given equation is a common scientific activity: it should be mastered and, however trivial, never despised.

The solving of problems leads to an appreciation of several important features:
  (a) the orders of magnitude of physical and chemical quantities;
  (b) the need for an understanding of units;
  (c) the value of checking dimensional homogeneity;
  (d) the sources of physical and chemical data;
  (e) the precision of the data and its transmission to the result.

Most problems involve algebraic manipulation. It is essential to obtain a clear picture of the chemistry and physics involved in the problem before embarking on a series of mathematical processes. It is often useful to obtain an explicit algebraic expression before inserting numerical values. There are several advantages in so doing:
  (f) the expression can be checked dimensionally;
  (g) the possible cancellation of terms may improve the precision of the result;
  (h) the chemical or physical significance of the result may be more important;

(i) similar problems with other magnitudes can be solved with little additional effort;

(j) if the result is erroneous, it is easy to check whether the error is in the deduction or in the arithmetic;

(k) in examinations, the derivation of a correct explicit expression will score marks, even though the arithmetic may be in error.

If the data are inserted into an expression in the form of numbers between 1 and 10, multiplied by the appropriate powers of 10, it is easy to estimate an approximate answer: calculators and other aids give wrong answers if manipulated incorrectly.

Suppose that we have for a relative permittivity ε_r

$$(\varepsilon_r - 1) = \mathcal{N}\mu^2/9\varepsilon_0 k_B T \qquad\qquad (A1.1)$$

$\mathcal{N}$ (numbers of molecules per unit volume)	$=$	$2.461 \times 10^{25} \text{ m}^{-3}$
μ (dipole moment)	$=$	$5.11 \times 10^{-30} \text{ C m}$
ε_0 (permittivity of a vacuum)	$=$	$8.8542 \times 10^{-12} \text{ F m}^{-1}$
k_B (Boltzmann constant)	$=$	$1.3807 \times 10^{-23} \text{ J K}^{-1}$
T (absolute temperature)	$=$	298.15 K

Inserting the parameters into equation (A1.1), we obtain

$$\varepsilon_r - 1 = \frac{2.461 \times 10^{25} \text{ m}^{-3} \times (5.11)^2 \times 10^{-60} \text{ C}^2 \text{ m}^2}{9 \times 8.8542 \times 10^{-12} \text{ F m}^{-1} \times 1.3807 \times 10^{-23} \text{ J K}^{-1} \times 298.15 \text{ K}} \qquad (A1.2)$$

Equation (A1.2) is dimensionally correct, as both sides are dimensionless. We can see that $(\varepsilon_r - 1) \approx 60/(100 \times 300)$, or 2×10^{-3}. Thus, when the expression is evaluated, we can write with confidence, $(\varepsilon_r - 1) = 1.96 \times 10^{-3}$.

A1.2.1 Approach to problems

There are different ways of tackling problems, so that these notes are offered only as a guide. Sometimes a recommended stage may be changed or bypassed. Elegant derivations are often concise: the converse is not necessarily true, and failure to justify a stage in a derivation may indicate a lack of judgement or a lack of confidence. On the other hand, over-elaboration of trivial detail or of arithmetic manipulation may be equally unacceptable in a polished answer to a problem. Some degree of subjective judgement is involved in the solution of problems, and in the marking of such solutions in examinations. Few examiners would give high marks for a completely correct numerical answer in the absence of satisfactory evidence of the method used.

A1.2.2 Suggested procedure

(a) Read the problem carefully. If you think that it contains an ambiguity (which can happen sometimes), assume the simplest interpretation of the ambiguity, and comment on it.

(b) Summarize the given information by appropriate means, such as:

(i) labelled drawings;

(ii) energy-level diagrams;

(iii) sketch-graphs, correctly labelled;

(iv) defining symbols used in diagrams and formulae;

(v) listing numerical values with units.

(c) State the answers required, defining the quantities involved, together with their units and symbols.
(d) Indicate relevant laws and equations which are to be used in developing the problem, at least initially.
(e) State the method to be used, for example 'take $\log_e$ of both sides of equation (1)'.
(f) Where appropriate, attempt to formulate an explicit equation before inserting numerical data. Look for cancellations of terms, and indicate any physical or functional approximations.
(g) Do *not* make needless numerical approximations, but state any approximations which are made, and include an estimate of the probable error as far as you are able.
(h) For convenience, substitute a new symbol for a complex group of symbols in deriving an expression.
(i) Check the dimensions of both quantities and expressions for consistency. Remember that exponents (and log terms) are dimensionless.
(j) Insert numerical values into expressions carefully. Determine an approximate result by 'gross cancellation', as in the example above.
(k) Think about the answer in terms of your knowledge of the physical sciences, and comment on it in the light of the question. If you feel doubtful about the validity of the result, check your arithmetic and deductions. If you still have some reservations about your answer, indicate their nature.
(l) Keep a neat format in your answer. In an examination, allocate a numerical question only its fair share of time.

A1.2.3   Solving an example problem

The diffusion coefficient (D) of carbon in α-iron, as a function of temperature, follows the equation

$$D = D_0 \exp(-E_d/\mathscr{R}T). \tag{A1.3}$$

Values of D are as follows:

T/K	300	500	700	900	1100
D/m^2 s^{-1}	4.73×10^{-21}	3.35×10^{-15}	1.08×10^{-12}	2.66×10^{-11}	2.05×10^{-10}

Verify the above equation, and find values for both the activation energy (E_d) for the diffusion process, and D_0. Comment briefly on the results. ($\mathscr{R} = 8.315$ J K^{-1} mol^{-1}.)
The following solution attempts to illustrate some of the points discussed above.

A1.2.4   Solution

The presence of $\mathscr{R}$ in the exponent and its units indicate that E_d is expected in J mol^{-1}. D_0 is the limiting value of D as $T \to \infty$. Taking ln of both sides of (A1.3) gives

$$\ln D = \ln D_0 - E_d/\mathscr{R}T \tag{A1.4}$$

If the graph of $\ln D$ against $1/T$ is a straight line, (A1.3) would be verified. The slope of the line, $\Delta(\ln D)/\Delta(1/T)$, is $-E_d/\mathscr{R}$, and the intercept, $\ln D$ at $1/T = 0$, is $\ln D_0$.

$D/\mathrm{m^2\ s^{-1}}$	$\ln D$	T/K	$(10^3/T)/\mathrm{K^{-1}}$
4.73×10^{-21}	-46.800	300	3.333
3.35×10^{-15}	-33.330	500	2.000
1.08×10^{-12}	-27.554	700	1.429
2.66×10^{-11}	-24.350	900	1.111
2.05×10^{-10}	-22.308	1100	0.909

Note: $\ln D$ does not have the units $\mathrm{m^2\ s^{-1}}$; $\ln(D/D_0)$ is dimensionless, and the separation of $\ln D$ from $\ln D_0$ is a mathematical convenience.

The straight-line graph (Fig. A1.1) verifies (A1.3) for the diffusion of carbon in α-iron. From a least-squares fit (see Appendix 9) to (A1.4), the slope is -10103 K, giving $E_\mathrm{d} = 84.006$ kJ mol^{-1}; $D_0 = 2.00 \times 10^{-6}$ m^2 ms^{-1}. (It is clear from the graph that no data point merits exclusion from the least-squares calculation.) The sign and magnitude of E_d seem reasonable, since work must be done on the system to cause diffusion, and the energies of such processes are usually of the order of 1 eV per atom. D increases with T and, if (A1.3) continues to hold, would tend to a limiting value of 2.00×10^{-6} m^2 s^{-1}. However, at such high temperatures the material would melt or even vaporize. The usefulness of D_0 here is mainly in connection with the evaluation of D from (A1.3).

The estimated standard deviation in the slope is, from Appendix 9, 0.75 K. This value is transmitted to the derived parameter E_d (see Section A9.1) as an estimated standard deviation of 0.006 kJ mol^{-1}, so that the final value for E_d is 84.006 ± 0.006 kJ mol^{-1}, usually written as $84.006(6)$ kJ mol^{-1}.

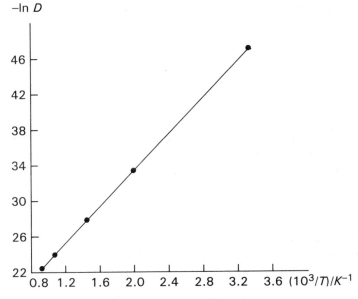

Fig. A1.1. Graph of $-\ln D$ against $10^3/T$; the slope is $10^{-3}E_\mathrm{d}/\mathscr{R}$.

A1.3   COMPUTER METHODS

Throughout this book, the reader is introduced to procedures, such as linear least squares, that often need to be invoked while gaining the desired familiarity with the subject matter. To this end a number of programs has been written that meet the requirements of problem solving in the context of this book; problems at the end of this appendix test the usefulness of the programs described.

A1.3.1   Program design and availability

The programs that have been written are self-contained, and mostly self-explanatory, or 'user-friendly' in the current jargon. They are executable on an IBM-compatible personal computer, and need only a monitor and a printer peripherals.

Some of the programs in the set could have been enhanced by graphical display, but it is the author's experience and belief that handling and assessing numerical data and plotting graphs gives an appreciation of and familiarity with the results that are not always obtained by an appraisal of the monitor output. Furthermore, not everyone will have a graph plotter available, and screen plots are only transient!

Most of the programs are provided with a facility for repeating the calculations with different values of the important input parameters. The availability of the programs can be ascertained from the author.

A1.3.2   Linear least squares

The mathematics of linear least squares is discussed in Appendix 9. The least-squares program LSLI solves the equation

$$y = ax + b \qquad\qquad (A1.5)$$

for the best values of a and b, assuming that the errors in x are negligible compared with those in y. The program also computes the estimated standard deviations in a and b, so that experimental errors reflected in a and b may be transmitted to derived parameters, according to Appendix 9.1.

A1.3.3   Gaussian quadrature

The process of numerical integration, or finding the area under a curve such as

$$y = \int_a^b f(x)\,\mathrm{d}x, \qquad\qquad (A1.6)$$

is addressed in Appendix 12. The Gaussian quadrature program QUAD has a greater precision, and the subdivisions in x need not be equally spaced, although they must be at least four in number, and input in increasing order of x. Ideally, they should be symmetrically spaced with respect to the centre of the range, and they must, of course, include the extrema. The value of the integral is given, together with its probable error.

A1.3.4   Madelung constant

One method of calculating Madelung constants is described in Appendix 13, and that method has been programmed. The Madelung constant depends on the given structure type, such as that of NaCl or CsCl. For a substance like MgO, which has the NaCl structure type, $\mathscr{A}$ is multiplied by q_+q_-, which is 4 for this substance.

The program MADC addresses the cubic, tetragonal, orthorhombic, hexagonal, trigonal and monoclinic crystal systems. For the cesium chloride structure, with $a = 0.4123$ nm and a summation limit of 3, $\mathscr{A}(\text{CsCl}) = 1.76260$. The summation limit can take the integer values 2, 3, 4 or 5; the higher the limit the better the precision, but the longer the computation time.

A1.3.5 Radial wave functions

The set of normalized radial distribution functions listed in Table 2.3 have been provided in the program RADL to calculate $R_{n,l}(r)$, $R_{n,l}^2(r)$ and $4\pi r^2 R_{n,l}^2(r)$ for values of r from zero to 0.8 nm, for 1s, 2s, $2p_z$, $2p_x$, $2p_y$, 3s, $3p_z$, $3p_x$, $3p_y$, $3d_{z^2}$, $3d_{xz}$, $3d_{yz}$, $3d_{x^2-y^2}$ and $3d_{xy}$. For numerical convenience, the results are listed in terms of r/a_0.

A1.3.6 Maxwell–Boltzmann distribution

The program GASD calculates the Maxwell–Boltzmann distribution of speeds, as given by equation (3.43), for v between 0 and 1500 m s^{-1}; provision is made for input of the relative mass of a species and of a temperature for the calculation. Additionally, one may determine the fraction ζ of species having speeds less than a chosen input value v_0. Since the area under the curve to ∞ is unity, it follows that $(1 - \zeta)$ is the fraction of species with speeds $\geqslant v_0$. This result is readily converted into an energy through the relationship $\frac{1}{2}mv^2 = k_B T$.

A1.3.7 Electron-in-a box

The program BOXS calculates the wave functions ψ_n and energies E_n for the first eight solutions of the one-dimensional 'electron-in-a-box' wave equation. It can be applied also to conjugated systems to predict spectral characteristics of the first excited electronic state, treating the π-electron system as a one-dimensional box.

The results for ψ_n are given between 0 and the box length a in integer steps from 0 to 24. The values of ψ_n are normalized to 99 for convenience of output. The results show the form of the function sufficiently well to indicate the number of nodes for each value of n, and a print of the results can be sketched satisfactorily.

A1.3.8 Monte Carlo calculations

The Monte Carlo calculations on liquids to which we referred in section 3.2.5 are too complex for the present context. However, the program MONC enables some of the principles of the Monte Carlo process to be examined.

Two calculations have been programmed: (1) the integration of a function between finite limits, using the example of Appendix 12; (2) the evaluation of an integral of the form $\int_0^a x^2 \exp(-\alpha x^2)\, dx$. The results with the first example can be compared with those given in Appendix 12, and as obtained by the use of Gaussian quadrature. The second calculation is used implicitly in the Maxwell–Boltzmann distribution. We consider briefly some aspects of Monte Carlo theory, through the evaluation of the integral

$$A = \int_a^b f(x)\, dx \tag{A1.7}$$

by calculating the average value for $f(x)$ in the interval a to b. Let x be a random number distributed uniformly between the limits a and b, with a probability $\phi(x)$

given by $1/(b-a)$ if x lies between a and b, and zero otherwise. The expectation value of $f(x)$ is given by

$$\mathscr{E}(x) = \int_a^b f(x)\phi(x)\,\mathrm{d}x = A/(b-a) \tag{A1.8}$$

Let N trials give the succession of random numbers $x_1, x_2, \ldots, x_N$. For large values of N, we have

$$\mathscr{E}(x) = (1/N)\sum_{i=1}^N f(x_i) \tag{A1.9}$$

Hence, an approximate value of the integral A is given by

$$A = (b-a)/N \sum_{i=1}^N f(x_i) \tag{A1.10}$$

In the program, values of x_i ($i = 1$ to N) are obtained by a random number generator, the lower limit a is fixed at zero, and A is determined from (A1.10). The program allows inspection of the progress of the calculation as N increases. The statistical nature of the calculation can be appreciated by running the program with successive numbers N of the same value.

In the first calculation, the given function (see Appendix 12) is

$$f(x) = 2x^2 + 3x + 2 \tag{A1.11}$$

and the second calculation uses the equation

$$f(v) = 4\pi(m/2\pi k_\mathrm{B} T)^{3/2} v^2 \exp(-mv^2/2k_\mathrm{B} T) \tag{A1.12}$$

It should be noted that at least 10 integration points must be supplied. Generally, larger numbers of points will be used and, preferably, multiples of 10. With the first problem, it is not too lengthy to use a maximum of 10^6 points.

A1.3.9   Angular wave functions

The program PLOT computes the wave functions listed in Table 2.4, and displays them in the xy plane so that the angular dependence of ψ is clearly shown. For any of the functions available, the program computes ψ and $|\psi|^2$. In the case of $|\psi|^2$, an alternative, symbolic plot is provided, wherein decreasing values of $|\psi|^2$ are represented by the character sequence $\& > 0 > \$ > \cdots > \cdot > $ 'space'. The output is scaled to the value of $\psi(1s)_{r=0}$, or $|\psi|^2(1s)_{r=0}$ in the case of $|\psi|^2$ plots, and normalized to a maximum of 99 for convenience in printing the results.

The nuclear charge Z (or Z_{eff}) can be varied to show its effect on the size of the orbital, and an appreciation of the three-dimensional structure of an orbital can be gained by calculating at varying levels of z. A scaling factor is given, so that the absolute magnitude of ψ or $|\psi|^2$ may be determined. By contouring the numerical output with lines of equal amplitude, or density, the shape and gradient of an orbital is shown clearly.

A1.3.10   Eigenvalue problems

Many procedures in theoretical chemistry, such as the Hückel molecular orbital calculations in section 2.10, involve eigenvalues and eigenvectors, and first we shall discuss some of the principles of this aspect of matrix algebra.

A1.3.10.1   Eigenvalues and eigenvectors

An eigenvector of a square matrix $\mathbf{A}$ is a column vector $\mathbf{x}$ which is non-zero and has the property

$$\mathbf{A} \cdot \mathbf{x} = \lambda \mathbf{x} \tag{A1.13}$$

where λ, a scalar quantity, is the corresponding eigenvalue, or characteristic value, of $\mathbf{A}$. Thus, in

$$\begin{pmatrix} 4 & 3 \\ 2 & 1 \end{pmatrix} \cdot \begin{pmatrix} 1 \\ 2 \end{pmatrix} = \begin{pmatrix} 10 \\ 4 \end{pmatrix} = 2\begin{pmatrix} 5 \\ 2 \end{pmatrix} \tag{A1.14}$$

$\begin{pmatrix} 10 \\ 4 \end{pmatrix}$ is an eigenvector of the matrix $\begin{pmatrix} 4 & 3 \\ 2 & 1 \end{pmatrix}$, with an eigenvalue of 2. If an eigenvector is multiplied by a scalar it remains an eigenvector, and the eigenvalue is unchanged. Thus, with a scalar α we have

$$\mathbf{A} \cdot (\alpha\mathbf{x}) = \alpha\mathbf{A} \cdot \mathbf{x} = \alpha\lambda\mathbf{x} = \lambda(\alpha\mathbf{x}) \tag{A1.15}$$

so that $\alpha\mathbf{x}$ is also an eigenvector, with the eigenvalue λ.

In order to find the eigenvalues and eigenvectors of a matrix $\mathbf{A}$, we begin by writing (A1.13) as

$$\mathbf{A} \cdot \mathbf{x} = \lambda\mathbf{I} \cdot \mathbf{x} \tag{A1.16}$$

where $\mathbf{I}$ is an identity matrix

$$\mathbf{I} = \begin{pmatrix} 1 & 0 & 0 & \cdots & 0 & 0 \\ 0 & 1 & 0 & \cdots & 0 & 0 \\ \cdot & \cdot & \cdot & \cdots & \cdot & \cdot \\ 0 & 0 & 0 & \cdots & 0 & 1 \end{pmatrix} \tag{A1.17}$$

whence

$$(\mathbf{A} - \lambda\mathbf{I}) \cdot \mathbf{x} = 0. \tag{A1.18}$$

A matrix $\mathbf{A}$ is said to be *singular* provided that there is a non-zero vector $\mathbf{x}$ that conforms to the condition

$$\mathbf{A} \cdot \mathbf{x} = 0. \tag{A1.19}$$

Thus, it follows that $(\mathbf{A} - \lambda\mathbf{I})$ in (A1.18) is singular, which means that the determinant

$$|\mathbf{A} - \lambda\mathbf{I}| = 0 \tag{A1.20}$$

Conversely, given (A1.20), there must be a non-zero vector $\mathbf{x}$ that conforms to (A1.18) which, in turn, means that (A1.16) holds, that is, λ is an eigenvalue of $\mathbf{A}$. In other words, and generally, λ is an eigenvalue of $\mathbf{A}$ if and only if $|\mathbf{A} - \lambda\mathbf{I}| = 0$. Equation (A1.20) is termed the *characteristic equation* of $\mathbf{A}$ in the variable λ, and its roots are the eigenvalues of $\mathbf{A}$.

In the matrix

$$A = \begin{pmatrix} 0 & 1 \\ 3 & 2 \end{pmatrix} \tag{A1.21}$$

$$A - \lambda I = \begin{pmatrix} 0 & 1 \\ 3 & 2 \end{pmatrix} - \lambda \begin{pmatrix} 1 & 0 \\ 0 & 1 \end{pmatrix} = \begin{pmatrix} -\lambda & 1 \\ 3 & 2 - \lambda \end{pmatrix} \tag{A1.22}$$

and the characteristic equation is formed by equating the *secular determinant* to zero:

$$\begin{vmatrix} -\lambda & 1 \\ 3 & 2 - \lambda \end{vmatrix} = 0 \tag{A1.23}$$

Expansion of the determinant gives the equation

$$\lambda^2 - 2\lambda - 3 = 0 \tag{A1.24}$$

whence the roots are $\lambda = 3, -1$. We now find the eigenvectors of **A**. For $\lambda = 3$

$$A - \lambda I = \begin{pmatrix} 0 & 1 \\ 3 & 2 \end{pmatrix} - 3\begin{pmatrix} 1 & 0 \\ 0 & 1 \end{pmatrix} = \begin{pmatrix} -3 & 1 \\ 3 & -1 \end{pmatrix} \tag{A1.25}$$

Solving

$$\begin{pmatrix} -3 & 1 \\ 3 & -1 \end{pmatrix} \cdot \begin{pmatrix} x \\ y \end{pmatrix} = \begin{pmatrix} 0 \\ 0 \end{pmatrix} \tag{A1.26}$$

we find that any vector of the form $\begin{pmatrix} p \\ 3p \end{pmatrix}$, such as $\begin{pmatrix} 1 \\ 3 \end{pmatrix}$, is an eigenvector, with an eigenvalue of 3. If the eigenvector is to be of unit length then, from $\sqrt{[p^2 + (3p)^2]} = 1$, the two possible values for the eigenvector are $\begin{pmatrix} \pm 1/\sqrt{10} \\ \pm 3/\sqrt{10} \end{pmatrix}$. For $\lambda = -1$, we can show in a similar manner that the eigenvector is $\begin{pmatrix} 1 \\ -1 \end{pmatrix}$, or $\begin{pmatrix} \pm 1/\sqrt{2} \\ \mp 1/\sqrt{2} \end{pmatrix}$ on a unit basis.

A1.3.10.2   *Diagonal matrices*
Consider again the matrix of (A1.21). We have found the eigenvalues 3 and -1, with the corresponding eigenvectors $\begin{pmatrix} 1 \\ 3 \end{pmatrix}$ and $\begin{pmatrix} 1 \\ -1 \end{pmatrix}$. Now consider a matrix **Q**, the columns of which are the eigenvectors of **A**:

$$Q = \begin{pmatrix} 1 & 1 \\ 3 & -1 \end{pmatrix} \tag{A1.27}$$

Forming $A \cdot Q$, we have

$$A \cdot Q = \begin{pmatrix} 0 & 1 \\ 3 & 2 \end{pmatrix} \cdot \begin{pmatrix} 1 & 1 \\ 3 & -1 \end{pmatrix} = \begin{pmatrix} 3 & -1 \\ 9 & 1 \end{pmatrix} \tag{A1.28}$$

each column of which is the corresponding column of **Q** multiplied by one of the eigenvalues. This result is general: if any non-singular $n \times n$ matrix, the columns of which are eigenvectors of another $n \times n$ matrix, has eigenvalues in the order $p, q,$

$r, \ldots$, then the diagonal matrix $\mathbf{D}$ is formed by

$$\mathbf{Q}^{-1} \cdot \mathbf{A} \cdot \mathbf{Q} = \mathbf{D} = \begin{pmatrix} p & 0 & 0 & \ldots & 0 & 0 \\ 0 & q & 0 & \ldots & 0 & 0 \\ 0 & 0 & r & \ldots & 0 & 0 \\ & & \cdot & \ldots & \cdot \end{pmatrix} \tag{A1.29}$$

where $\mathbf{Q}$ is an orthogonal matrix,[1] and $\mathbf{Q}^{-1} \cdot \mathbf{A} \cdot \mathbf{Q}$ is a *similarity transformation* on $\mathbf{A}$. Similarity transformations are encountered in several areas of theoretical chemistry.[2] In the matrix

$$\mathbf{A} = \begin{pmatrix} 1 & 2 \\ 2 & 1 \end{pmatrix} \tag{A1.30}$$

the foregoing shows that the eigenvectors are $\begin{pmatrix} \sqrt{2} \\ \sqrt{2} \end{pmatrix}$ and $\begin{pmatrix} \sqrt{2} \\ -\sqrt{2} \end{pmatrix}$ for the eigenvalues 3 and -1, respectively.

It follows that $\mathbf{Q}$ and $\mathbf{Q}^{-1}$ are given by

$$\mathbf{Q} = \begin{pmatrix} 1 & 1 \\ 1 & -1 \end{pmatrix}; \quad \mathbf{Q}^{-1} = \begin{pmatrix} 1/2 & 1/2 \\ 1/2 & -1/2 \end{pmatrix} \tag{A1.31}$$

Since $|\mathbf{Q}| = -2$, it is non-singular and will diagonalize $\mathbf{A}$. Hence,

$$\mathbf{D} = \mathbf{Q}^{-1} \cdot \mathbf{A} \cdot \mathbf{Q} = \begin{pmatrix} 3 & 0 \\ 0 & -1 \end{pmatrix} \tag{A1.32}$$

where 3 and -1 are the eigenvalues of $\mathbf{A}$, in the order corresponding to the columns of $\mathbf{Q}$.

This technique is straightforward with a 2×2 matrix, or even a 3×3, but it becomes very cumbersome with larger matrices and a more convenient method is needed. One such procedure is the Jacobi diagonalization; we consider it here and use it in the program of the next section.

A1.3.10.3   Jacobi diagonalization
This iterative procedure for selecting $\mathbf{Q}$ is based on the transformation of the reference axes x, y to a new frame x', y'. A clockwise rotation θ (Fig. A1.2) leads to the equation

$$\begin{pmatrix} x' \\ y' \end{pmatrix} = \begin{pmatrix} \cos\theta & -\sin\theta \\ \sin\theta & \cos\theta \end{pmatrix} \cdot \begin{pmatrix} x \\ y \end{pmatrix} = \mathbf{Q} \cdot \begin{pmatrix} x \\ y \end{pmatrix} \tag{A1.33}$$

In this particular case, the transpose matrix $\mathbf{Q}^{\mathrm{T}}$ is also the inverse $\mathbf{Q}^{-1}$; $\mathbf{Q}$ is an orthogonal matrix—orthogonal transformations do not change the angle between the axes.

Since we have in $\mathbf{A}$ a symmetric, non-singular matrix, application of (A1.29) leads to

$$\begin{pmatrix} \cos\theta & \sin\theta \\ -\sin\theta & \cos\theta \end{pmatrix} \cdot \begin{pmatrix} 1 & 2 \\ 2 & 1 \end{pmatrix} \cdot \begin{pmatrix} \cos\theta & -\sin\theta \\ \sin\theta & \cos\theta \end{pmatrix} = \begin{pmatrix} \lambda_1 & 0 \\ 0 & \lambda_2 \end{pmatrix} \tag{A1.34}$$

---

[1] See, for example, C-E. Fröberg (1969) *Introduction to Numerical Analysis*, 2nd edition, Addison-Wesley.
[2] See, for example, M. F. C. Ladd, (1989; reprinted 1992) *Symmetry in Molecules and Crystals*, Ellis Horwood Limited.

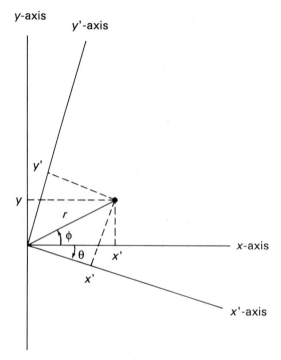

Fig. A1.2. Transformation of axes: the clockwise rotation θ leads to the matrix in (1.33).

Expanding (A1.34), we obtain

$$\begin{pmatrix} [\cos^2 \theta + 4 \cos \theta \sin \theta + \sin^2 \theta] & [2 \cos^2 \theta - 2 \sin^2 \theta] \\ [2 \cos^2 \theta - 2 \sin^2 \theta] & [\cos^2 \theta - 4 \cos \theta \sin \theta + \sin^2 \theta] \end{pmatrix} = \begin{pmatrix} \lambda_1 & 0 \\ 0 & \lambda_2 \end{pmatrix}$$

(A1.35)

This equation is satisfied if the off-diagonal elements equal zero, that is, the desired value of θ is given by

$$\theta = 45°.$$

(A1.36)

Substituting this value of θ in the diagonal elements gives

$$\mathbf{D} = \begin{pmatrix} 3 & 0 \\ 0 & -1 \end{pmatrix}.$$

(A1.37)

In general, for a matrix

$$\mathbf{A} = \begin{pmatrix} a_{11} & a_{12} \\ a_{21} & a_{22} \end{pmatrix}$$

(A1.38)

$$\tan 2\theta = 2a_{12}/(a_{22} - a_{11})$$

(A1.39)

Now consider an $n \times n$ matrix $\mathbf{A}$, and let the elements a_{ii}, a_{ij}, a_{ji} and a_{jj} be isolated and operated on by the Jacobi procedure, such that $a_{ij} = a_{ji} = 0$. The appropriate θ rotation is given by (A1.39), with a transformation $\mathbf{Q}^{-1} \cdot \mathbf{A} \cdot \mathbf{Q}$ operating only on those elements at the intersection of the i, j rows and columns. Generally, the largest off-diagonal terms are then located and the procedure iterated until each off-diagonal term is less than a sufficiently low prescribed quantity.

A1.3.10.4   General matrix diagonalization

The program EIGN performs this diagonalization, and produces the eigenvalues and the corresponding eigenvectors. It can be incorporated into programs for other purposes, such as the Hückel molecular orbital calculations considered in the next program of this set.

In some problems that involve degeneracy among the eigenvalues, the Jacobi method may lead to ill-conditioning, and an alternative procedure is the Householder method of tridiagonalization. This technique is more complex, and the reader is referred to a specialized text.[1]

It should be noted that although the Jacobi procedure will operate on a non-symmetric matrix, the results will not be correct. Try, for example, the matrix $\begin{pmatrix} -32 & 66 \\ 15 & 31 \end{pmatrix}$ by the program and by the procedure of (A1.21) *et seq.*

A1.3.11   Hückel molecular orbital calculations

Hückel molecular orbital theory has been discussed sufficiently for our present purposes in section 2.10. The program HUCK has been built around the matrix diagonalization program, increased in matrix size, and is more convenient with problems on molecules.

It can be used for conjugated hydrocarbons by choosing $\beta_{ij} = 1$ where atoms i and j are bonded directly, but zero otherwise; all S_{ij} are taken to be zero (section 2.10). The program can be used with molecules containing heteroatoms by making changes to α and β; the following values have been recommended:[2]

X	$h(X)$	$h(X^-)$	$h(X^+)$	$k(C—X)$
C	0.0	—	—	1.0
N	0.5	1.5	2.0	1.0
O	1.0	2.0	2.5	0.8^a
Cl	2.0	—	—	0.4
CH$_3$	−0.5	—	—	1.0

$^a k(C{=}O) = 1.0$

The diagonal term of (2.160) may be written, generally, as $(\alpha + h\beta - E)/k\beta$, where $k\beta$ represents the off-diagonal non-zero terms; it reduces to the form of (2.162) for $h = 0$, $k = 1$.

It is possible for a program such as this to be enhanced by including calculations of bond order, charge density and so on, but to do so here would reduce the need of practising this type of calculation.

[1]See, for example, J. P. Lowe (1978) *Quantum Chemistry*, Academic Press.
[2]See A. Streitweiser (1961) *Molecular Orbital Theory for Organic Chemists*, Wiley.

A1.3.12   Point-group recognition

It has not been possible to include a discussion on symmetry within this book, although it is an important topic and we have referred to it from time to time. Nevertheless, it seems opportune to include in this appendix a point-group recognition program, developed by the author and used with success in teaching this subject over many years.[1]

Symmetry and its application in chemistry has been discussed by the author in another publication,[2] which includesa detailed treatment of symmetry, symmetry notations and the other information that would be needed in order to use this program profitably—it is essentially a teaching aid.

We consider a point-group recognition scheme based on a division of the 32 crystallographic point groups into four types. We use the Hermann–Mauguin point-group notation, and give a table to show the equivalent Schönflies symbols.

The first step in the scheme is to search the given model for the presence of a centre of symmetry or a mirror plane. They are easy to recognize: if a model with a centre of symmetry $\bar{1}$ (a cube, or a model of SF_6) is placed on a flat surface, it will have a similar face or plane of atoms uppermost and parallel to the supporting surface. If a model has a mirror plane m (a tetrahedron or a model of CH_4), a search is made for the object/mirror-image (right-hand/left-hand) relationship in the model. Some models will have both of these symmetry elements and others will have none of them. Thus, we can divide the 32 crystallographic point groups into four types, as follow:

> I: No m and no $\bar{1}$
>   1, 2, 222, 3, 32, 4, $\bar{4}$, 422, 6, 622, 23, 432
> II: m present, but no $\bar{1}$
> m, $mm2$, $3m$, $4mm$, $\bar{4}2m$, $\bar{6}$, $6mm$, $\bar{6}m2$, $\bar{4}3m$
> III: $\bar{1}$ present, but no m
> $\bar{1}$, $\bar{3}$
> IV: m and $\bar{1}$ both present
> $\dfrac{2}{m}$, mmm, $\bar{3}m\,\dfrac{4}{m}$, $\dfrac{4}{m}mm$, $\dfrac{6}{m}$, $\dfrac{6}{m}mm$, $m3$, $m3m$

The model should then be examined carefully, so as to identify the other symmetry elements present, and their relative orientations.

The identification proceeds systematically by means of the block diagram scheme in Fig. A1.3. Here R refers to the maximum degree of pure rotational symmetry in the model, and N is the number of such axes. Where diads (twofold rotation axes) are present with $R \geqslant 3$, there will be R of them. It should be noted that the program reports $\bar{R}$ as $-R$.

In the figure, questions are given in 'rectangovals', point groups in squares and error paths in diamonds. It will be noted that in types I, II and IV, the first three questions are mostly similar. The cubic point groups evolve from question 2 in I, II and IV.

[1]M. F. C. Ladd (1976) *International Journal of Mathematical Education in Science and Technology*, 7, 395.
[2]M. F. C. Ladd (1989; paperback 1992) *Symmetry in Molecules and Crystals* Ellis Horwood.

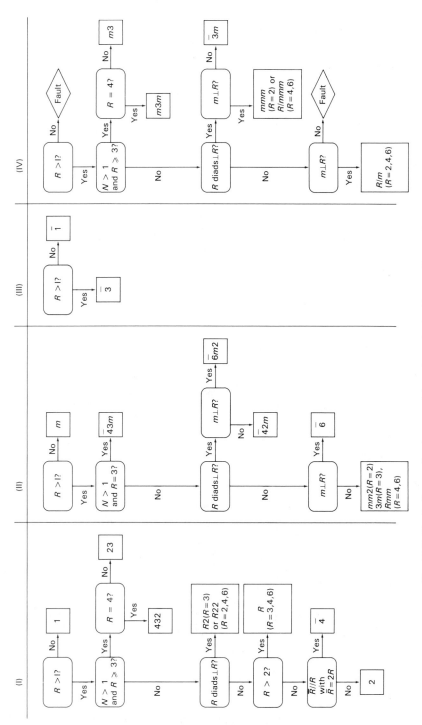

Fig. A1.3. Block diagram for the recognition scheme in the 32 crystallographic point groups.

Finally, we give in Table A1.1 a list of point groups in both the Hermann–Mauguin and Schönflies notations. Students of the physical sciences will inevitably encounter both of them.

Table   A1.1. Hermann–Mauguin   and   Schönflies   symmetry   notations

Schöenflies	Hermann–Mauguin[a]	Schöenflies	Hermann–Mauguin[a]
C_1	1	D_3	32
C_2	2	D_4	422
C_3	3	D_6	622
C_4	4	D_{2h}	mmm
C_6	6	D_{3h}	$\bar{6}m2$
C_i, S_2	$\bar{1}$	D_{4h}	$\dfrac{4}{m}\,mm$
C_s, S_1	$m\ (\bar{2})$		
S_6	$\bar{3}$		
S_4	$\bar{4}$	D_{6h}	$\dfrac{6}{m}\,mm$
C_{3h}	$\bar{6}$	D_{2d}	$\bar{4}2m$
C_{2h}	$2/m$	D_{3d}	$\bar{3}m$
C_{4h}	$4/m$	T	23
C_{6h}	$6/m$	T_h	$m3$
C_{2v}	$mm2$	O	432
C_{3v}	$3m$	T_d	$\bar{4}3m$
C_{4v}	$4mm$	O_h	$m3m$
C_{6v}	$6mm$	$C_{\infty v}$	∞m
D_2	222	$D_{\infty h}$	$\infty/m\ (\bar{\infty})$

[a] $2/m$ is an acceptable way of writing $\dfrac{2}{m}$, but $4/mmm$ is not as satisfactory as $\dfrac{4}{m}\,mm$.

A1.3.14   Curve fitting and interpolation

Another useful program, code name INTP, permits the fitting of a curve to a series of data points, and interpolation of function values lying within the extrema of the data.

Unique quadratic functions are fitted to n sample data points s_i, s_{i+1}, s_{i+2} ($i = 1, 2, 3, \ldots, n-2$), and first and second derivatives calculated for each function. Unique quintic functions are developed for the points s_i ($i = 1, 2, 3, \ldots, n - 1$), using the data $f_i, f_{i+1}, f'_i, f'_{i+1}, f''_i$ and f''_{i+1}. In this way, the separate curves are matched in slope and curvature. Sturm sequences are calculated to determine the number and position of any turning points and inflexion points in each polynomial. In the first and final ranges of the data, the end points are obtained perforce from the first and $(n-2)$th quadratic functions, and the quintic functions are degraded to quadratic status in these ranges.

PROBLEMS FOR APPENDIX 1

These problems have been designed around the programs described in this Appendix, but they can be solved by any other programs that perform the appropriate functions.

A1.1 In X-ray crystallography, the unit-cell dimension a for cubic crystals is given by

$$\sin^2 \theta_{hkl} = \lambda^2/4a^2(h^2 + k^2 + l^2) = \lambda^2 N/4a^2$$

where λ is the wavelength of the X-radiation, h, k and l are the (integer) Miller indices of a crystal plane, θ_{hkl} is the Bragg angle for that plane and N is an integer equal to $(h^2 + k^2 + l^2)$.

The following values of $\sin^2 \theta_{hkl}$ were obtained from the lines on an X-ray powder photograph of a cubic oxide, using Cu $K\alpha$ X-radiation, $\lambda = 0.15418$ nm:

$$0.1028 \quad 0.2735 \quad 0.3756 \quad 0.4102 \quad 0.5467 \quad 0.8205 \quad 0.9227$$

It follows from the equation above that these data have an integer relationship. By inspection, determine a common factor for the $\sin^2 \theta_{hkl}$ values and, hence, the value of N for each line. Obtain a best value for a by the method of least squares, and determine the estimated standard deviation $\sigma(a)$.

A1.2 Re-evaluate the standard molar entropy of nickel by Gaussian quadrature, using the data in problem 4.19. Apply the Debye T^3 approximation as before, and assume that between 283.0 K and 298.15 K, C_p remains at the value 26.09 J mol^{-1} K^{-1}.

A1.3 Magnesium fluoride has the Rutile structure type (see problem 4.11) with $a = 0.4623$ nm, $c = 0.3052$ nm and $x = 0.3030$. Determine the shortest Mg—F distance and, hence, evaluate the Madelung constant for the MgF_2 structure; use a summation limit of 3 with the calculation described in Appendix 13. Assuming $\rho/r_e = 0.15$ (larger than 0.1 for AX_2 structure types), where r_e is the shortest Mg—F distance, calculate the lattice energy for MgF_2. Evaluate, for comparison, the thermodynamic lattice energy of MgF_2, using the appropriate data from problem 4.4 and Tables 4.5 and 4.6, and with ΔH_f (MgF_2, s) $= -1102.5$ kJ mol^{-1}.

A1.4 (a) Calculate and plot the 1s radial functions R, R^2 and $4\pi r^2 R^2$ for hydrogen. At what values of r/a_0 do maxima occur in R, R^2 and $4\pi r^2 R^2$? Confirm your results from the radial wave function given in Table 2.3.

(b) Repeat for the hydrogen 3s R function. At what values of r/a_0 does R cross the abscissa? Confirm your results from the appropriate radial function.

A1.5 Calculate and plot the Maxwell-Boltzmann distribution of speeds $\Phi(v)$ for argon as a function of v, at 5 K, 50 K, 25 °C, 500 °C and 1000 °C. At each temperature, note the change in both the form of the curve and the values of v_{max}, $\bar{v}$ and $\sqrt{\bar{v^2}}$. Determine an equation relating v_{max} and T.

A1.6 Problem 2.25 can also be solved with the program BOXS, by a correct choice of input parameters and with a little manipulation of the output results. Carry out this calculation.

A1.7 Run the program MONC: for calculation (1) let N take the values 10, 100, 100 000. For (2), choose, say, argon ($M_r = 39.948$) and run for increasing N at different values of T.

A1.8 Determine the extent ($\sim 99\%$) of the $\psi^2(2p_z)$ atomic density function for carbon, outwards along its axis. To be precise, the value of Z should be modified by Slater's screening constant to give Z_{eff} as an input parameter. Evaluate the result also from the wave function itself, and compare the results with each other and with the van der Waal's radius for this species.

A1.9 (a) Solve the following matrices for their eigenvalues and eigenvectors:

$$(i) \begin{pmatrix} 1 & 3 \\ 3 & -2 \end{pmatrix} \quad (ii) \begin{pmatrix} 3 & 2 \\ -1 & 0 \end{pmatrix} \quad (iii) \begin{pmatrix} 1 & 2 & 0 \\ 2 & -1 & 2 \\ 0 & 2 & 2 \end{pmatrix}.$$

(b) Input the matrix for butadiene (section 2.10) to the program EIGN, and confirm the results given there for the eigenvalues (energies) and eigenvectors (coefficients of the wave functions).

A1.10 Solve methylene cyclopropene by the HMO method:

$$\underset{C_2}{\overset{C_3}{\underset{\diagup}{\big\|}}}\diagdown C_1 {=} C_4$$

From the results, calculate the total π energy, the delocalization π energy, the bond orders, the charge densities and the free-valence parameters. What can be deduced from the results?

A1.11 (a) Rework problem 2.28 on bicyclobutadiene, to obtain the eigenvalues, and calculate the delocalization energy. Examine also cyclobutadiene in the same way, and decide which of the two molecules is more stable.

(b) Carry out an HMO calculation for naphthalene, and determine the total energy, the delocalization energy and the bond orders. Construct a graph of bond order against bond length (see Fig. 2.40), and estimate the bond lengths in naphthalene.

(c) What would the delocalization energy and predicted bond lengths become in the naphthalene anion $C_{10}H_8^-$? Would it be more stable than naphthalene?

A1.12 Use the program SYMM on the following models, after having studied them carefully and identified the symmetry elements present:

Model number	Model
18	Tetrahedron, or CH_4
2	Cube, or SF_6
81	Beer mug (with handle), or CH_2FCl

A1.13 The following data were obtained experimentally for y as a function of x:

x	0	1	2	3	4	5	6	7	8
y	0.00	1.70	2.50	2.51	2.00	-0.85	-1.00	-0.70	-0.15

x	9	10	11	12	13	14	15	16
y	0.60	1.85	2.15	2.85	3.85	3.75	3.55	3.00

Fit a curve to the data points and, hence, determine the values of the function at $x = 2.5, 6.5$ and 11.5.

Appendix 2 Stereoviewing

The representation of crystal and molecular structures by stereoscopic pairs of drawings has become commonplace in recent years. Indeed, some very sophisticated computer programs have been written that draw stereoviews from crystallographic data. Several illustrations herein are presented as stereoviews. Two diagrams of a given object are necessary, and they must correspond to the views seen by the eyes in normal vision. Correct viewing requires that each eye sees only the appropriate drawing, and there are two ways in which it can be accomplished.

1. Two suppliers of stereoviewers are:

(a) C. F. Casella and Company Limited, Regent House, Britannia Walk, London N1 7ND, England. This maker supplies two grades of stereoscope.

(b) Taylor–Merchant Corporation, 212-T West 35th Street, New York, NY 10001, U.S.A.

A stereoviewer may be obtained by completing the card at the front of the book, and with it stereoscopic pairs of drawings may be viewed directly.

2. The unaided eyes can be trained to defocus, so that each eye sees only the appropriate diagram. The eyes must be relaxed, and look straight ahead. This process may be aided by placing a white card edgeways between the drawings so as to act as an optical barrier. When viewed correctly, a third (stereoscopic) image is seen in the centre of the given two views. It may be found helpful to open the eyes wide and then allow them to relax, without making any attempt to focus on the views.

Appendix 3 The hypsometric formula— an example of the Boltzmann distribution

Consider a rectangular column of an ideal gas of cross-sectional area A and height z, with reference to an origin O at ground level, at a uniform temperature T (Fig. A3.1). The mass of a gas molecule is m and, at the height z, let the pressure of the gas be p and let there be N molecules in a volume V. We need to determine how N varies with z. Since the gas is assumed to be ideal, no intermolecular attractions need be considered. Hence, from the gas laws,

$$pV = n\mathscr{R}T. \tag{A3.1}$$

The gas constant $\mathscr{R}$ is Lk_B, where L is the Avogadro constant and k_B is the Boltzmann constant. Since $nL = N$

$$p = Nk_B T/V. \tag{A3.2}$$

At a height $(z + dz)$ the pressure is $(p + dp)$. The gravitational force on the segment of width dz is $ADg\,dz$, where D is the density of the gas and g is the gravitational acceleration. The pressure difference across the segment is $(-dp/dz)\,dz$, the negative sign indicating that the gas pressure decreases in the positive direction of z. The hypsometric force on the segment is $-A(dp/dz)\,dz$, and at equilibrium the two forces are balanced:

$$-A(dp/dz)\,dz = ADg\,dz \tag{A3.3}$$

or

$$-dp = Dg\,dz. \tag{A3.4}$$

Equation (A3.4) might be obtained also from a definition of pressure.
From (A3.2), at a constant T and V

$$dp = k_B T/V\,dN \tag{A3.5}$$

and using the fact that $D = mN/V$, we have

$$\frac{dN}{N} = -mg\,dz/k_B T. \tag{A3.6}$$

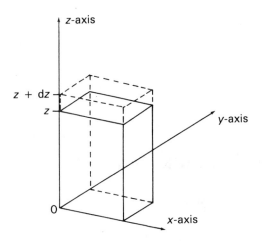

Fig. A3.1. Construction for the hypsometric formula.

On integration, we obtain

$$\ln N = -mgz/k_{\mathrm{B}} T + \text{constant}. \tag{A3.7}$$

At $z = 0$ let $N = N_0$. Thus the constant becomes $\ln N_0$ and

$$N = N_0 \exp(-mgz/k_{\mathrm{B}} T). \tag{A3.8}$$

Now, mgz is the gravitational potential energy per molecule of gas at the height z. Let U represent this potential energy per mole of gas. Then $U = Lmgz$, and

$$N = N_0 \exp(-U/\mathscr{R}T). \tag{A3.9}$$

As an example, consider air, of mean molar mass $0.0288 \text{ kg mol}^{-1}$, at a height of 10 km above ground level (the height of a transatlantic air liner) at 298 K. From (A3.9), N/N_0 is approximately 0.32 times its value at ground level; at 100 km it is only about 1.1×10^{-5} times its value at ground level.

Appendix 4 Gamma function

The gamma function is useful in handling integrals of the type

$$\int_0^\infty x^n \exp(-ax^2)\, dx \tag{A4.1}$$

where a is a constant; they occur in several areas of chemistry and chemical physics. The gamma function $\Gamma(n)$ may be represented by the integral equation

$$\Gamma(n) = \int_0^\infty t^{n-1} \exp(-t)\, dt \tag{A4.2}$$

The following particular results are important[1]:

(a) For $n > 0$ and integral,

$$\Gamma(n) = (n-1)! \tag{A4.3}$$

(b) For $n > 0$,

$$\Gamma(n+1) = n\Gamma(n) \tag{A4.4}$$

and if n is also integral,

$$\Gamma(n+1) = n! \tag{A4.5}$$

(c) $$\Gamma(1/2) = \sqrt{\pi}. \tag{A4.6}$$

As an example, we shall consider the solution of the integral

$$I = \int_0^\infty x^4 \exp(-x^2/2)\, dx. \tag{A4.7}$$

Let $x^2/2 = t$, so that $x = (2t)^{1/2}$ and $dx = (2t)^{-1/2}\, dt$. Then,

$$I = 2\sqrt{2} \int_0^\infty t^{3/2} \exp(-t)\, dt \tag{A4.8}$$

Hence,

$$I = 2\sqrt{2}\,\Gamma(\tfrac{5}{2}), \qquad \text{or } 3\sqrt{\pi/2}. \tag{A4.9}$$

---

[1] See, for example, H. Margenau and G. M. Murphy (1943) *The Mathematics of Physics and Chemistry*, van Nostrand.

Appendix 5 Spherical polar coordinates

A5.1   COORDINATES

The polar coordinates, r, θ and ϕ are defined by

$$x = r \sin \theta \cos \phi$$
$$y = r \sin \theta \sin \phi \qquad \text{(A5.1)}$$
$$z = r \cos \theta$$

where

$$r^2 = x^2 + y^2 + z^2. \qquad \text{(A5.2)}$$

A5.2   VOLUME ELEMENT, dτ

In normalization problems, we may need to express a volume element $d\tau$, $dx\, dy\, dz$ in Cartesian coordinates, in polar coordinates. Consider the volume element $d\tau$ shown in Fig. A5.1; it corresponds to the quantity $dx\, dy\, dz$. From the diagram, it is a straightforward matter to determine the magnitudes of the sides of the volume element, which may be taken to be parallelepipedal. Hence,

$$d\tau = r^2\, dr \sin \theta\, d\theta\, d\phi. \qquad \text{(A5.3)}$$

The integration limits of the variables, which correspond to x, y and z each between $-\infty$ and $+\infty$, are

$$0 \leqslant r \leqslant \infty$$
$$0 \leqslant \theta \leqslant \pi \qquad \text{(A5.4)}$$
$$0 \leqslant \phi \leqslant 2\pi.$$

It is seen easily that these limits define a sphere of infinite radius.

A5.3   LAPLACIAN OPERATOR ∇²

From the rules of partial differentiation, for any function $f(r)$, where $r = g(x, y, z)$, we can transform variables in the following manner:

$$\frac{\partial f}{\partial x} = \frac{\partial f / \partial r}{\partial x / \partial r} \qquad \text{(A5.5)}$$

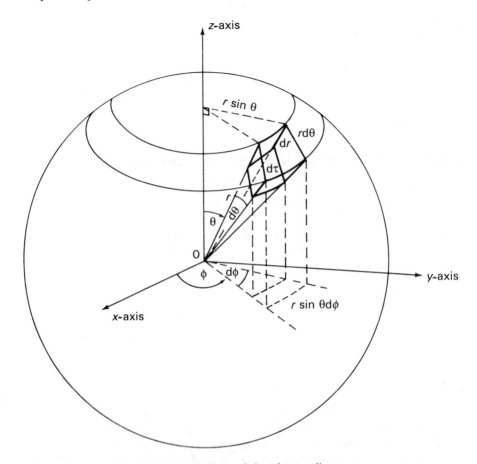

Fig. A5.1. Volume element $d\tau$ in polar coordinates.

and

$$\frac{\partial^2 f}{\partial x^2} = \left[\frac{1}{\partial x/\partial r}\right]\frac{\partial}{\partial r}\left[\frac{\partial f/\partial r}{\partial x/\partial r}\right]$$

$$= \frac{(\partial^2 f/\partial r^2)(\partial x/\partial r) - (\partial^2 x/\partial r^2)(\partial f/\partial r)}{(\partial x/\partial r)^3}. \tag{A5.6}$$

If $g(x, y, z)$ is given by (A5.2), $\partial x/\partial r = r/x$, and $\partial^2 x/\partial r^2 = (x^2 - r^2)/x^3$. Thus

$$\frac{\partial^2}{\partial x^2} = \frac{r}{x}\left(\frac{\partial^2}{\partial r^2}\right) + (r^2 - x^2)\left(\frac{\partial}{\partial r}\right). \tag{A5.7}$$

Equations for $\partial^2/\partial y^2$ and $\partial^2/\partial z^2$ are symmetrical; hence,

$$\frac{\partial^2}{\partial x^2} + \frac{\partial^2}{\partial y^2} + \frac{\partial^2}{\partial z^2} = r^{-3}\left[r^3\frac{\partial^2}{\partial r^2} + 2r^2\frac{\partial}{\partial r}\right] \tag{A5.8}$$

or

$$\nabla_r^2 = \frac{\partial^2}{\partial r^2} + \frac{2}{r}\frac{\partial}{\partial r} = \frac{1}{r^2}\frac{\partial}{\partial r}\left(r^2\frac{\partial}{\partial r}\right). \tag{A5.9}$$

In a similar manner, using (A5.1), we can show that

$$\frac{\partial x}{\partial\phi} = -r\frac{\sin\theta}{\sin\phi}$$

$$\frac{\partial^2 x}{\partial\phi^2} = r\frac{\sin\theta\cos\phi}{\sin^2\phi}$$

$$\frac{\partial y}{\partial\phi} = r\frac{\sin\theta}{\cos\phi} \tag{A5.10}$$

$$\frac{\partial^2 y}{\partial\phi^2} = r\frac{\sin\theta\sin\phi}{\cos^2\phi}$$

$$\frac{\partial z}{\partial\phi} = \frac{\partial^2 z}{\partial\phi^2} = 0$$

and, following (4.6),

$$\nabla_\phi^2 = \frac{1}{r^2\sin^2\theta}\frac{\partial^2}{\partial\phi^2}. \tag{A5.11}$$

It is left as an exercise to the reader to show that

$$\nabla_\theta^2 = \frac{1}{r^2}\frac{\partial^2}{\partial\theta^2} + \frac{\cos\theta}{r^2\sin\theta}\frac{\partial}{\partial\theta} \tag{A5.12}$$

whence

$$\nabla^2 = \nabla_r^2 + \nabla_\theta^2 + \nabla_\phi^2 = r^{-2}\left\{\frac{\partial}{\partial r}\left(r^2\frac{\partial}{\partial r}\right) + \frac{1}{\sin^2\theta}\frac{\partial^2}{\partial\phi^2} + \frac{1}{\sin\theta}\frac{\partial}{\partial\theta}\left(\sin\theta\frac{\partial}{\partial\theta}\right)\right\}. \tag{A5.13}$$

Appendix 6 Reduced mass

The calculation of a reduced mass is a central force problem in mechanics. We shall consider a two-particle problem in which the potential energy of the system is determined by the distance between the two particles.

Let two particles of masses m_1 and m_2 be vibrating along a line joining their centres. The speeds of the particles at any instant are v_1 and v_2, respectively, and the distance between them is r (Fig. A6.1); C is the centre of mass of the system.

The kinetic energy E_K of the system is given by

$$E_K = \tfrac{1}{2}m_1 v_1^2 + \tfrac{1}{2}m_2 v_2^2 \tag{A6.1}$$

and the centre of mass is defined by

$$m_1 r_1 + m_2 r_2 = 0. \tag{A6.2}$$

From Fig. A6.1, noting the positive direction of r

$$r = r_2 - r_1. \tag{A6.3}$$

Therefore,

$$m_1(r_2 - r) = -m_2 r_2 \tag{A6.4}$$

or

$$r_2 = m_1 r/(m_1 + m_2). \tag{A6.5}$$

Hence from (A6.3) and (A6.5),

$$r_1 = -m_2 r/(m_1 + m_2). \tag{A6.6}$$

Since

$$v_i = \frac{dr_i}{dt} = \dot{r}_i, \tag{A6.7}$$

$$E_K = \tfrac{1}{2}m_1 \dot{r}_1^2 + \tfrac{1}{2}m_2 \dot{r}_2^2 \tag{A6.8}$$

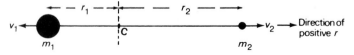

Fig. A6.1. Two-particle system vibrating about a centre of mass, C.

or

$$E_K = \tfrac{1}{2}m_1m_2r^2/(m_1 + m_2) + \tfrac{1}{2}m_1m_2r^2/(m_1 + m_2).$$ (A6.9)

Equation (A6.9) may be written

$$E_K = \mu\dot{r}^2$$ (A6.10)

where μ, given by,

$$\mu = m_1m_2/(m_1 + m_2)$$ (A6.11)

is the reduced mass of the system. Evidently, (A6.10) represents the kinetic energy of two particles of equal mass, μ, separated by a distance r, and vibrating about their centre of mass midway between them. Thus,

$$E_K = \tfrac{1}{2}\mu(\dot{r}/2)^2 + \tfrac{1}{2}\mu(\dot{r}/2)^2 = \mu\dot{r}^2.$$ (A6.12)

Appendix 7 Solution of a second-order differential equation

Consider the equation

$$\frac{d^2y}{dx^2} + k^2y = 0 \tag{A7.1}$$

where k^2 is a constant. Let d/dx be represented by D. Then

$$D^2y + k^2y = 0. \tag{A7.2}$$

Consider next the equation

$$[(D - p_1)(D - p_2)]y = 0 \tag{A7.3}$$

where p_1 and p_2 are constants. Expanding (A7.3)

$$D^2y - (p_1 + p_2)\, Dy + p_1 p_2 y = 0. \tag{A7.4}$$

Comparing (A7.2) and (A7.4), we see that they will be equivalent provided that $p_2 = -p_1$. Hence

$$p_1^2 = -k^2 \tag{A7.5}$$

and the two roots of (A7.5) are

$$p_1 = \pm ik. \tag{A7.6}$$

Taking the terms in (A7.3) in turn

$$(D - p_1)y = 0 \tag{A7.7}$$

or

$$dy/y = p_1\, dx. \tag{A7.8}$$

On integrating (A7.8) we obtain

$$\ln y = p_1 x + A \tag{A7.9}$$

where A is a constant; using (A7.6)

$$y = A\, \exp(ikx). \tag{A7.10}$$

In a similar manner, we have for the second root of (A7.5)

$$y = B \exp(-ikx) \tag{A7.11}$$

and the complete solution is then written as

$$y = A \exp(ikx) + B \exp(-ikx). \tag{A7.12}$$

It is a simple matter to show that double differentiation of (A7.12) leads to (A7.1).

Appendix 8 Separation of variables

Certain partial differential equations are termed *separable*, and may be solved by the following type of procedure. Consider a function

$$\Psi(x, y) = \psi(x)\psi(y). \tag{A8.1}$$

It is a simple matter to show that

$$\frac{\partial^2 \Psi(x, y)}{\partial x^2} = \psi(y)\frac{\partial^2 \psi(x)}{\partial x^2} \tag{A8.2}$$

and

$$\frac{\partial^2 \Psi(x, y)}{\partial y^2} = \psi(x)\frac{\partial^2 \psi(y)}{\partial y^2}. \tag{A8.3}$$

Following the form of (2.54), we may write

$$\psi(y)\frac{\partial^2 \psi(x)}{\partial x^2} + \psi(x)\frac{\partial^2 \psi(y)}{\partial y^2} = E\psi(x)\psi(y). \tag{A8.4}$$

Dividing by $\psi(x)\psi(y)$, we obtain

$$\frac{1}{\psi(x)}\frac{\partial^2 \psi(x)}{\partial x^2} + \frac{1}{\psi(y)}\frac{\partial^2 \psi(y)}{\partial y^2} = E. \tag{A8.5}$$

Since E is a constant, both terms on the left-hand side of (A8.5) are independently constant. Hence, we may write

$$\frac{\partial^2 \psi(x)}{\partial x^2} = E\psi(x) \tag{A8.6}$$

and

$$\frac{\partial^2 \psi(y)}{\partial y^2} = E\psi(y) \tag{A8.7}$$

whence

$$E\psi(x) + E\psi(y) = E\Psi(x, y). \tag{A8.8}$$

It is straightforward to introduce $-(h^2/8\pi m)$ into (A8.4) and show that the results of (2.54)–(2.57) can be so obtained.

Appendix 9 Least-squares line and propagation of errors

If it is desired to fit a straight-line relationship to a number of observations in excess of two, it is often appropriate to use the method of least squares. Let the equation be of the form

$$y = ax + b, \tag{A9.1}$$

where a and b are constants which have to be determined. For any observation i,

$$ax_i + b - y_i = e_i, \tag{A9.2}$$

where e_i is an error that will be assumed both to be random and to reside in the value of the dependent variable y_i, the error in the independent variable x_i being relatively negligible. According to the principle of least squares, the best values of a and b are chosen such that the sum of the squares of the errors e_i is a minimum. Thus

$$\mathrm{Min}\left(\sum_i e_i^2 \right) = \mathrm{Min}\left(\sum_i (ax_i + b - y_i)^2 \right). \tag{A9.3}$$

The required minimum value may be found by differentiating the right-hand side of (A9.3) partially with respect to both a and b, and setting each of the derivatives equal to zero. Hence

$$\frac{\partial \left(\sum_i e_i^2 \right)}{\partial a} = 2 \sum_i (ax_i^2 + bx_i - x_i y_i) = 0 \tag{A9.4}$$

and

$$\frac{\partial \left(\sum_i e_i^2 \right)}{\partial b} = 2 \sum_i (ax_i + b - y_i) = 0. \tag{A9.5}$$

Thus, we may derive

$$a[x^2] + b[x] - [xy] = 0, \tag{A9.6}$$

and

$$a[x] + bN - [y] = 0. \tag{A9.7}$$

Equations (A9.6) and (A9.7) are known as the normal equations and $[x]$, for example, means $\sum_i x_i$ over the number N of observations. If each observation has a weight w then the normal equations become

$$a[wx^2] + b[wx] - [wxy] = 0, \tag{A9.8}$$

and

$$a[wx] + b[w] - [wy] = 0. \tag{A9.9}$$

Solving for a and b,

$$a = ([w][wxy] - [wx][wy])/\Delta, \tag{A9.10}$$

and

$$b = ([wx^2][wy] - [wx][wxy])/\Delta, \tag{A9.11}$$

where Δ is given by

$$\Delta = [w][wx^2] - [wx][wx]. \tag{A9.12}$$

If all of the weights are unity, $[w] = N$.

The standard deviations in a and b may be estimated by the following procedure. From (A9.2),

$$[e^2] = \sum_i w_i(ax_i + b - y_i)^2. \tag{A9.13}$$

Then, we write, without proof here[1],

$$\sigma^2(a) = \{[e^2]/(N - 2)\}[w]/\Delta \tag{A9.14}$$

and

$$\sigma^2(b) = \{[e^2]/(N - 2)\}[wx^2]/\Delta, \tag{A9.15}$$

σ being an estimated standard deviation and σ^2 the corresponding variance.

It is recommended that the least-squares line be compared, where feasible, with a plot of the experimental x, y values. In the light of this inspection, certain observations may be reasoned to be unreliable. It must be remembered that a least-squares procedure will always give the best fit to the observations, including the bad ones.

A9.1　PROPAGATION OF ERRORS

The number of significant figures in a result is not necessarily similar to the number of significant figures in the data. Consider $y = p^n$, where $p = 2.0 \pm 0.1$. For $n = 0.1$, y lies between 1.066 and 1.077, whereas for $n = 4$, y lies between 13.0 and 19.4.

Consider any function $y = f(p)$ (Fig. A9.1). In the small interval δp, the change δy in y is given with good accuracy by

$$\delta y = \left(\frac{dy}{dp}\right)\delta p. \tag{A9.16}$$

---

[1] See, for example, E. T. Whittaker and G. Robinson (1949) *The Calculus of Observations*, Blackie.

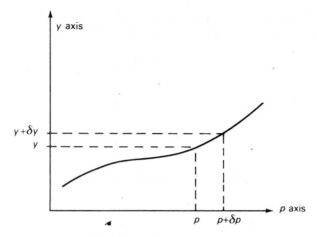

Fig. A9.1. A function $y = f(p)$.

Consider next any function $y = f(p_1, p_2)$ where p_1 and p_2 are independent variables. For two small independent changes δp_1 and δp_2, the changes in y are given by analogy with (A9.16) by

$$(\delta y)_{p_1} = \left(\frac{\partial y}{\partial p_1}\right)\delta p_1 \qquad\qquad (\text{A9.17})$$

and

$$(\delta y)_{p_2} = \left(\frac{\partial y}{\partial p_2}\right)\delta p_2. \qquad\qquad (\text{A9.18})$$

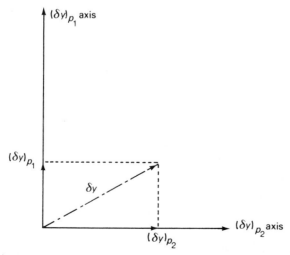

Fig. A9.2. Representation of the uncorrelated errors $(\delta y)_{p_1}$ and $(\delta y)_{p_2}$.

Since we have assumed that these two variations in y are uncorrelated, they can be represented along two rectangular axes (Fig. A9.2). Hence

$$(\delta y)^2 = (\delta y)_{p_1}^2 + (\delta y)_{p_2}^2 = \left(\frac{\partial y}{\partial p_1}\right)^2 (\delta_{p_1})^2 + \left(\frac{\partial y}{\partial p_2}\right)^2 (\delta_{p_2})^2. \qquad \text{(A9.19)}$$

Generalizing for a function $y = f(p_j)(j = 1, 2, 3, \ldots, n)$:

$$(\delta y)^2 = \sum_{j=1}^{n} \left(\frac{\partial y}{\partial p_j}\right)^2 (\delta p_j)^2. \qquad \text{(A9.20)}$$

The quantity δy can be equated to the standard deviation in y, $\sigma(y)$.

Appendix 10 Overlap integrals

In applying the variation principle (section 2.7), we encountered the overlap integral $\int \psi_i \psi_j \, d\tau$. In this appendix, we will carry out the evaluation of this quantity in a fairly simple case. Consider the overlap between two hydrogen 1s atomic orbitals, 1 and 2. Taking each AO to be of the form $\psi = 1/\sqrt{(\pi a_0^3)}\exp(-r/a_0)$, the overlap integral $\int \psi_1 \psi_2 \, d\tau$ is given by

$$S = \frac{1}{\pi a_0^3} \int_0^\infty \int_0^\pi \int_0^{2\pi} \exp[-(r_1 + r_2)/a_0] r^2 \, dr \sin \theta \, d\theta \, d\phi. \qquad (A10.1)$$

The overlap system has cylindrical symmetry, and it is convenient to transform the integral to spheroidal coordinates μ, v, ϕ given by

$$\mu = \frac{r_1 + r_2}{R} \qquad v = \frac{r_1 - r_2}{R} \qquad \phi = \phi \qquad (A10.2)$$

where R is the internuclear distance in units of a_0, that is, $R = r/a_0$; μ ranges from 1 to ∞, v ranges from -1 to $+1$ and ϕ ranges from 0 to 2π. The volume element $d\tau$ becomes $(R^3/8)(\mu^2 - v^2) \, d\mu \, dv \, d\phi$.[1] Then (A10.1) may be rewritten, taking $Z = 1$, as

$$S = \frac{R^3}{8\pi} \int_0^{2\pi} d\phi \int_{-1}^1 \int_1^\infty \exp(-R\mu)(\mu^2 - v^2) \, dv \, d\mu. \qquad (A10.3)$$

Integrating first over ϕ leads to:

$$S = \frac{R^3}{4} \int_{-1}^1 dv \int_1^\infty \exp(-R\mu)(\mu^2 - v^2) \, d\mu. \qquad (A10.4)$$

Expanding (A10.4):

$$S = \frac{R^3}{4} \int_1^\infty 2\mu^2 \exp(-R\mu) \, d\mu - \int_{-1}^1 v^2 \, dv \int_1^\infty \exp(-R\mu) \, d\mu \qquad (A10.5)$$

which reduces to

$$S = \frac{R^3}{4} \int_1^\infty 2\mu^2 \exp(-R\mu) \, d\mu - \frac{2}{3} \int_1^\infty \exp(-R\mu) \, d\mu. \qquad (A10.6)$$

Using the integration formula given in the solution to problem 2.12 now leads readily to the result

$$S = \exp(-R)(1 + R + R^2/3). \qquad (A10.7)$$

---

[1]The formulation of $d\tau$ is discussed exhaustively in H. Eyring, J. Walter and G. E. Kimball (1963) *Quantum Chemistry*, Wiley and, with a slight change in notation, in J. C. Slater (1963) *Quantum Theory of Molecules and Solids*, Volume 1, McGraw-Hill.

Appendix 11 Equation of state for a solid

The Helmholtz free energy A is defined by

$$A = U - TS. \tag{A11.1}$$

Since A is an extensive property,

$$dA = dU - T\,dS - S\,dT. \tag{A11.2}$$

At constant temperature, and since $T\,dS = dU + p\,dV$, we find

$$dA = -p\,dV \tag{A11.3}$$

or

$$\left(\frac{\partial A}{\partial V}\right)_T = -p. \tag{A11.4}$$

From (A11.2), we obtain

$$\left(\frac{\partial A}{\partial V}\right)_T = \left(\frac{\partial U}{\partial V}\right)_T - T\left(\frac{\partial S}{\partial V}\right)_T. \tag{A11.5}$$

Using Maxwell's relations[1]

$$\left(\frac{\partial S}{\partial V}\right)_T = \left(\frac{\partial p}{\partial T}\right)_V \tag{A11.6}$$

and furthermore,

$$\left(\frac{\partial p}{\partial T}\right)_V \left(\frac{\partial T}{\partial V}\right)_p \left(\frac{\partial V}{\partial p}\right)_T = -1 \tag{A11.7}$$

or

$$\left(\frac{\partial p}{\partial T}\right)_V = -\left(\frac{\partial V}{\partial T}\right)_p \div \left(\frac{\partial V}{\partial p}\right)_T. \tag{A11.8}$$

From the thermodynamic definitions of expansivity α and of isothermal compressibility κ, it follows that the right-hand side of (A1.18) is α/κ, and we obtain now,

$$\left(\frac{\partial A}{\partial V}\right)_T = \left(\frac{\partial U}{\partial V}\right)_T - \frac{T\alpha}{\kappa} \tag{A11.9}$$

[1]See, for example. K. Denbigh (1966) *The Principles of Chemical Equilibrium*, CUP.

and, with (A11.4),

$$\left(\frac{\partial U}{\partial V}\right)_T = -p + \frac{T\alpha}{\kappa} \tag{A11.10}$$

which is a thermodynamic equation of state for a solid.

Appendix 12   Numerical integration

A numerical integration procedure computes the value of a definite integral from a set of values of the integrand. When these values are ordinates of a curve, the integration defines the area under the curve; we meet numerical integration often in this form.

It may happen that a curve can be simulated by a least-squares fit to an appropriate function. Then there is no problem, because we can obtain an analytical solution from the fitted function. Suppose that the curve in Fig. A12.1 can be fitted by the function

$$y = 2x^2 + 3x + 2. \tag{A12.1}$$

Then

$$\int_0^{0.8} y \, dx = 2x^3/3 + 3x^2/2 + 2x \Big|_0^{0.8} = 2.9013 \ldots \tag{A12.2}$$

A12.1   NUMERICAL AND OTHER METHODS

It may not be possible to deduce a function which will represent satisfactorily a set of experimental measurements, and it becomes necessary to adopt another procedure. Two rather different methods will be described here.

Some numerical integration techniques depend on the relationship

$$\int_a^c f(x) \, dx = \int_a^b f(x) \, dx + \int_b^c f(x) \, dx. \tag{A12.3}$$

One such method is embodied in Simpson's rule.

A12.1.1   Simpson's rule
In this procedure a curve such as that in Fig. A12.1 is divided into an EVEN number n of intervals of equal width h. Then Simpson's rule, given without proof here, states that for any function $y = f(x)$

$$\int_{x_0}^{x_n} y \, dx = \int_{x_0}^{x_0 + nh} y \, dx$$

$$= (h/3)[y_0 + y_n + 2(y_2 + y_4 + y_6 + \cdots + y_{n-2})$$
$$+ 4(y_1 + y_3 + y_5 + \cdots + y_{n-1})]. \tag{A12.4}$$

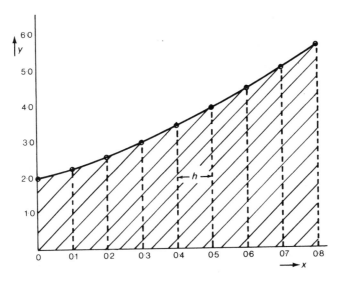

Fig. A12.1. A function, $y = f(x)$.

Example
Let the curve in Fig. A12.1 be divided into 8 intervals of width 0.1 units. Then, from the curve, we have the following data:

x	0.0	0.1	0.2	0.3	0.4	0.5	0.6	0.7	0.8
y	2.0	2.3	2.7	3.1	3.5	4.0	4.5	5.1	5.7

Using (A12.4),

$$\int_0^{0.8} y \, dx = (0.1/3)[2.0 + 5.7 + 2(2.7 + 3.5 + 4.5)$$

$$+ \, 4(2.3 + 3.1 + 4.0 + 5.1)] = 2.903, \qquad \text{(A12.5)}$$

which agrees with the analytical result to within 0.06%.

A12.1.2   Direct weighing (analogue) method
The curve in Fig. A12.1 is drawn on good-quality graph paper. The shaded area is cut out carefully and weighed; let its mass be 0.2759 g. A certain known number of squares are cut out and weighed so as to provide a calibration relationship between mass and area. Let 500 squares have a mass of 4.7563 g and, using the graph scales, let 1 square be equal to 0.1 units of area. Then

$$\frac{\text{Mass of curve}}{\text{Area under curve}} = \frac{\text{Mass of 500 squares}}{500 \times \text{scale}}. \qquad \text{(A12.6)}$$

Hence,

$$\text{Area under curve} = 50 \times 0.2759/4.7563 = 2.90, \qquad (A12.7)$$

which is within 0.05% of the analytical result.

Problems requiring numerical integration arise, for example, in evaluating thermodynamic functions such as ΔH or S (see also Appendix A1.3.3).

Appendix 13 Calculation of the Madelung constant

We have discussed the Madelung constant in Chapter 4, and it is useful to have a simple method of calculating it for any lattice array, to a precision that is sufficiently high for most purposes. The method given here[1] has been applied successfully to a wide range of compounds, and has been programmed for IBM-compatible personal computers. It assumes spherical charge distributions, with densities that are a linear function of the radial coordinate in reciprocal space.

The Madelung constant $\mathscr{A}$ is given by

$$\mathscr{A} = \frac{(g - Q)D}{RZ} \sum_j q_j^2 - \frac{\pi R^2 D}{ZV} \sum_{\mathbf{h}} |F_{\mathbf{h}}|^2\, \phi(\mathbf{h}) \tag{A13.1}$$

where the terms have the following meanings:

g $= 26/35$

Q a correction for termination of the $\mathbf{h}$ series (as below)

D a standard distance in the structure, often the nearest neighbour distance

R an arbitrary distance less than half (0.495) the nearest neighbour distance

Z the number of formula-entities in the unit cell

q_j the charge, including sign, of the jth atom species

V the unit-cell volume

h the magnitude of the reciprocal lattice vector $\mathbf{h}$

$\phi(\mathbf{h})$ $= 288(\alpha \sin \alpha + 2 \cos \alpha - 2)^2\, \alpha^{-10}$

α $= 2\pi h R$

$F_{\mathbf{h}}$ $= \sum_j q_j \exp(i2\pi\, \mathbf{h} . \mathbf{r}_j)$—it is the crystallographic structure factor for point atoms of form factors q_j.

The sums over j include all atoms in the unit cell, and the sum over $\mathbf{h}$ includes all reciprocal lattice vectors hkl in a sphere of radius α; h, k and l are the components of $\mathbf{h}$ with respect to the reciprocal lattice axes. The series termination correction Q depends

[1] F. Bertaut (1952) *Journal of Physics Radium* **13**, 499; D. H. Templeton (1955) *Journal of Chemical Physics*, **23**, 1629; R. E. Jones and D. H. Templeton (1956) *Journal of Chemical Physics*, **25**, 1062.

on the radius α according to the following table:

α	Q
2π	0.00030
3π	0.000090
4π	0.000012
5π	0.0000057

Termination of the series at $\alpha = 2\pi$ including the correction term Q gives results better than 0.02%, which is sufficient to match other terms in many lattice energy calculations.

Appendix 14   Some thermodynamics of solutions

In describing the state of a solution, or indeed that of any other system, two kinds of property are used. There are intensive properties such as density, viscosity and refractive index, that are independent of the amount of substance considered, and there are extensive properties like volume, energy and entropy, which do depend on the amount of substance under consideration. It may be noted that while volume, for example, is an extensive property, volume per unit mass (density) is an intensive property. The extensive properties of a given amount of solution can be measured, often directly, and in our thermodynamic investigation of solubility we need to know how some of these properties vary with composition.

A14.1 PARTIAL MOLAR VOLUME

We wish to study open systems, those in which composition is a variable. Introductory courses in thermodynamics are concerned largely with closed systems, those in which the composition of the system is constant, so that the thermodynamics of partial molar properties is perhaps the less familiar.

Consider a solution consisting of n_A mole of a solvent species A and n_B mole of a solute species B. On addition of an amount dn_A mole of A the volume increase will be dV; then we may write

$$\frac{dV}{dn_A} = \bar{V}_A \qquad\qquad (A14.1)$$

and $\bar{V}_A$ is the partial molar volume of the solvent in the solution.

Similarly, we write

$$\frac{dV}{dn_B} = \bar{V}_B, \qquad\qquad (A14.2)$$

for the partial molar volume of the solute in the solution.

A partial molar quantity is a property of the solution as a whole and not solely of the component in question. Thus $\bar{V}_B$ is the change in total volume when solute molecules are added. An important part of this change is concerned with the packing of adjacent solvent molecules and therefore with the forces between the species in solution.

In order to assist in the appreciation of a partial molar property, Fig. A14.1 is given as an illustration of the partial molar volume of $CaCl_2$ in water, $\bar{V}(CaCl_2)$, as a function of the number of moles of $CaCl_2$, $n(CaCl_2)$, at 298.15 K.

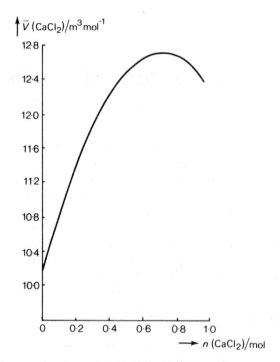

Fig. A14.1. Partial molar volume of $CaCl_2$ in water at 298.15 K as a function of moles of $CaCl_2$.

A14.2 PARTIAL MOLAR ENTROPY

In studying solubility in Chapter 4, we are concerned with the change in free energy per mole of solute $\Delta G_d^{\ominus}$ in establishing an equilibrium state, starting with the solid and hydrated ions in their standard states. We show how $\Delta G_d^{\ominus}$ is divided between the enthalpy of solution to infinite dilution $\Delta H_d^{\ominus}$ and the corresponding entropy change, which includes the sum of the partial molar entropies of the hydrated ions; the reader may wish to refer to Fig. 4.21 and (4.82). The process of dissolution involves changes in the solute, the solvent and the solution: we shall find that the term $\sum_i \bar{S}_i^{\ominus}$ may be either positive or negative for different pairs of ions.

A.14.3 MEASUREMENT OF THE PARTIAL MOLAR ENTROPY OF AN ION

The partial molar entropy of hydrated ions may be determined from measurements of the temperature variation of emf in a suitable galvanic cell. We shall consider $ZnCl_2$ as a solute, and the following cell could be set up:

$$Zn \,|\, ZnCl_2(c) \,|\, HCl(a_{\pm} = 1) \,|\, H_2(1\ atm),\ Pt, \qquad\qquad (A14.3)$$

for which the spontaneous cell reaction is

$$Zn(s) + 2H^+(aq) \rightarrow Zn^{2+}(aq) + H_2\ (g). \qquad\qquad (A14.4)$$

The emf is measured at several values of the concentration c and extrapolated, conveniently against $\sqrt{c}$, to zero concentration (infinite dilution). These measurements are repeated at three or four temperatures between about 288 and 308 K so as to obtain $dE^{\ominus}/dT$; since

$$\Delta S^{\ominus} = n\mathscr{F}\,\frac{dE^{\ominus}}{dT}, \qquad\qquad (A14.5)$$

we can obtain $\Delta S^{\ominus}$ for reaction (A14.4): n is the number of electrons involved in the reaction and $\mathscr{F}$ is the Faraday constant. But $\Delta S^{\ominus}$ is given also by

$$\Delta S^{\ominus} = \{\bar{S}^{\ominus}(Zn^{2+}) + S^{\ominus}(H_2)\} - \{S^{\ominus}(H^+) + \bar{S}^{\ominus}(Zn)\}. \qquad\qquad (A14.6)$$

where the thermodynamic states given above are assumed. By convention, we put $\bar{S}^{\ominus}(H^+) = 0$. Hence

$$\bar{S}^{\ominus}(Zn^{2+}) = \Delta S^{\ominus} + S^{\ominus}(Zn) - S^{\ominus}(H_2). \qquad\qquad (A14.7)$$

In a typical experiment, $dE^{\ominus}/dT$ (at 298.15 K) was found to be $- 1.00 \times 10^{-4}$ V K^{-1}; $S^{\ominus}(Zn) = 41.6$ J mol^{-1} K^{-1} and $S^{\ominus}(H_2) = 130.6$ J mol^{-1} K^{-1}. From (A14.5), $\Delta S^{\ominus} = - 19.30$ J mol^{-1} K^{-1} and, thus, $\bar{S}^{\ominus}(Zn^{2+}) = 108.3$ J mol^{-1} K^{-1}. This result is, of course, relative to the chosen value of $\bar{S}^{\ominus}(H^+)$ as zero.

A14.4   GENERALIZED DESCRIPTION OF PARTIAL MOLAR
QUANTITIES

Any extensive property, X, is determined by both the state of the system, and the amounts of substances present. Thus

$$X = f(T, p, n_i) \qquad (i = 1, 2, \ldots, N\,). \qquad\qquad (A14.8)$$

Hence,

$$dX = \left(\frac{\partial X}{\partial T}\right)_{p,\,n_i} dT + \left(\frac{\partial X}{\partial p}\right)_{T,\,n_i} dp + \sum_{i=1}^{N} \left(\frac{\partial X}{\partial n_i}\right)_{T,\,p,\,n_j} dn_i\ (i = 1, 2, \ldots, N; j \neq i).$$

$$(A14.9)$$

The derivative $(\partial X/\partial n_i)_{T,\,p,\,n_j}$ is the partial molar property $\bar{X}_i$ for the component i in the whole system.

If the extensive property is the Gibbs free energy, the partial molar property is identical with the chemical potential. Thus

$$\left(\frac{\partial G}{\partial n_i}\right)_{T,\,p,\,n_j} = \bar{G}_i = \mu_i. \qquad\qquad (A14.10)$$

Differentiating (A14.10) with respect of T, we write

$$\left(\frac{\partial^2 G}{\partial n_i\, \partial T}\right)_{p,\,n_j} = \left(\frac{\partial \mu_i}{\partial T}\right)_{p,\,n_j}. \tag{A14.11}$$

Since we have

$$\left(\frac{\partial G}{\partial T}\right)_p = -S, \tag{A14.12}$$

differentiating with respect to n_i gives

$$\left(\frac{\partial^2 G}{\partial T\, \partial n_i}\right)_{p,\,n_j} = -(\partial S/\partial n_i)_{p,\,n_j} = -\bar{S}_i. \tag{A14.13}$$

Since the order of differentiation in (A14.11) and (A14.13) is immaterial,

$$\left(\frac{\partial \mu_i}{\partial T}\right)_{p,\,n_j} = \bar{S}_i. \tag{A14.14}$$

Using $G = H - TS$ and differentiating with respect to n_i we have

$$\bar{G}_i = \bar{H}_i - T\bar{S}_i = \mu_i. \tag{A14.15}$$

Using (A14.14) and rearranging, we obtain

$$\mu_i - T\left(\frac{\partial \mu_i}{\partial T}\right)_{p,\,n_j} = \bar{H}_i. \tag{A14.16}$$

We have now developed general expressions for the partial molar entropy, partial molar free energy and partial molar enthalpy.

A14.5   PARTIAL MOLAR ENTHALPY IN SOLUTIONS

In an equilibrium between a solute and its solution, the solid and the solute in the saturated solution are at the same chemical potential. Hence, for any species i,

$$\mu_i = \mu_i^{\ominus} + \mathscr{R}T \ln a_i \tag{A14.17}$$

or

$$\mathscr{R} \ln a_i = \mu_i/T - \mu_i^{\ominus}/T. \tag{A14.18}$$

We should remember that quantities such as a_i, the activity of a single ionic species, are not measurable experimentally; nevertheless, it is convenient to discuss them as though they were. Under the conditions of constant pressure and composition

$$\mathscr{R}\left\{\frac{\partial(\ln a_i)}{\partial T}\right\}_{p,\,n_j} = \left\{\frac{\partial(\mu_i/T)}{\partial T}\right\}_{p,\,n_j} - \left\{\frac{\partial(\mu_i^{\ominus}/T)}{\partial T}\right\}_{p,\,n_j}. \tag{A14.19}$$

From (A14.16), dividing by T^2,

$$\frac{\mu_i}{T^2} - \frac{1}{T}\left(\frac{\partial u_i}{\partial T}\right)_{p,\,n_j} = \frac{\bar{H}_i}{T^2}, \tag{A14.20}$$

or

$$\left\{\frac{\partial(\mu_i/T)}{\partial T}\right\}_{p, n_j} = -\frac{\bar{H}_i}{T^2}.$$ (A14.21)

Hence from (A14.19)

$$\left\{\frac{\partial(\ln a_j)}{\partial T}\right\}_{p, n_j} = \frac{H_i^\ominus - \bar{H}_i}{\mathscr{R}T^2};$$ (A14.22)

$\bar{H}_i$ is the partial molar enthalpy of the ith constituent in the solution, and $H_i^\ominus$, equivalent here to $\bar{H}_i^\ominus$, is the corresponding value in the pure state of the ith constituent. We know that

$$a_i = c_i f_i,$$ (A14.23)

where f_i is the molar activity coefficient for the ith species; but concentrations c_i in mol dm^{-3}, the usual units of solubility, are not independent of temperature. So we may write, from (A14.23),

$$\left\{\frac{\partial(\ln f_i)}{\partial T}\right\}_{p, n_j} = \frac{H_i^\ominus - \bar{H}_i}{\mathscr{R}T^2} + \left\{\frac{\partial[\ln(D_0/D)]}{\partial T}\right\}_{p, n_j},$$ (A14.24)

where D_0 and D are the densities of the solvent and solution, respectively; c_i is constant under differentiation with respect to T.

A14.6   STANDARD STATE FOR SOLUTIONS

The standard state of a solution is based on the concept of unit activity, and there is freedom in its choice. The standard state should be convenient for the given application, and capable of forming a basis for comparison of different electrolyte solutions. It is permissible to choose different standard states for a given substance in two phases at equilibrium with each other. The equilibrium constant will then be altered in value, but not in its constancy, at a given temperature. However, when different standard states are chosen for two such substances, their activities are not equal, although their chemical potentials must be identical because they are in equilibrium. Thus, for any species in phase I,

$$\mu_I = \mu_I^\ominus + \mathscr{R}T \ln a_I,$$ (A14.25)

and for phase II in equilibrium with phase I

$$\mu_{II} = \mu_{II}^\ominus + \mathscr{R}T \ln a_{II}.$$ (A14.26)

Now $\mu_I = \mu_{II}$; hence

$$\mu_I^\ominus + \mathscr{R}T \ln a_I = \mu_{II}^\ominus + \mathscr{R}T \ln a_{II}.$$ (A14.27)

If the standard states are one and the same, $\mu_I^\ominus = \mu_{II}^\ominus$ and $a_I = a_{II}$.

The standard state for solutions is the infinitely dilute solution, the activity a of the solute in solution being defined such that the ratio $a/c \to 1$ as $c \to 0$, c being the concentration of the solution in mol dm^{-3} (see also Appendix 15). The standard state is hypothetical: it corresponds to a solution of concentration 1 mol dm^{-3} in which, from (A14.24) since $D \to D_0$ as $c \to 0$ and $f_i \to 1$, the partial molar enthalpy of the solute in the standard state, $\overline{H}_i$, has the same value as in the infinitely dilute solution, $H_i^{\ominus}$. Thus, in discussing solubility, it is convenient to refer $\Delta H_d^{\ominus}$ to an infinitely dilute solution, and $\Delta G_d^{\ominus}$ to a hypothetical solution of unit activity, knowing that these descriptions apply, in our context, to one and the same standard state.

In practice, we are concerned with the measurable quantity, the mean activity $a_{\pm}$, but we may use a similar definition of the standard state. The reader may wish to refer to Appendix 15 for the definition of mean activity properties.

Consider next the equilibrium

$$\text{NaCl(s)} \; \rightleftharpoons \; \text{Na}^+ \text{(aq)} + \text{Cl}^- \text{(aq)} \qquad \text{(A14.28)}$$
$$\text{(saturated solution)}$$

The standard state is a hypothetical solution of unit mean concentration $c_{\pm}$ and unit mean activity coefficient $f_{\pm}$. This choice has the required property that

$$a_{\pm}(\text{NaCl}) = 1 = c^2 f_{\pm}^2, \qquad \text{(A14.29)}$$

where c, the stoichiometric concentration, has, in this example, the same value as $c_{\pm}$. The concentration of each ionic species is also unity. Hence, the condition

$$f_{\pm} = f_+ = f_- \qquad \text{(A14.30)}$$

holds, and permits us to write

$$\mu^{\ominus}(\text{NaCl}) = \mu^{\ominus}(\text{Na}^+) + \mu^{\ominus}(\text{Cl}^-). \qquad \text{(A14.31)}$$

Now consider an unsymmetrical electrolyte, such as MgCl_2, in saturated solution of stoichiometric concentration c:

$$\text{MgCl}_2\text{(s)} \; \rightleftharpoons \; \text{Mg}^{2+}\text{(aq)} + 2\text{Cl}^-\text{(aq)} \qquad \text{(A14.32)}$$
$$\text{(saturated solution)}$$

The equation

$$\mu(\text{MgCl}_2) = \mu^{\ominus}(\text{MgCl}_2) + \mathscr{R}T \ln a(\text{MgCl}_2) \qquad \text{(A14.33)}$$

requires that, in the standard state,

$$a_{\pm}(\text{MgCl}_2) = 1 = 4c^3 f_{\pm}^3. \qquad \text{(A14.34)}$$

The standard state refers to unit mean concentration $4c^3$ and unit mean activity coefficient. The concentration of MgCl_2 in the standard state is $4^{-1/3}$, that of Mg^{2+} being $4^{-1/3}$ and that of Cl^-, $2 \times 4^{-1/3}$. It is clear that $c_{\pm}^3 = (4^{-1/3}) \times (2 \times 4^{-1/3})^2 = 1$. Apparently the standard state for Cl^- is different in the NaCl and MgCl_2 solutions in the same concentration terms. The following argument may be used in order to combat this apparent inconsistency.

Let 1 mol of Mg^{2+} be concentrated from the hypothetical solution of concentration $4^{-1/3}$ to a new solution of unit concentration, while 2 mol of Cl^- are diluted from the hypothetical solution of concentration $(2 \times 4^{-1/3})$ to a new solution of unit concentration. Both of the new solutions will be deemed to obey the requirement that $f(\text{Mg}^{2+}) = f(\text{Cl}^-) = 1$. Hence,

$$\Delta G(\text{Mg}^{2+}) = -\mathscr{R}T \ln(4^{-1/3}) = \tfrac{1}{3}\mathscr{R}T \ln(4), \qquad \text{(A14.35)}$$

and

$$\Delta G(\text{Cl}^-) = -2\mathcal{R}T \ln(2 \times 4^{-1/3}) = \tfrac{2}{3}\mathcal{R}T \ln(4) - 2\mathcal{R}T \ln(2). \quad \text{(A14.36)}$$

The total free energy change is evidently zero, and we can write

$$\mu^{\ominus}(\text{MgCl}_2) = \mu^{\ominus}(\text{Mg}^{2+}) + 2\mu^{\ominus}(\text{Cl}^-) \quad \text{(A14.37)}$$

to compare with (A14.31). As long as we refer to the standard state of unit concentration *and* unit activity coefficient, we can compare electrolyte solutions on a common thermodynamic basis.

Appendix 15 Debye–Hückel limiting law

A strong electrolyte in aqueous solution, while fully dissociated, may not behave as though the concentration of free ions is equal to the corresponding stoichiometric concentration. On dissolution in water the ions in an electrolyte become hydrated: they become attached, albeit loosely, to a number of water molecules in a hydration sphere, and the ionic charge is, to some extent, distributed over this sphere. Positive and negative hydrated ions attract one another electrostatically, and every hydrated ion may be regarded as being surrounded by oppositely charged species, so forming an ionic 'atmosphere'. The hydrated ions cluster and disperse dynamically, but over a period of time which is long in comparison with the lifetime of any cluster, there will be a certain fraction of the total stoichiometric ionic concentration that is unavailable as free ions. This effect is expressed by the activity a of a species, defined such that

$$a_i = c_i f_i, \tag{A15.1}$$

where c_i and f_i are the concentration and activity coefficient of the ith species. It is further defined that

$$\underset{c_i \to 0}{\text{Limit}}\; f_i = 1. \tag{A15.2}$$

In many thermodynamic arguments, particularly those involving strong electrolytes, it is necessary to know the value of a or f. Although single-ion activities cannot be measured, the Debye–Hückel theory of strong electrolytes leads to an approximate equation for the calculation of the activity coefficient of a single ion. Without proof here, we write

$$\ln f_i = -A q_i^2 \sqrt{I}; \tag{A15.3}$$

A is given by

$$A = 1.8247 \times 10^6/(\varepsilon_r T)^{3/2}, \tag{A15.4}$$

and is equal to 1.186 at 298.15 K; ε_r is the relative permittivity of the solvent, T is the absolute temperature, q_i is the numerical charge on the ion i, and I is the total ionic strength of the solution given by

$$I = \tfrac{1}{2} \sum_i c_i q_i^2 . \tag{A15.5}$$

Equation (A15.3) provides satisfactory values of f_i provided that $I \leqslant 0.02$, for which reason it is referred to as the Debye–Hückel limiting law.

The measurable activity properties are the mean activity $a_\pm$ and the mean activity coefficient $f_\pm$. For a generalized electrolyte $A_{v+}^{q+} B_{v-}^{q-}$, $f_\pm$ is given by

$$f_\pm^v = (f_+^{v+} f_-^{v-}),\tag{A15.6}$$

where

$$v = v_+ + v_-;\tag{A15.7}$$

$a_\pm$ and $c_\pm$ are defined in a similar manner. Then (A15.3) becomes

$$\ln f_\pm = -Aq_+q_-\sqrt{I}.\tag{A15.8}$$

Extensions of the Debye–Hückel equation for higher ionic strengths have been proposed, one of the most satisfactory being a modified Davies equation:

$$\ln f_\pm = -Aq_+q_-\left\{\frac{\sqrt{I}}{(1+\sqrt{I})} - 0.3I/q_+q_-\right\}.\tag{A15.9}$$

The following data for aqueous K_2SO_4 at 25°C exemplifies these equations:

c	I	A^a	$f_\pm$(A15.8)	$f_\pm$(A15.9)	$f_\pm$(Experimental)
0.001	0.003	1.186	0.88	0.89	0.88
0.005	0.015	1.186	0.75	0.78	0.77
0.01	0.030	1.186	0.66	0.71	0.69
0.05	0.15	1.186	0.40	0.54	0.51
0.10	0.30	1.186	0.27	0.48	0.42

[a]The units of A must reciprocate those of $\sqrt{I}$, so that if I is measured in mol dm^{-3}, A is in mol$^{-1/2}$ dm$^{3/2}$.

Appendix 16 Average classical thermal energies

A16.1 AVERAGE KINETIC ENERGY

Consider a system of classical particles, each of mass m but with different speeds v. The kinetic energy ε_K of any particle is $\frac{1}{2}mv^2$. We shall assume that the energies of the particles follow a Boltzmann distribution; then, from (3.35), we have for the average kinetic energy

$$\bar{\varepsilon}_K = \overline{mv^2/2} = \frac{\int_{-\infty}^{\infty}\int_{-\infty}^{\infty}\int_{-\infty}^{\infty} (mv^2/2)\exp(-mv^2/2k_BT)\,dv_x\,dv_y\,dv_z}{\int_{-\infty}^{\infty}\int_{-\infty}^{\infty}\int_{-\infty}^{\infty} \exp(-mv^2/2k_BT)\,dv_x\,dv_y\,dv_z}. \tag{A16.1}$$

Following Appendix 5, and since the integrands are symmetrical functions, we can write

$$\bar{\varepsilon}_K = \frac{2\int_0^{\infty} (mv^2/2)\exp(-mv^2/2k_BT)v^2\,dv \int_0^{\pi}\sin\theta\,d\theta \int_0^{2\pi} d\phi}{2\int_0^{\infty}\exp(-mv^2/2k_BT)v^2\,dv \int_0^{\pi}\sin\theta\,d\theta \int_0^{2\pi} d\phi} \tag{A16.2}$$

which simplifies to

$$\bar{\varepsilon}_K = \frac{m/2 \int_0^{\infty} v^4\exp(-mv^2/2k_BT)\,dv}{\int_0^{\infty} v^2\exp(-mv^2/2k_BT)\,dv}. \tag{A16.3}$$

Following Appendix 4, it is straightforward to show that the numerator and denominator in (A16.3) evaluate to $(m/4)t^{-5/2}\Gamma(\frac{5}{2})$ and $\frac{1}{2}t^{-3/2}\Gamma(\frac{3}{2})$, respectively, where $t = m/2k_BT$. Hence,

$$\bar{\varepsilon}_K = \tfrac{3}{2}k_BT. \tag{A16.4}$$

A16.2 AVERAGE VIBRATIONAL ENERGY

In section 5.2.2.2, we treated the vibrational energy of a monatomic solid in terms of that of a one-dimensional simple harmonic oscillator. Using (5.2), and following the treatment in section A16.1, we can write the average vibrational energy as

$$\overline{\varepsilon}_v = \frac{\dfrac{m/2 \displaystyle\int_{-\infty}^{\infty} \int_{-\infty}^{\infty} (v^2 + 4\pi^2 v^2 x^2)\exp\{-m(v^2 + 4\pi^2 v^2 x^2)/2k_BT\}\, dv\, dx}{}}{\displaystyle\int_{-\infty}^{\infty} \int_{-\infty}^{\infty} \exp\{-m(v^2 + 4\pi^2 v^2 x^2)/2k_BT\}\, dv\, dx}. \qquad \text{(A16.5)}$$

It is readily confirmed that (A16.5) is equivalent to

$$\overline{\varepsilon}_v = \frac{m \displaystyle\int_0^{\infty} v^2 \exp(-mv^2/2k_BT)\, dv}{2 \displaystyle\int_0^{\infty} \exp(-mv^2/2k_BT)\, dv} + \frac{4\pi^2 v^2 m \displaystyle\int_0^{\infty} x^2 \exp(-4\pi^2 v^2 mx^2/2k_BT)\, dx}{2 \displaystyle\int_0^{\infty} \exp(-4\pi^2 v^2 mx^2/2k_BT)\, dx}.$$

$$\text{(A16.6)}$$

Each term on the right-hand side of (A16.6) solves to $k_BT/2$, whence

$$\overline{\varepsilon}_v = k_BT. \qquad \text{(A16.7)}$$

The reader may care to compare this derivation with the less rigorous discussion in section 1.3.1. It may be noted that both terms on the right-hand side of (5.22) are 'squared terms'.

Appendix 17 Fermi–Dirac statistics

Each energy state g_i, of energy E_i, in a system of electrons is determined by the four quantum numbers n_x, n_y, n_z and m_s. Electrons are indistinguishable, but any electron of given n_x, n_y, n_z and m_s occupies the corresponding energy state; thus, each state can be either empty or completely filled by one electron. Two states of the same values of n_x, n_y and n_z, occupied by a pair of electrons with $m_s = \pm\frac{1}{2}$, constitute an electron pair, or fully occupied atomic orbital.

In classical (Boltzmann) statistics, the state of each particle in a system is determined solely by the energy that it possesses, that is to say, its energy states are non-degenerate. In quantum statistics, we are usually dealing with degenerate energy states, as the following calculation shows.

Consider an electron at 300 K. From Appendix 16 the mean kinetic energy is $\frac{3}{2}k_B T$, or 6.21×10^{-21} J. From (5.45) and (5.49) we have

$$E = \frac{h^2}{2m_e a^2} (n_x^2 + n_y^2 + n_z^2). \tag{A17.1}$$

The mass (m_e) of an electron is 9.10×10^{-31} kg, and if we confine the electron to a cubical box of side (a) 1 mm then, by equating E to 6.21×10^{-21} J, we find

$$n_x^2 + n_y^2 + n_z^2 \approx 2.6 \times 10^{10}. \tag{A17.2}$$

Thus an electron possesses the mean classical kinetic energy, $\frac{3}{2}k_B T$, if it selects any quantum numbers satisfying (A17.2): in other words the energy levels are highly degenerate.

Consider a set of energy states or cells, g_i ($i = 1, 2, \ldots, s$), of similar energy E_s, given by (5.50), and let the number of electrons in the set be N_s, so that N_s cells are occupied, and $g_s - N_s$ are empty. Following section 4.10.3 and in particular (4.88), we can write for the number of different ways (W_s) in which the N_s electrons can occupy g_s cells, allowing for the indistinguishability of electrons, as

$$W_s = \frac{g_s!}{(g_s - N_s)!N_s!}. \tag{A17.3}$$

To a given range of energies, there corresponds for each range an equation like (A17.3). Hence the total number of distinguishable arrangements (W) for an entire system of p sets is given by

$$W = \prod_{i=1}^{p} W_i = \prod_{i=1}^{p} \frac{g_i!}{(g_i - N_i)!N_i!}. \tag{A17.4}$$

The most probable distribution is that which maximizes W. However the maximization is subject to the constraints

$$\sum_{i=1}^{p} N_i = N \qquad (A17.5)$$

where N is the number of electrons in the entire system, and

$$\sum_{i=1}^{p} E_i N_i = E \qquad (A17.6)$$

where E is the total energy of the entire system. Using Stirling's approximation, (4.93), with (A17.4) gives

$$\ln W = \sum_{i=1}^{p} [g_i \ln g_i - (g_i - N_i) \ln(g_i - N_i) - N_i \ln N_i]. \qquad (A17.7)$$

For a maximum,[1] $d \ln W = 0$; thus

$$d \ln W = \frac{\partial \ln W}{\partial N_1} dN_1 + \frac{\partial \ln W}{\partial N_2} dN_2 + \cdots + \frac{\partial \ln W}{\partial N_i} dN_i + \cdots + \frac{\partial \ln W}{\partial N_p} dN_p = 0. \qquad (A.17.8)$$

From (A17.5) and (A17.6), we have

$$\sum_{i=1}^{p} dN_i = 0, \qquad (A17.9)$$

and

$$\sum_{i=1}^{p} E_i \, dN_i = 0. \qquad (A17.10)$$

To solve (A17.8) subject to the constraints (A17.9) and (A17.10), we use Lagrange's method of undetemined multipliers.[2] Using (A 17.7) in (A 17.8), we obtain

$$d \ln W = \sum_{i=1}^{p} [\ln (g_i - N_i) - \ln N_i] \, dN_i = 0. \qquad (A17.11)$$

We assign multiplers, α and β, such that

$$\sum_{i=1}^{p} [\ln (g_i - N_i) - \ln N_i + \alpha + \beta E_i] \, dN_i = 0; \qquad (A17.12)$$

only two multipliers are needed since there are only two constraining equations. Equation (A17.12) must hold for any small change in N_i, and we shall consider the simplest general variation. More than one term of N_i must be involved, so as to conform to (A17.9). If we vary only two different terms N_i, and N_j, then since

---

[1] A consideration of entropy, through (4.90), will convince us that we are discussing a maximim rather than a minimum.
[2] See, for example, B. J. McLelland (1973) *Statistical Thermodynamics*, Wiley.

$dN_i = -dN_j$ to satisfy (A17.9), (A17.10) cannot be satisfied simultaneously, since $E_i \neq E_j$. We conclude that the simplest general variation involves three N_i terms.

Let dN_i ($i = 1, 2, 3$) be non-zero. Then, the coefficients of these quantities must be zero, from (A17.12). Hence

$$\ln (g_1 - N_1) - \ln N_1 + \alpha + \beta E_1 = 0, \tag{A17.13}$$

and

$$\ln (g_2 - N_2) - \ln N_2 + \alpha + \beta E_2 = 0. \tag{A17.14}$$

Since we have chosen $dN_3 \neq 0$ too,

$$\ln (g_3 - N_3) - \ln N_3 + \alpha + \beta E_3 = 0. \tag{A17.15}$$

We can repeat this procedure for $dN_i \neq 0$ ($i = 2, 3, 4$), giving

$$\ln (g_4 - N_4) - \ln N_4 + \alpha + \beta E_4 = 0 \tag{A17.16}$$

and so on. In general, for the ith state, we have

$$\ln (g_i - N_i) - \ln N_i + \alpha + \beta E_i = 0. \tag{A17.17}$$

This equation may be written in the form

$$\frac{g_i}{N_i} - 1 = \exp(\alpha)\exp(\beta E_i). \tag{A17.18}$$

(A17.18) is one form of the Fermi–Dirac distribution function. We need next to identify α and β: α can be determined through (A17.5), but for the moment we shall write $\exp(\alpha) = A$. At high temperatures, the number of states which are energetically accessible is very large; then $g_i/N_i \gg 1$, and

$$N_i \approx g_i A^{-1} \exp(-\beta E_i). \tag{A17.19}$$

In the high-temperature limit, Boltzmann statistics apply, and from (1.4) we can equate β to $1/k_B T$. We have now

$$N_i/g_i = \frac{1}{A \exp(E_i/k_B T) + 1}. \tag{A17.20}$$

It is convenient to define an energy, E_F, such that

$$A = \exp(-E_F/k_B T). \tag{A17.21}$$

Hence,

$$\frac{N_i}{g_i} = \frac{1}{\exp(E - E_F)/k_B T + 1}. \tag{A17.22}$$

We write $f(E) = N_i/g_i$, where $f(E)$ gives the probability that a state of energy E is occupied. Thus,

$$f(E) = \frac{1}{\exp(E - E_F)/k_B T + 1} \tag{A17.23}$$

which is a convenient form of the Fermi–Dirac distribution function.

Solutions to problems

CHAPTER 1

1.1 Extrapolation of a graph of melting temperature against relative molar mass, excluding HF (for obvious reasons), indicates a melting temperature of ~ 280 K for HAt.

1.2 From (1.2), the frequency of oscillation is 225 mHz; the stored energy is 2.5×10^{-5} J.

1.3 Using (1.2), $\ell = 517$ N m.

1.4 The energy gap $h\nu$ corresponds to 25.1 kJ mol^{-1} (4.17×10^{-20} J). From (1.5) or (1.4), $n_1/n_0 \sim 4 \times 10^{-5}$, so that most molecules are in the ground state at 298 K.

1.5 There are 4 NaCl entities per unit cell (remember to consider sharing); the density is 2164 kg m^{-3}.

1.6 Relative molar mass 596.67; absolute mass 9.9079×10^{-25} kg.

CHAPTER 2

```
        H           O
        ..         . .
    H : C : C
        ..         .      O : C : C : H
        H                 ..    ..
                          H     H
```

(There are two 'lone pairs' on each oxygen atom.)

```
        H   H
        ..  ..
    H : C : N :
        ..  ..
        H   H
```

2.1 It may be noted that H has the configuration (He), and C, N and O that of (Ne).

2.2 Use (2.1) and either (2.2) or (2.3): $E = 4.114 \times 10^{-16}$ J; $p = 2.745 \times 10^{-23}$ N s. If you did not use the relativistic mass m' of the electron $\{m' = m_e/[1 - (v/c)^2]^{1/2}\}$, the results would be 4.093×10^{-16} J and 2.731×10^{-23} N s, respectively.

2.3 Use (2.14); note that (2.11) does not give a maximum. Differentiating and equating the derivative to zero gives $\lambda_{max} = 0.2z/[1 - \exp(-z/\lambda_{max})]$, where $z = hc/k_B T$. Solving by successive approximations gives $\lambda_{max} = 5.795\ \mu m$. The following algorithm may be helpful. Choose a starting value for λ_{max}, say, y, and a precision parameter ε, which may be set at 10^{-8}. Then:

```
→ x = 0.2z/[1 − exp(−z/y)]
  if |y − x| < ε then
    print x
  else
    y = x
```

This algorithm is easily programmed.

2.4 Use (2.14), with $\lambda = 790$ nm and $d\lambda = 20$ nm, $E(\lambda)\ d\lambda = 1.123 \times 10^{-5}\ J\ m^{-3}$ With the Rayleigh–Jeans equation (2.11), $E(\lambda)\ d\lambda = 3.159 \times 10^{-2}\ J\ m^{-3}$, so that the Planck equation is essential.

2.5 By extrapolation of a graph of θ against V to $\theta = 0$:

λ/nm	546.1	365.0	312.6
$-V_0$/V	2.05	0.92	0.32

The least squares line of V_0 against ν has a slope of -4.19×10^{-15} V s. Multiplying this value by $-e$ gives 6.71×10^{-34} J s (J Hz^{-1}). Thus, k here is the Planck constant, and this experiment led to an early evaluation of it.

2.6 $\delta\lambda = \lambda_C (1 - \cos 45°) = 2.426$ pm $\times\ 0.293 = 0.71$ pm; the scattered wavelength is 100.71 pm.

2.7 Using (2.28) and (2.2), $v = 1.45 \times 10^6\ m\ s^{-1}$. (A relativistic correction is unnecessary here.)

2.8 By rearranging, $\bar{\nu}(= 1/\lambda) = 4/K[1/2^2 - 1/n^2]$; hence, $K = 4/R_H$. The red line corresponds to $n = 3$; hence, the required energy is 3.026×10^{-19} J.

2.9 From (2.44), $\Delta p_x \approx 5 \times 10^{-20}$ J; hence, the mean kinetic energy is $3\Delta p_x^2/2m_e \approx 4 \times 10^{-9}$ J. The potential energy is $-e^2/4\pi\varepsilon_0 r \approx 2 \times 10^{-13}$ J. Thus, the potential energy does not balance the kinetic energy associated with the uncertainty principle, and the system would be unstable.

2.10 From (2.49) and (2.52), form ψ_n^2, and integrate between the limits for a given n. Thus, the probabilities are:

n	5–15 nm	9–11 nm
1	0.818	0.198
2	$\frac{1}{2}$	0.00645

2.11 Using (2.64), $E = 1.81 \times 10^{-7}$ J; from (2.83), this energy may be written as $4.15 \times 10^{10}\ E_H$ (Hartree).

2.12 We need to evaluate the triple integral

$$\frac{1}{\pi a_0^3} \int_{1.10a_0}^{1.11a_0} r^2 \exp(-2r/a_0)\ dr \int_{0.20\pi}^{0.21\pi} \sin\theta\ d\theta \int_{0.60\pi}^{0.61\pi} d\phi.$$

The θ and ϕ integrals are straightforward. For the r integral, use

$$\int x^n \exp(ax)\,dx = \left[\frac{x^n}{a}\right]\exp(ax) = \left[\frac{n}{a}\right]\int x^{n-1}\exp(ax)\,dx.$$

Probability $= 2.53 \times 10^{-7}$.

2.13 N $(1s)^2(2s)^2(2p)^3$ or $(He)(2s)^2(2p)^2$

      Al $(1s)^2(2s)^2(2p)^6(3s)^2(3p)^1$ or $(Ne)(3s)^2(3s)^1$

      Cl^- $(1s)^2(2s)^2(2p)^2(3s^2)(3p)^6$ or (Ar)

      K $(1s)^2(2s)^2(2p)^6(3s)^2(3p)^6(4s)^1$ or $(Ar)(4s)^1$

2.14 He^+ is a hydrogen-like species, and its ground state energy is given by (2.82), with $n = 1$, multiplied by Z_{eff}, which is implicitly unity in (2.82). For He^+ there is no screening to take into account, and $Z_{eff} = 2$. Hence, $E = -E_H = -4.36 \times 10^{-18}$ J.

2.15 $\mathscr{H} = -(h^2/8\pi^2 m_e)\{\nabla_1^2 + \nabla_2^2\} - V(r)$

where the subscripts refer to electrons 1 and 2. $V(r)$ is given by

$$V(r) = -2e^2/4\pi\varepsilon_0 r_1 - 2e^2/4\pi\varepsilon_0 r_2 + e^2/4\pi\varepsilon_0 r_{12}$$

where r_1 and r_2 are distances of electrons 1 and 2, respectively, from the nucleus, and r_{12} is the distance between the two electrons. Hence, we have

$$-\frac{h^2}{8\pi^2 m_e}\{\nabla_1^2 + \nabla_2^2\}\psi - \frac{e^2}{4\pi\varepsilon_0}\left\{\frac{2}{r_1} + \frac{2}{r_2} - \frac{1}{r_{12}}\right\}\psi = E\psi.$$

We could improve this equation by replacing the value of 2, the atomic number, by 1.70, the effective atomic number, which takes account of screening.

2.16   (a) $S \propto \int_0^\infty R_{1,0,0}R_{2,0,0}\,r^2\,dr \int_0^\pi \sin\theta\,d\theta \int_0^{2\pi} d\phi$; the angular functions integrate to 4π and, since the radial functions are non-zero except at infinite distance. S is non-zero.

    (b) $S \propto \int_0^\infty R_{1,0,0}R_{2,1,\pm1}\,r^2\,dr \int_0^\pi \sin\theta\cos\theta\,d\theta \int_0^{2\pi}\exp(i\phi)d\phi$. The integral with respect to ϕ is zero and, hence, S is zero. When $R_{2,1,\pm1}$ is replaced by $R_{2,1,0}$, S is no longer zero. Hence, bonding MOs can be formed between 1s, 2s and 1s, $2p_z$, but not between 1s, $2p_x$ or 1s, $2p_y$, if the internuclear axis is along z.

2.17   From the 1s, 1s graph, (a) $S(H_2^+) = 0.57$, (b) $S(H_2) = 0.75$; from the 1s, 2p graph, $S(HF) = 1.15$; (d) the maximum on the 1s, 2p graph occurs at $\rho = 2.10$ Differentiating the 1s, 2p overlap function with respect to ρ and equating the derivative to zero gives $\rho^3 - 1 = \rho$. Solving this equation by successive approximations gives $\rho_{max} = 2.104$.

2.18   Be is $(1s)^2(2s)^2$, so Be_2 is $(1s\sigma^2)(1s\sigma^*)^2(2s\sigma^2)(2s\sigma^*)^2$, which is overall antibonding, $\kappa = 0$.

    C is $(1s)^2(2s)^2(2p)^2$, so C_2 is $(1s\sigma)^2\,(1s\sigma^*)^2\,(2s\sigma)^2\,(2s\sigma^*)^2\,(2p\pi)^4$, which is bonding; $\kappa = 2$.

2.19   Refer to the MO energy level diagram of Fig. 2.30 (for N_2), but without any arrows. Electrons are fed in according to the aufbau principle, so that the following configurations obtain:

NO $(1s\sigma)^2(1s\sigma^*)^2(2s\sigma)^2(2s\sigma^*)^2(2p\pi)^4(2p\sigma)^2(2p\pi^*)^1$, $\kappa = 2.5$.

CN $(1s\sigma)^2(1s\sigma^*)^2(2s\sigma)^2(2s\sigma^*)^2(2p\pi)^4(2p\sigma)^1$, $\kappa = 2.5$.

NO^+ would be stabilized with respect to NO by loss of the unpaired electron in the antibonding $2p\pi^*$ MO; $\kappa = 3.0$. CN^- would be stabilized with respect to CN by the addition of an electron to the bonding $2p\sigma$ MO; $\kappa = 3.0$.

2.20   The two O—H bond moments combine to give a vector of effective length $2(0.096)$ nm $\times \cos(104.4/2)°) = 0.118$ nm. Hence, $q(O) = -0.32$ and $q(H) = +0.16$.

2.21   Follow the treatment around (2.149) to (2.152). The s:p electron density ratio in the sp^2 hybrid is 1:2. Hence, $\lambda = 1/\sqrt{2}$, and $\theta_\sigma = 120°$.

2.22   Let $\psi(sp^x) = N\{\psi(s) + \lambda\psi(p)\}$. The ratio of s:p is $1:x^2$, so that $\lambda = \sqrt{x}$; hence

$$\int \psi^2(sp^x)\, d\tau = N^2\left\{ \int \psi^2(s)\, d\tau + x \int \psi^2(p)\, d\tau + 2\sqrt{x} \int \psi(s)\psi(p)\, d\tau \right\} = 1$$
$$= N^2\{1 + x + 0\}.$$

Hence, $\psi(sp^x) = (1 + x)^{-1/2}\{\psi(s) + x^{1/2}\psi(p)\}$.

2.23   (a) The required result follows from solution 2.22, with $x = 3$.

(b) $p_x = p \cos(109.47/2)° = p/\sqrt{3}$; similarly for p_y and p_z. Then:

$$\psi(sp^3)_{1,1,1} = \tfrac{1}{2}\{s + p_x + p_y + p_z\}$$
$$\psi(sp^3)_{-1,-1,1} = \tfrac{1}{2}\{s - p_x - p_y + p_z\}$$
$$\psi(sp^3)_{1,-1,-1} = \tfrac{1}{2}\{s + p_x - p_y - p_z\}.$$
$$\psi(sp^3)_{-1,1,-1} = \tfrac{1}{2}\{s - p_x + p_y - p_z\}.$$

$\int \psi^2(sp^3)\, d\tau = \tfrac{1}{4}\{\int [s^2 + p_x^2 + p_y^2 + p_z^2]\} = 1$, and so on, for the other three hybrids, because for the mixed integrals $\int s\, p\, d\tau = \delta_K$, which is zero. Hence, all hybrids are separately normalized.

$\int \psi(sp^3)_{1,1,1}\psi(sp^3)_{-1,-1,1}\, d\tau = \tfrac{1}{4}\{\int [s^2 - p_x^2 - p_y^2 + p_z^2]\} = 0$, and so on, for the other possible pairs. Hence, all pairs of sp^3 wave functions are mutually orthogonal.

2.24   Following through the Hückel MO procedure, with the assumptions as before, leads to the secular determinant

$$\begin{vmatrix} y & 1 & 0 & 1 \\ 1 & y & 1 & 0 \\ 0 & 1 & y & 1 \\ 1 & 0 & 1 & y \end{vmatrix}$$

where y $(\alpha - E)/\beta$. The determinant solves to $y^4 - 4y = 0$, and $y = 0, 0, \pm 2$; hence, $E = \alpha, \alpha, \alpha \pm \beta$. Two electrons occupy the lowest MO $(\alpha + 2\beta)$, and two occupy a doubly degenerate MO (α); the other MOs are antibonding and unoccupied. $E_\pi = 2(\alpha + 2\beta) + 2\alpha = 4\alpha + 4\beta$. Two ethnic double bonds have energy $4\alpha + 4\beta$; hence, the delocalization energy of cyclobutadiene is zero. The frontier MOs are the two with the same energy, α.

We interject here a note on the Hückel $4n + 2$ rule for *cyclic* conjugated molecules containing N π electrons. If N is expressed in terms of a smaller integer n, then:

$N = 4n + 2$ molecule is stable;
$N = 4n + 1$ molecule forms a free radical;
$N = 4n$ molecule is unstable.

For cyclobutadiene, $N = 4n$, and the molecule is highly unstable. (Compare the corresponding situation for benzene.)

2.25   This problem is an example of the so-called free-electron molecular orbital theory. Using the particle-in-a-box result for energy, we have

$$\Delta E = \{(N + 2)^2 - (N + 1)^2\}h^2/(8m_e a^2).$$

Since $E = hc/\lambda$, we obtain

$$\lambda = \frac{8cm_e a^2}{h(2N + 3)} = \frac{3.297 \times 10^{12} \text{ nm}^{-2} \times a^2}{2N + 3}.$$

Since $N = 3$ and the box length $a = 1.12$ nm, $\lambda = 459.5$ nm, which corresponds to the orange region of the visible spectrum.

2.26 Ne_2 VB: there are no electrons for pairing and, hence, no canonical structures, so Ne_2 is not a stable molecule.

$$MO: (1\sigma_g)^2(1\sigma_u)^2(2\sigma_g)^2(2\sigma_u)^2(3\sigma_g)^2(1\pi_u)^4(1\pi_g)^4(3\sigma_u)^2$$

which is overall an antibonding configuration, so Ne_2 is unstable; $\kappa = 0$.

VB: canonical form is Li—H, with ionic forms Li^+H^-, Li^-H^+; Li^-H^+ is of minimal significance. The structure is a resonance hybrid between the three forms.

$MO: [(1s\sigma)^2](2s\sigma)^2$; a σ bond is formed between H(1s) and Li(2p), with the Li 1s electrons remaining relatively undisturbed in 'core'; $\kappa = 1$.

2.27   (a) $\psi_{cov} = c_1\psi_H(1)\psi_F(2) + c_2\psi_H(2)\psi_F(1)$

(b) $\psi_{ion} = c_1\psi_F(1)\psi_F(2) + c_2\psi_H(1)\psi_H(2)$

the latter term in (b) will be negligible.

(c) $\psi_{res} = \psi_{cov} + \lambda\psi_{ion}$

where ψ_{cov} and ψ_{ion} have meanings as above.

2.28   Following through the HMO procedure, with $S = 0$, $H_{ii} = \alpha$ and $H_{ij} = \beta$, leads to the secular determinant:

$$\begin{vmatrix} \alpha - E & \beta & 0 & \beta \\ \beta & \alpha - E & \beta & \beta \\ 0 & \beta & \alpha - E & \beta \\ \beta & \beta & \beta & \alpha - E \end{vmatrix} = 0.$$

Dividing through by β and setting $y = (\alpha - E)/\beta$ gives

$$\begin{vmatrix} y & 1 & 0 & 1 \\ 1 & y & 1 & 1 \\ 0 & 1 & y & 1 \\ 1 & 1 & 1 & y \end{vmatrix} = 0.$$

Expanding by cofactors leads to the equation

$$y^4 - 5y^2 + 4y = y(y^3 - 5y + 4) = y(y - 1)(y^2 + y - 4) = 0$$

for which the roots are $y = 0$, 1, 1.5616, -2.5616. The MO energies E_π are $\alpha + 2.5616\beta$, α, $\alpha - \beta$, $\alpha - 1.5616\beta$, and their total is $4\alpha + 5.123\beta$; only Ψ_1 and Ψ_2 are occupied. Two ethenic double bonds correspond to $E_\pi = 4\alpha + 4\beta$. Hence, D_π for bicyclobutadiene is 1.123β, or approximately 79 kJ mol^{-1}. Proceeding as for butadiene, we can obtain the coefficients c_n, and hence the complete wave functions:

$$\Psi_1 = 0.4352\psi_1 + 0.5573\psi_2 + 0.4352\psi_3 + 0.5573\psi_4$$

$$\begin{aligned}\Psi_2 &= 0.7071\psi_1 && - 0.7071\psi_3 \\ \Psi_3 &= && 0.7071\psi_2 && - 0.7071\psi_4\end{aligned}$$

$$\Psi_4 = 0.5573\psi_1 - 0.4352\psi_2 + 0.5573\psi_3 - 0.4352\psi_4$$

$p_{12} = p_{23} = p_{34} = p_{41} = 0.485$; $P_{12} = 1.485$. $p_{24} = 0.621$; $P_{24} = 1.621$. A useful check on the bond orders in hydrocarbons is given by the equation $E_\pi = N\alpha + 2\beta \sum p_{ij}$, where N is the number of π bonds. Thus, $E_\pi = 4\alpha + 5.122\beta$, which differs from that previously calculated only by round-off errors.

$$\mathscr{F}_1 = \mathscr{F}_3 = 4.732 - \sum_j P_{1j} = 4.732 - 2 \times 1.485 - 1_{(\sigma, \text{C—H})} = 0.762$$

$$\mathscr{F}_2 = \mathscr{F}_4 = 4.732 - 2 \times 1.485 - 1.621 = 0.141$$

$$q_1 = q_3 = 1 - 2 \times 0.4352^2 - 2 \times 0.7071^2 = -0.379$$

$$q_2 = q_4 = 1 - 2 \times 0.5573^2 \qquad\qquad = +0.379$$

2.29 Using $\mu = qed$, $q = 0.123$; hence the fractional ionic character λ is 0.37

2.30

n in d^n	Weak field (High spin)		Strong field (Low spin)	
	Config	N	Config	N
1	t^1	1	t^1	1
2	t^2	2	t^2	2
3	t^3	3	t^3	3
4	t^3e^1	4	t^4	2
5	t^3e^2	5	t^5	1
6	t^4e^2	4	t^6	0
7	t^5e^2	3	t^6e^1	1
8	t^6e^2	2	t^6e^2	2
9	t^6e^3	1	t^6e^3	1

2.31 Refer to Fig. S2.1. In tetrahedral symmetry, the axes of the d_{z^2} and $d_{x^2-y^2}$ orbitals lie midway between the ligand positions. Consequently, the two e_g levels now lie *below* the three t_{2g} levels, the opposite of the octahedral case. Since zinc (Zn^{2+}) is d^{10} there are no unpaired electrons, and the complex would be diamagnetic. (Note that the same is true of the Cu^+ ion.)

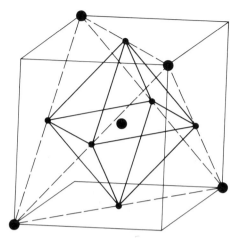

Fig. S.2.1. Octahedron and tetrahedron inscribed in a cube.

2.32 (a) Linear on the sphere, but lone-pair repulsion causes the molecule to be bent. Using domains, we can see that the F—O—F angle should be less than that in H_2O because F is more electronegative than H (actually, F—O—F is 103°). (b) Tetrahedral on the sphere; the lone pair on N repels the F atoms so that a pyramidal shape is obtained, with the lone pair at the apex. Using domains, we expect F—N—F to be less than H—N—H in ammonia; actually, F—N—F = 102°. (c) Tetrahedral on the sphere; since there are no non-bonding pairs, the tetrahedral symmetry is not destroyed. The same picture is given by domains, because each is of the same size and electronegativity. (d) Molecules with five electron pairs have a basic trigonal bipyramidal arrangement, with the lone pair/s occupying the equatorial position. This situation arises because the axial positions have three closest contacts at 90°, but the equatorial positions have only two and are less crowded. Thus, SF_4 will have the axial F—S—F bond bent away from the lone pair to less than 180° (actually 173°); again, because of repulsion by the equatorial lone pair, the equatorial F—S—F will be less than the trigonal planar value of 120° (actually 101°).

CHAPTER 3

3.1 (a) $N = 3pV/m\overline{v^2}$. For T constant, kinetic energy $\frac{1}{2}mv^2$ is constant. If p and V are also constant, N is constant.

(b) $pV = \frac{1}{3}Nm\overline{v^2}$. For T constant, $\overline{v^2}$ is constant. If N is also constant, pV is constant.

(c) $V = 3Nm\overline{v^2}/p$. For constant p and N, $V \propto m\overline{v^2} \propto T$.

(d) $\overline{v^2} = 3pV/Nm = 3p/D$ (D = bulk density). Rate of diffusion $\propto \sqrt{\overline{v^2}}$. For constant p, rate $\propto \sqrt{D}$.

3.2 (a) $l = 1/(\sqrt{2}\sigma N/V) = (k_B T/p)(1/\sqrt{2}\sigma)$, so $p = k_B T/\sqrt{2}\sigma l$. For $l = d$, $p = k_B T/[\sqrt{2}\sigma\sqrt{(\sigma/\pi)}]$ (257 atm).

(b) $z = \sqrt{2}\sigma\overline{v}N/V = 4\sigma p(\pi m k_B T)^{-1/2}$ (2.4×10^{10} s^{-1}).

3.3 $l = (k_B T/p)(1/\sqrt{2}\sigma) = (k_B T/p)[1/(\sqrt{2}\pi d^2)]$ (1150 nm).

3.4 From section 3.1.1.5 $v_{max} = (2k_B T/m)^{1/2}$; $\overline{v} = (8k_B T/\pi m)^{1/2}$, and from section 3.1.1.4 $\sqrt{\overline{v^2}} = (3k_B T/m)^{1/2}$. The latter result may also be obtained from $\overline{v^2} = \int_{-\infty}^{\infty} v^2\Phi(v)\,dv/\int_{-\infty}^{\infty} \Phi(v)\,dv$. Hence, $v_{max}: \overline{v}: \sqrt{\overline{v^2}} = \sqrt{2}: \sqrt{(8/\pi)}: \sqrt{3} = 1 : 1.128 : 1.225$.

3.5   In one dimension, we have $\varepsilon = \frac{1}{2}mv_x^2$. Then the probability of a molecule having a velocity between v_x and $v_x + dv_x$ is given by $\phi(v_x)\,dv_x = \alpha\,\exp(-mv_x^2/2k_BT)\,dv_x$, where α is a constant. Since a velocity must lie between $\pm\infty$, $\int_{-\infty}^{\infty}\phi(v_x)\,dv_x = 1$. Solving the integral as in section 3.1.1.4, $\alpha = (m/2\pi k_BT)^{1/2}$. Hence, $\phi(v_x) = (m/2\pi k_BT)^{1/2}\exp(-mv_x^2/2k_BT)$, which is the Maxwell–Boltzmann one-dimensional velocity distribution.

3.6   For potassium at 900 K:

(a) $\bar{v}_x = (2k_BT/\pi m)^{1/2}$ (349 m s^{-1}).

(b) $\bar{v} = (8k_BT/\pi m)^{1/2}$ (698 m s^{-1}).

(c) $z = \sqrt{2}\sigma\bar{v}p/k_BT = \sqrt{2}(\pi d^2)p \times 698$ m s$^{-1} \div k_BT$ (7.75 $\times$ 10^8 s^{-1}).

(d) $Z_{KK} = \frac{1}{2}z(N/V) = \frac{1}{2}zp/k_BT$ (4.78 $\times$ 10^{32} s^{-1}).

(e) $l = \bar{v}/z = 698$ m s$^{-1}/(7.75 \times 10^8$ s$^{-1})$ (901 nm).

3.7 $p = n\mathcal{R}T/(V-nb) - an^2/V^2 = (n\mathcal{R}T/V)[1/(1-nb/V)] - an^2/V^2$. Using the expansion $1/(1-x) = 1 + x + x^2 + \cdots$, where $x = nb/V$: $p = (n\mathcal{R}T/V)[1 + nb/V + (nb/V)^2 + \cdots] - an^2/V^2 = (n\mathcal{R}T/V)\{1 + [nb - (an^2/\mathcal{R}T)]/V + (nb/V)^2 + \cdots\}$. From (3.45), $p = (n\mathcal{R}T/V)\{1 + B(T)/V + \cdots\}$. Comparing coefficients, $B(T) = nb - (an^2/\mathcal{R}T)$ or, for 1 mole of gas, $B(T) = b - (a/\mathcal{R}T)$. At T_B, $B(T) = 0$; hence, $b = a/\mathcal{R}T_B$, or $T_B = a/b\mathcal{R}$.

For CO_2,

$$T_B = \frac{3.592 \text{ dm}^6 \text{ atm mol}^{-2}}{0.08207 \text{ dm}^3 \text{ atm K}^{-1} \text{ mol}^{-1} \times 0.04267 \text{ dm}^3 \text{ mol}^{-1}} = 1026 \text{ K}.$$

3.8   From (3.54), $p_c = 49.8$ atm; $V_{m,c} = 95.4$ cm^3 mol^{-1}; $T_c = 154.4$ K.

3.9   Using vector addition:

μ(1,2-dichlorobenzene) = 2.94 D

μ(1,3-dichlorobenzene) = 1.70 D

μ(1,4-dichlorobenzene) = 0.

The third result is precise, because of the molecular symmetry.

3.10   Flux of colliding molecules may be taken as $\frac{1}{4}(N/V\bar{v} = \frac{1}{4}(p/k_BT)(8k_BT/\pi m)^{1/2} = p(2\pi mk_BT)^{-1/2}$ (2.88 $\times$ 10^{27} m^{-2} s^{-1}). Hence, total number of collisions = 2.88 $\times$ 10^{27} m^{-2} s$^{-1} \times 0.025$ m^2 = 6.9 $\times$ 10^{25} s^{-1}.

3.11   Differentiate (3.90) with respect to r, and equate the derivate to zero. For $r_e = 0.37 \pm 0.02$ nm, $\delta = 0.33 \pm 0.02$ nm.

3.12   From (3.57) and (3.99), $\eta = \frac{1}{3}m\bar{v}\,l/(\sqrt{2}\delta)$, where $\bar{v}$ is given by $(8k_BT/\pi m)^{1/2}$; hence, $\eta = \frac{2}{3}\sigma(k_BTm/\pi)$.

At 0 °C, $\eta = 133 \times 10^{-5}$ kg m^{-1} s^{-1} = 133 μpoise;

At 400 °C, $\eta = 209 \times 10^{-5}$ kg m^{-1} s^{-1} = 209 μpoise.

Temperature coefficient of viscosity = 0.19 μpoise deg^{-1}. (The experimental values are 202 μpoise, 369 μpoise and 0.42 μpoise deg^{-1}, respectively.)

3.13   From (3.47) with $n = 1$, $p = \mathcal{R}T/(V-b) - a/V^2$; hence $(\partial p/\partial V)_T = -\mathcal{R}T/(V-b)^2 + 2a/V^3$. The compressibility κ is given by $\kappa = -(1/V)[1/(\partial p/\partial V)_T] = V^2(V-b)/[\mathcal{R}TV^3 - 2a(V-b)^2]$. Inserting data from Table 3.1, with $\mathcal{R}$ in dm^3 atm K^{-1} mol^{-1}, $\kappa = 0.037$ atm^{-1}.

3.14   A graph of $g(r)$ against r (independent variable) is drawn, and the (first) major peak extended to the abscissa. By counting squares (a) total peak from 0.25 nm to 0.50 nm gives 10.8; (b) half-peak from 0.25 nm to 0.37 nm doubled gives 9.6. Thus, the average coordination number is 10.2, with an average nearest neighbour distance of 0.37 nm. The radial distribution function $g(r) \to 1$ as r becomes very large because the distribution tends to uniformity at large distances.

3.15   Let the cube side be a. Spheres, radius r, are in contact along any face of the cube, so that the diagonal $a\sqrt{2} = 4r$. Thus, the volume occupied per sphere is $a^3 = 4r^3\sqrt{2}$, and the packing efficiency is $(4\pi r^3/3)/(4r^3\sqrt{2}) = 0.74$.

3.16   By analogy with the sodium chloride structure, gold is face-centred cubic, space group $Fm3m$, with Au at 0, 0, 0. The space group symmetry relates all other atoms in the crystal.

3.17   In dicyanoethyne, overlap of the π orbitals is facilitated by the packing adopted; the C...N non-bonded distance is less than the sum of the van der Waals' radii by ~ 0.01 nm.

3.18   From the equation given, a plot of $\ln \eta$ against $1/T$ should be linear. A graph confirms this view and, by least squares, $E_{a,\eta} = 4.7$ kJ mol^{-1}.

3.19   We can treat the sample of gas as an array of particles in a cubic box of side a. We know from section 2.4.2 that any particle of mass m in the box has an energy ε given by $\varepsilon = (h^2/8ma^2)(n_x^2 + n_y^2 + n_z^2)$. It follows that the average energy $\bar{\varepsilon}$ of the gas is proportional to $1/V^{2/3}$, where V is the volume of the box. If the distribution of energies follows the Boltzmann equation, the probability of any state is proportional to $\exp(-\bar{\varepsilon}/k_\mathrm{B}T)$. Since it is reasonable that this probability remains constant under compression, then the exponential term must be constant. Thus, $T \propto 1/V^{2/3}$, which shows that T increases as V decreases (with increased pressure).

3.20   (a) The three-dimensional particle-in-a-box equation may be written as $n_x^2 + n_y^2 + n_z^2 = (8mV^{2/3}/h^2)\varepsilon = R^2$, where m is the mass of any particle and V is the volume of the (cubic) box of side a. The eigenstates for n_x, n_y, $n_z \geqslant 0$ may be represented by a cubic lattice of spacing a. The values of n_x, n_y, and n_z that correspond to a set of degenerate energy levels lie on the surface of a sphere of radius R drawn with its origin at the origin of the lattice; there is one lattice point (eigenstate) per unit volume a^3 (a primitive unit-cell condition). The total number $N(\varepsilon)$ of eigenstates with energy $\leqslant \varepsilon$ is the number of lattice points in the positive octant of the sphere, or $\frac{1}{8}(4\pi R^3/3)$, or $\pi R^3/6$. Hence,

$$N(\varepsilon) = \pi/6(8mV^{2/3}/h^2)^{3/2}\varepsilon^{3/2}.$$

If, further, we let $N(\varepsilon)$ increase to $N(\varepsilon) + dN(\varepsilon)$ as ε increases to $\varepsilon + d\varepsilon$, and take $D(\varepsilon)$ to represent the number of quantum states per unit energy range local to ε, then

$$dN(\varepsilon) = D(\varepsilon)\, d\varepsilon = \frac{dN(\varepsilon)}{d\varepsilon}\, d\varepsilon = \frac{\pi}{4}\left(\frac{8mV^{2/3}}{h^2}\right)^{3/2} \varepsilon^{1/2}\, d\varepsilon.$$

$D(\varepsilon)$ is the required density of states function, similar to that discussed in this chapter. The above equations apply to particles of zero spin. For spin states s other than zero, the right-hand sides of the above equations should be multiplied by the spin degeneracy of $2s + 1$.

(b) At any temperature, the probability of (translational) energies follows the Boltzmann equation given in section 1.3.2. The energy ε of any particle of mass m confined to a cubic box of side a is given by $\varepsilon = (h^2/8ma^2)(n_x^2 + n_y^2 + n_z^2)$, as shown in section 2.4.2. Hence, for a given value of ε, there can exist more than one quantum state; the energy is degenerate. These quantum states exist in families; a given energy, say $\varepsilon_{1,2,3}$ has the same magnitude as $\varepsilon_{1,3,2}, \varepsilon_{2,3,1}, \varepsilon_{2,1,3}, \varepsilon_{3,1,2}$, and $\varepsilon_{3,2,1}$; a degeneracy of six. The exponential factor in the Maxwell–Boltzmann distribution of speeds favours small speeds; the factor is proportional to the density of lattice points in velocity space, and is greatest at the origin. However, the volume of a spherical shell decreases with speed v as $4\pi v^2\,dv$, and very small speeds have a low probability of occurrence. In high states the degeneracy is large, but the exponential factor gives large values of the speed a low probability. In between are the states in which most species are found, so that a maximum exists in the distribution.

CHAPTER 4

4.1 $I_1(\text{Tl}) = 589.15\ \text{kJ mol}^{-1}$.

4.2   The slope of the graph of $\ln p$ against $1/T$ is $22723\ \text{J mol}^{-1}\ \text{K}$ (by least squares), and $\sigma(\text{slope}) = 822\ \text{J mol}^{-1}\ \text{K}$. Hence, $\Delta H_e = 189(7)\ \text{kJ mol}^{-1}$.

4.3   From the thermochemical cycle below

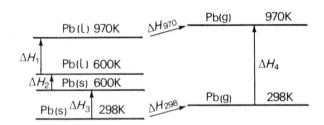

$$\Delta H_{298} = \Delta H_s^{\ominus}(\text{Pb}) = \Delta H_{970} + \Delta H_1 + \Delta H_2 + \Delta H_3 - \Delta H_4$$

$$= (189 + 11.1 + 4.8 + 8.4 - 14.0)\ \text{kJ mol}^{-1}$$

$$= 199\ \text{kJ mol}^{-1}.$$

4.4 $\Delta U_c(\text{MgCl, s}) = -726\ \text{kJ mol}^{-1}$; $\Delta U_c(\text{MgCl}_2, \text{s}) = -2057\ \text{kJ mol}^{-1}$. Although more energy is expended in producing $\text{Mg}^{2+}(\text{g})$ compared with $\text{Mg}^+(\text{g})$, it is more than balanced by the lower lattice energy of $\text{MgCl}_2(\text{s})$, so that this compound is formed in preference to MgCl(s).

4.5   From the cycle below, $\Delta H_f^{\ominus}(\text{NH}_4\text{Cl, s}) = -315.5\ \text{kJ mol}^{-1}$.

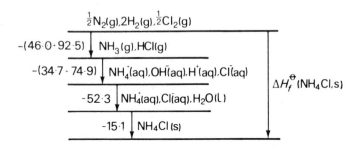

4.6   The central ion may be regarded as surrounded by a sphere of radius $r_{\rm e}$. From (4.8), the energy w lost per ion is $0.806r^2/4\pi\varepsilon_0 V^{1/3}$. The volume V occupied per ion in the NaCl structure type is $r_{\rm e}^3$. Hence, $w = 0.806e^2/4\pi\varepsilon_0 r_{\rm e}$, and the Madelung constant is 1.612.

4.7   Using (4.24) and (4.13), $\rho/r_{\rm e} = 0.162$ and $U(r_{\rm e}) = -3384\,{\rm kJ\,mol^{-1}}$; hence, $E({\rm O}\to{\rm O}^{2-}) = 609\,{\rm kJ\,mol^{-1}}$.

4.8   Average $\Delta H_{\rm h}^{\ominus}({\rm Sr}^{2+}, {\rm g}) = -1437\,{\rm kJ\,mol^{-1}}$.

4.9 $U_{\rm s} = e^2/4\pi\varepsilon_0\{6(1.0\times1.0)/(0.212\,{\rm nm}) - 4\qquad(1.0\times1.0)/(0.130\,{\rm nm})\}\times10^9 = -5.69\times10^{-19}\,{\rm J\,ion^{-1}}$, or $-343\,{\rm kJ\,mol^{-1}}$.

4.10 $a = 0.456\,{\rm nm}$, whence the density of cesium iodide is $4550\,{\rm kg\,m^{-3}}$.

4.11   In the diagrams, the unique bond lengths are labelled p, q, ϕ and ξ, and have the values 0.1945 nm, 0.1985 nm 80.96° and 90.01°, respectively.

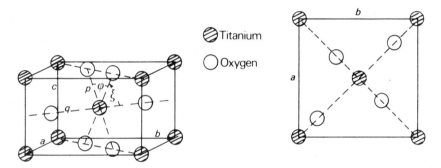

4.12   From the diagram, when the atoms are in maximum contact, $a\sqrt{2} = 2r_-$ and $a\sqrt{3} = 2r_+ + 2r_-$. Hence, $\mathbb{R} = \sqrt{3/2} - 1 = 0.225$.

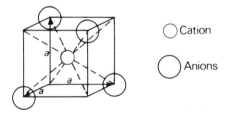

Cation

Anions

4.13

	$\Delta H/{\rm kJ\,mol^{-1}}$
${\rm AgI(s)} + {\rm K^+(aq)} + {\rm I^-(aq)} \to {\rm K^+(aq)} + {\rm [AgI_2]^-(aq)}$	−9.54
${\rm AgI(s)} + {\rm Na^+(aq)} + {\rm I^-(aq)} \to {\rm Na^+(aq)} + {\rm [AgI_2]^-(aq)}$	−7.46
${\rm Ag(s)} + {\rm I_2(s)}\quad + {\rm K^+(aq)}$	
$\qquad\qquad\qquad + {\rm I^-(aq)} \to {\rm K^+(aq)} + {\rm [AgI_2]^-(aq)}$	−72.0
${\rm Ag(s)} + {\rm I_2(s)}\quad + {\rm Na^+(aq)}$	
$\qquad\qquad\qquad + {\rm I^-(aq)} \to {\rm Na^+(aq)} + {\rm [AgI_2]^-(aq)}$	−70.3

Average $\Delta H_{\rm f}^{\ominus}({\rm AgI, s}) = 62.7\,{\rm kJ\,mol^{-1}}$.

4.14 $\Delta G_{\rm d}^{\ominus}({\rm MgF_2}) = 39.6\,{\rm kJ\,mol^{-1}}$. The negative $\Delta H_{\rm a}$ promotes solubility; this effect is more than negated by the large negative $\Delta S_{\rm d}$. The entropy of ${\rm MgF_2}$ has a normal value, so it is the $\bar{S}_i^{\ominus}$ terms for the ions that have a controlling effect here. The small highly charged Mg^{2+} ion, and the F^- ion, are strongly structure-making in water, aided in the case of F^- by $F\ldots H\ldots F$ hydrogen bonds, so that the entropy of the (ions + water) system is significantly less than that of the sum of the separate components.

4.15 Following (4.88), W_1 (for the number of defects) is $N!/(N - n)!\,n!$, and W_2 (for the number of interstices) is $N'!/(N' - n)!\,n!$. The total probability W is $W_1 W_2$. By analogy with (4.89) through to (4.94), $\Delta u - \ln[(N - n)/n] - \ln[(N' - n)/n]$. Assuming that $n \ll N$ and $n \ll N'$, then $n = \sqrt{NN'}\,\exp(-\Delta u/2k_B T)$. For AgCl at 500 K, $n/\sqrt{NN'} = 9.1 \times 10^{-6}$.

4.16 W in (4.88) is cubed, leading to $n/N = \exp(-\Delta u/3k_B T)$.

4.17 From the equation given, $\ln \beta_t = \ln A - x^2/Dt$, which is confirmed by a linear plot of $\ln \beta_t$ against x^2. By least squares, $D = 9.70 \times 10^{-7}\,\text{mm}^2\,\text{s}^{-1}$.

4.18 From the equation given, $\ln D = \ln D_0 = E_d/\mathscr{R}T$. By least squares, $E_d = 191\,\text{kJ mol}^{-1}$ and $D_0 = 3.42 \times 10^{-7}\,\text{m}^2\,\text{s}^{-1}$. The meaning of D_0 might be said to be the diffusion coefficient at infinite temperature, or the diffusion coefficient for zero E_d. Its practical value is in calculating D from the given expression.

4.19 A graph of C_p/T against T was drawn from $T = 15.05\,\text{K}$ to $T = 283.0\,\text{K}$, and extrapolated to 298 K. By cutting and weighing the relevant portion of the graph against a standard area of the same paper, the area between 15.05 K and 298 K was found to be 3726 squares. The calibration gave 1 square $= 8.04 \times 10^{-3}\,\text{J mol}^{-1}\,\text{K}^{-1}$; hence, $S(15.05\,\text{K to }298\,\text{K}) = 29.96\,\text{J mol}^{-1}\,\text{K}^{-1}$. From the Debye approximation, $S(0\,\text{K to }15.05\,\text{K}) = 0.06\,\text{J mol}^{-1}\,\text{K}^{-1}$. Thus, the standard entropy of nickel is $30.0\,\text{J mol}^{-1}\,\text{K}^{-1}$.

4.20

Side	Madelung constant
a	1.2929
$2a$	1.6069
$3a$	1.6105
$4a$	1.6135

By constant second differences over the last *three* results (because of the large change from a to $2a$), the probable value for the extended array is 1.615.

CHAPTER 5

5.1 From the density, relative atomic mass, and number of atoms per unit cell (BCC), $a = 0.350\,\text{nm}$; hence, $\mathscr{N} = 4.66 \times 10^{28}\,\text{m}^3$. From (5.14), $l = 1.16 \times 10^{-8}\,\text{m}$; from (5.6), $\mu = 1.57 \times 10^{-3}\,\text{V}^{-1}\,\text{m s}^{-1}$.

5.2 From (5.16), with the correct value of $\mathscr{K}$, $\kappa = 422\,\text{W m}^{-1}\,\text{K}^{-1}$.

5.3 From (5.34), $C_{V,1}(10\,\text{K}) = 0.0248\,\text{J mol}^{-1}\,\text{K}^{-1}$. From (5.66), with $\pi^2/2$ replacing 3, $C_{V,\text{el}}(10\,\text{K}) = 9.14 \times 10^{-3}\,\text{J mol}^{-1}\,\text{K}^{-1}$ (each atom of aluminium provides 3 valence electrons). By equating the two contributions to C_V, they are shown to be equal at 6.06 K.

5.4   Using (5.54), and remembering that $N/V = 2/a^3$, $E_F = 7.57 \times 10^{-19}$ J, or 4.73 eV, whence $T_F = 5.48 \times 10^4$ K. For $g(E_F)$, we can use (5.54), remembering that V is the volume associated with 1 valence electron, which leads to (5.65). Thus, $(\mathrm{d}N/\mathrm{d}E)_{E=E_F} = g(E)_{E=E_F} = \frac{3}{2}N/E_F = 2.0 \times 10^{18}$ J^{-1}.

5.5   (a)

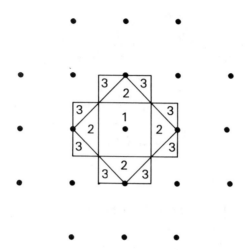

(b) A cube of side $2\pi/a$.

5.6   The Brillouin zone is a cube of side $2\pi/a$. Hence, the sphere radius k is equal to π/a, or $\pi/V^{1/3}$. Using (5.52) and (5.53), $n_e = \pi/3$.

5.7   From the geometry of the structure, we have the construction shown below. Thus, $c = 2r\sqrt{(8/3)}$, where r is the radius of the sphere. Since $a = 2r$, the volume occupied per sphere is $2\sqrt{8}\, r^3$, and the packing efficiency is 0.74.

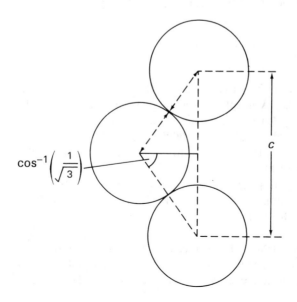

5.8   Eight tetrahedral holes:

$$\tfrac{1}{4},\tfrac{1}{4},\tfrac{1}{4}; \qquad \tfrac{1}{4},\tfrac{3}{4},\tfrac{1}{4}; \qquad \tfrac{3}{4},\tfrac{1}{4},\tfrac{1}{4}; \qquad \tfrac{3}{4},\tfrac{3}{4},\tfrac{1}{4}$$

$$\tfrac{1}{4},\tfrac{1}{4},\tfrac{3}{4}; \qquad \tfrac{1}{4},\tfrac{3}{4},\tfrac{3}{4}; \qquad \tfrac{3}{4},\tfrac{1}{4},\tfrac{3}{4}; \qquad \tfrac{3}{4},\tfrac{3}{4},\tfrac{3}{4}$$

Four octahedral holes:

$$\tfrac{1}{2},0,0; \qquad 0,\tfrac{1}{2},0; \qquad 0,0,\tfrac{1}{2}; \qquad \tfrac{1}{2},\tfrac{1}{2},\tfrac{1}{2}$$

5.9   Using $\varepsilon = hc/\lambda_{max}$, $\lambda_{max} = 512$ nm. Following (5.77), taking the band gap ΔE as 2.42 eV, $\rho_{300}/\rho_{400} = \exp(\Delta E/600 \text{ K } k_B)/\exp(\Delta E/800 \text{ K } k_B) = 1.2 \times 10^5$.

5.10   From (3.35),

$$\bar{r} = \frac{\int_{-\infty}^{\infty} r \exp\{\beta(-ar^2 + br^3)\}\, dr}{\int_{-\infty}^{\infty} \exp\{\beta(-ar^2 + br^3)\}\, dr}$$

where $\beta = 1/kT$. Expanding:

$$\bar{r} = \frac{\int_{-\infty}^{\infty} \{r \exp(-\beta ar^2) + \beta br^4 \exp(-\beta ar^2)\}\, dr}{\int_{-\infty}^{\infty} \{\exp(-\beta ar^2) + \beta br^3 \exp(-\beta ar^2)\}\, dr}.$$

The integrals are, in order, $1/(2\beta a)$, $3\beta b\sqrt{\pi}/8(\beta a)^{5/2}$, $\sqrt{\pi}/2(\beta a)^{1/2}$ and $\beta b/2(\beta a)^2$, whence $r = 3/4\ \beta a$ or $3k_B T/4a$, showing that $\bar{r}$ is directly proportional to T.
showing that $\bar{r}$ is directly proportional to T.

5.11   Use (5.32) for C_V; Θ(Cu) from Table 5.1 is 343 K; hence, $\Theta/T = 343/298.15 = 1.150$. The integration from $x = 0$ to $x = 1.150$ may be carried out by quadrature. Using x values 0 (0.1) 1.0, 1.150, the integral is 0.475. The term $9\mathcal{R}(T/\Theta)^3$ is 49.20 J mol^{-1} K^{-1}, whence $C_V(298.15 \text{ K}) = 23.37$ J mol^{-1} K^{-1} (experimental, 23.3 J K^{-1} mol^{-1}).

APPENDIX 1

A1.1   The common factor is ~ 0.0343, whence the values of N are, in order, 3, 8, 11, 12, 16, 24 and 27. A graph of a against N showed that no data point should be excluded, and by least squares the value of the slope $(\lambda^2/4a^2)$ is 0.03417, with an estimated standard deviation of 0.00001. From Appendix 9 it follows that $\sigma(a) = 0.00006$, and we report $a = 0.4170(6)$ nm. The correlation coefficient is 1, confirming an excellent linear fit.

A1.2   The integral $\int C_p/T \, dt$ between 15.05 K and 298.15 K, from the Gaussian quadrature program, is 31.53 J mol^{-1} K^{-1}. From 0 K to 15.05 K, the entropy is $C_p/3$, or 0.06 J mol^{-1} K^{-1}. Thus $S^{\ominus}$(Ni) $= 31.59$ J mol^{-1} K^{-1}, with a probable error of -0.16 J mol^{-1} K^{-1}.

A1.3   The shortest Mg—F distance is 0.1997 nm. Using this value in the Madelung constant program leads to $\mathscr{A} = 4.8187$ (this value includes the ionic charges q_+ and q_-). Hence, the lattice energy is -2850 kJ mol^{-1}. The thermodynamic (preferred) value is -2801 kJ mol^{-1}, which indicates the approximate nature of this particular electrostatic calculation.

A1.4   (a) From the graph, the maximum in R (and R^2) occurs at $r/a_0 = 2.0$. From the function, it is confirmed at exactly 2. For $4\pi r^2 R^2$, the maximum occurs at $r/a_0 = 4.0$, and from the function it is exactly 4.

(b) From the graph of $R(3s)$, the function crosses the abscissa at r/a_0 values of 2.0 and 7.0. From the function, we have to solve the equation $\rho^2 - 6\rho - 6 = 0$. The roots are $3 \pm \sqrt{3}$, or 1.268 and 4.732. Since $\rho = \frac{2}{3}(\rho/a_0)$ for $n = 3$, the values of r/a_0 for $R(3s) = 0$ are 1.902 and 7.098.

A1.5   As T increases, the curves become progressively less peaked, and with longer Maxwellian tails. The parameters v_{max}, $\bar{v}$ and $\sqrt{\overline{v^2}}$ move to higher v with increasing T, as the following results show:

T/K	$v_{max}/$m s^{-1}	$\bar{v}/$m s^{-1}	$\sqrt{\overline{v^2}}/$m s^{-1}
5	45.6	51.5	55.9
50	144.3	162.8	176.7
298	352.2	397.4	431.4
777	567.3	640.1	694.8
1273	728.0	821.4	891.6

For argon, the following relationship holds; ideally, the curve should pass through the origin, and the constant term indicates merely computational error.

$$v_{max} = 20.40(0)\sqrt{T} - 0.0016(7).$$

A1.6   The length of the 'box' is 0.14×8 nm, or 1.12 nm. Use this value and 0 double bonds. The first four energy levels are filled, so the transition energy required is $E_5 - E_4 = 4.322 \times 10^{-19}$ J, and $\lambda = 495.5$ nm^{-1}. Using the conjugated system option with a mean bond length of 0.14 and three double bonds would ignore the two π electrons on the nitrogen. Using four 'double bonds' is not satisfactory because it would produce a box length of 9×0.14 nm.

A1.7   Calculation (1). With values of 10 increasing by factors of 10 up to 100 000 for N, the result converges towards the true value of $2.90133\dots$. Repeating the calculation several times with the same value of N, say 1000, will not produce identical results because of the statistical nature of the Monte Carlo process. Calculation (2). If we choose argon (39.948) and a temperature of, say, 500 K, the calculation converges towards the value of 0.507. Note that these results can be checked through the programs QUAD and GASD, respectively.

A1.8   From a plot, 99% of the $2p_z$ electron density is enclosed by an orbital of radial coordinate $r = 0.18$ nm. By calculation from the $\psi^2(2p_z)$ wave function, $r(99\%) = 0.189$ nm. The van der Waals' radius for carbon is 0.185 nm, which would correspond to 98.9% included density. Thus, there is a very close correspondence between the 99% density contour of the outer orbital and the van der Waals' radius for carbon. Integrating the appropriate density equation

$$\frac{1}{32\pi}\left(\frac{Z_{eff}}{a_0}\right)^3 \int_0^R \rho^2 \exp(-\rho) r^2 \, dr \int_0^\pi \cos^2\theta \sin\theta \, d\theta \int_0^{2\pi} d\phi = 0.99$$

leads to the equation

$$\frac{1}{24}\left(\frac{Z_{eff}}{a_0}\right)^5\left\{e^{tR}\left[\frac{R^4}{t}-\frac{4R^3}{t^2}+\frac{12R^2}{t^3}-\frac{24R}{t^4}+\frac{24}{t^5}\right]-\frac{24}{t^5}\right\}=0.99$$

where $\rho = 2Z_{eff}r/na_0$, with $Z_{eff} = 3.25$ from Slater's rules, $n = 2$, $t = -Z_{eff}/a_0$ and R is that value of the radial coordinate r that corresponds to the 0.99 fraction. This equation can be solved for R by successive approximation ($R = 0.189$ nm).

A1.9   (a)

(i) $$2.8541\begin{pmatrix}0.8507\\0.5257\end{pmatrix}-3.8541\begin{pmatrix}0.5257\\-0.8507\end{pmatrix}$$

(ii)   (α) by program—incorrect

$$3.1926\begin{pmatrix}0.9820\\0.1891\end{pmatrix}-2.1926\begin{pmatrix}0.1891\\-0.9820\end{pmatrix}$$

(β) by (6.17) *et seq.*—correct

$$2\begin{pmatrix}0.8944\\-0.4472\end{pmatrix}\quad 1\begin{pmatrix}0.7071\\-0.7071\end{pmatrix}$$

(iii) $$3.4337\begin{pmatrix}0.4318\\0.5255\\0.7331\end{pmatrix}1.4259\begin{pmatrix}0.7917\\0.1686\\-0.5872\end{pmatrix}-2.8595\begin{pmatrix}0.4321\\-0.8339\\0.3432\end{pmatrix}$$

(b) The program confirms the eigenvalues and eigenvectors given for buta-1,3-diene in (2.163) and (2.169).

A1.10   Methylene cyclopropene

$$E_1 = \alpha + 2.1701\beta$$

$$\Psi_1 = 0.6116\psi_1 + 0.5227\psi_2 + 0.5227\psi_3 + 0.2818\psi_4$$

$$E_2 = \alpha + 0.3111\beta$$

$$\Psi_2 = 0.2536\psi_1 - 0.3682\psi_2 - 0.3682\psi_3 + 0.8152\psi_4$$

$$E_3 = \alpha - 1.0000\beta$$

$$\Psi_3 = 0.0000\psi_1 - 0.7071\psi_2 + 0.7071\psi_3 + 0.0000\psi_4$$

$$E_4 = \alpha - 1.4812\beta$$

$$\Psi_4 = 0.7494\psi_1 - 0.3020\psi_2 - 0.3020\psi_3 - 0.5059\psi_4$$

$$E_\pi = \alpha + 4.962\beta \qquad\qquad\qquad D_\pi = 0.962\beta$$

$$P_{12} = P_{13} = 0.4526 \qquad P_{14} = 0.7582 \qquad P_{23} = 0.8176.$$

We check the calculations with the relation

$$E_\pi = 4\alpha + 2\beta\sum P_{ij}$$
$$= 4\alpha + 4.962\beta$$

$$q_1 = 0.877 \qquad q_2 = q_3 = 0.818 \qquad q_4 = 1.489.$$

A further check is $\sum q_i = 4.002$, which is 4 within round-off error.

Since the formal charges on the atoms are each unity, the residual charges on atoms 1, 2, 3 and 4 are $+0.123$, $+0.182$, $+0.182$ and -0.489, respectively.

In determining the free-valence parameter, we can, instead of adding 1 for the σ C—C bond order and 1 for the σ C—H and then subtracting the total from the maximum bonding power of 4.732, just form the sum

$$\mathscr{F}_i = \sqrt{3} \sum_j p_{ij}$$

where the sum is taken over the j neighbours of atom i. Hence, completing the problem

$$\mathscr{F}_1 = 0.0686 \qquad \mathscr{F}_2 = \mathscr{F}_3 = 0.462 \qquad \mathscr{F}_4 = 0.974.$$

We can see from the results that the charges are unequally distributed. The three-membered ring carries a positive charge, and atom 4 carries a nearly unit negative charge. Consequently this molecule will have a dipole moment μ, which we can calculate approximately from the following scheme:

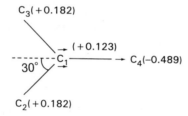

Then, taking a mean bond length of 0.14 nm, we have

$$\mu = (2 \times 0.182 \cos 30° + 0.489)0.14 \times 10^{-9} \,\mathrm{m} \times 1.6022 \times 10^{-19} \,\mathrm{C}$$

$$= 0.1804 \times 10^{-29} \,\mathrm{C}\,\mathrm{m}, \text{ or } 5.41 \text{ D.}$$

(The arrows indicate the bond moment vectors.)

A1.11 (a) Bicyclobutadiene

$$E_1 = \alpha + 2.5616\beta$$

$$\Psi_1 = 0.4352\psi_1 + 0.5573\psi_2 + 0.4352\psi_3 + 0.5573\psi_4$$

$$E_2 = \alpha$$

$$\Psi_2 = 0.7071\psi_1 + 0.0000\psi_2 - 0.7071\psi_3 + 0.0000\psi_4$$

$$E_3 = \alpha - 1.0000\beta$$

$$\Psi_3 = 0.0000\psi_1 + 0.7071\psi_2 + 0.0000\psi_3 - 0.7071\psi_4$$

$$E_4 = \alpha - 1.5616\beta$$

$$\Psi_4 = 0.5573\psi_1 - 0.4352\psi_2 + 0.5573\psi_3 - 0.4352\psi_4$$

$$E_\pi = \alpha + 5.123\beta$$

$$D_\pi = 1.123\beta.$$

Cyclobutadiene. We studied this molecule in problem 2.24, but these results are given here for immediate comparison with bicyclobutadiene:

$$E_1 = \alpha + 2.0000\beta$$

$$\Psi_1 = 0.5000\psi_1 + 0.5000\psi_2 + 0.5000\psi_3 + 0.5000\psi_4$$

$$E_2 = \alpha$$

$$\Psi_2 = 0.5000\psi_1 + 0.5000\psi_2 - 0.5000\psi_3 - 0.5000\psi_4$$

$$E_3 = \alpha$$

$$\Psi_3 = 0.5000\psi_1 - 0.5000\psi_2 - 0.5000\psi_3 + 0.5000\psi_4$$

$$E_4 = \alpha - 2.0000\beta$$

$$\Psi_4 = 0.5000\psi_1 - 0.5000\psi_2 + 0.5000\psi_3 - 0.5000\psi_4$$

$$E_\pi = 4\alpha + 4\beta$$

$$D_\pi = 0.$$

We can conclude that, by virtue of its greater delocalization energy, bicyclobutadiene is more stable than cyclobutadiene.

(b) Naphthalene

$$E_1 = \alpha + 2.3028\beta$$

$$\Psi_1 = 0.3006\psi_1 + 0.2307\psi_2 + 0.2307\psi_3 + 0.3006\psi_4 + 0.3006\psi_5 +$$
$$0.2307\psi_6 + 0.2307\psi_7 + 0.3006\psi_8 + 0.4614\psi_9 + 0.4614\psi_{10}$$

$$E_2 = \alpha + 1.6180\beta$$

$$\Psi_2 = 0.2629\psi_1 + 0.4253\psi_2 + 0.4253\psi_3 + 0.2629\psi_4 - 0.2629\psi_5 -$$
$$0.4253\psi_6 - 0.4253\psi_7 - 0.2629\psi_8 + 0.0000\psi_9 + 0.0000\psi_{10}$$

$$E_3 = \alpha + 1.3028\beta$$

$$\Psi_3 = 0.3996\psi_1 + 0.1735\psi_2 - 0.1735\psi_3 - 0.3996\psi_4 - 0.3996\psi_5 -$$
$$0.1735\psi_6 + 0.1735\psi_7 + 0.3996\psi_8 + 0.3470\psi_9 - 0.3470\psi_{10}$$

$$E_4 = \alpha + 1.0000\beta$$

$$\Psi_4 = 0.0000\psi_1 + 0.4082\psi_2 + 0.4082\psi_3 + 0.0000\psi_4 + 0.0000\psi_5 +$$
$$0.4082\psi_6 + 0.4082\psi_7 + 0.0000\psi_8 - 0.4082\psi_9 - 0.4082\psi_{10}$$

$$E_5 = \alpha + 0.6180\beta$$

$$\Psi_5 = 0.4253\psi_1 + 0.2629\psi_2 - 0.2629\psi_3 - 0.4253\psi_4 + 0.4253\psi_5 +$$
$$0.2629\psi_6 - 0.2629\psi_7 - 0.4253\psi_8 + 0.0000\psi_9 + 0.0000\psi_{10}$$

$$E_\pi = 10\alpha + 13.684\beta$$

$$D_\pi = 3.684\beta$$

The program will return another five unoccupied molecular orbitals.

From symmetry, we can see that

$$p_{12} = p_{34} = p_{56} = p_{78}; \qquad p_{23} = p_{67}; \qquad p_{19} = p_{4,10} = p_{5,10} = p_{89}$$

and from the wave functions

$$p_{12} = 0.725 \qquad p_{23} = 0.603 \qquad p_{19} = 0.555 \qquad p_{9,10} = 0.518.$$

From a graph of total bond order (Fig. 2.40; $P_{ij} = p_{ij} + 1$), the predicted bond lengths are $d_{12} = 0.139$ nm, $d_{23} = 0.141$ nm, $d_{19} = 0.142$ nm and $d_{9,10} = 0.143$ nm. Experimental values (X-ray crystallography) have been given as 0.1364 nm, 0.1415 nm, 0.1421 nm and 0.1418 nm, respectively.

(c) The eigenvalues and eigenvectors are the same as with naphthalene. However, there is now one electron in the non-bonding MO Ψ_6. From the output of the program, we see that only p_{12} and p_{23} will change: $p_{12} = 0.613$; $p_{23} = 0.672$. Hence, the revised bond lengths would be $d_{12} = 0.141$ nm and $d_{23} = 0.140$ nm. The total energy E_π now is $11\alpha + 13.066\beta$, so that $D_\pi = 4.044\beta$, so that the anion is stabilized with respect to the neutral compound by 0.36 β, or ~ 25 kJ mol^{-1}.

A1.12

Model	Point group
18	$\bar{4}3m$
2	$m3m$
81	m

A1.13   The x, y results are 2.5, 2.59;   6.5, -0.89;   11.5, 2.46.

Index

absorption spectra
  of atoms, 187
  of ionic species, 216
  of molecules, 188, 190
activity, 138, 211, 311, 314
activity coefficient, 211, 311, 314
additivity of ionic radii, 200ff
  departure from, 208
alkali-metal halides, interionic distances in, 199
allotrope, 3, 160
alloys, 259ff
  Hume-Rothery rules for, 262
  copper–gold, 259
    random solid solution in, 259
    superlattice structures in, 260
  silver–cadmium, 261
amorphous solid, 154
angular momentum, 40ff, 50
  orbital, 42
  quantization of, 40ff
  spin, 42
angular wave function, 41, 42, 46ff, 272
  program for, 272
anharmonic vibration, 9, 265
aniline, charge distribution on, 90
antibonding atomic orbital, 66, 69, 71, 82ff
argon, solid, structure of, 4
aromatic molecules, 88ff
asymmetric unit, 156
atomic mass unit, 12
atomic number, 44, 51
  effective, 51, 55
atomic orbital, 1, 44ff, 53
  energy of, 43, 52, 55
atomic orbitals, terminology for, 48
atomic spectra, 27, 50
atoms, many-electron, 51ff
attractive forces, 3, 177
Aufbau principle, 53
average energy, classical, evaluation of, 316
average value, general equation for, 113
axes rotation of, 275ff

Balmer spectral series, 27
band theory, 241ff, 247
  and MO theory, 245
band, energy, 235ff, 241ff
basis set, 62
benzene molecule
  HMO theory of, 88ff
  VB theory of, 91
black-body radiation, 20ff
  Planck theory of, 22
  Rayleigh–Jeans theory of, 21
Bloch theory, 242, 245
Bohr atomic theory, 28
Bohr radius, 30, 44, 62
Boltzmann distribution, 8, 21, 23, 111, 127, 220, 232, 240, 248, 284, 319
bond
  double, 18, 60
  multiple, 18, 80
  pi (π), 66, 71, 80ff
  sigma (σ), 66, 71, 80ff
  single, 18, 79
bond angle, 163ff
bond dissociation energy, 69, 183
bond dissociation enthalpy, 81, 183
bond length, 48, 79, 163ff
bond order, 85, 326
  and bond length, 87
  total, 86
bonding
  covalent, 3, 18ff
  hydrogen-, 4, 17, 161ff
  interatomic, 3ff
  ionic, 3, 177ff
  metallic, 5, 227ff
  van der Waals', 4, 109ff
bonding orbital, 65, 69, 71, 82ff
Born interpretation, 30
Born–Haber–Fajans cycle, 184
Born-Oppenheimer approximation, 60, 65
boundary conditions, 32ff
Boyle temperature, 117
Bravais lattice, 156
bridging bond, 101

Brillouin zone, 242ff, 264
buta-1,3-diene
  HMO theory of, 83ff
  bond lengths in, 83
  bond orders in, 86
  delocalization energy of, 84
  free-valence parameters of, 86
  wave functions for, 84

calcium sulphate dihydrate, crystal struc-
    ture of, 5
canonical form, 91
carbon (diamond), crystal structure of, 3,
    104, 254
carbon (graphite), crystal structure of, 159
cesium chloride, crystal structure of, 180
characteristic equation, 273
charge distributions, 88
  in polyatomic ions, 195ff
charge-transfer compound, 161, 216
chemical potential, 138, 212, 309, 311
cholesteric phase, 16
classical mechanics, 1, 7, 19ff
clathrate compound, 161
Clausius–Clapeyron equation, 14, 186
close-packing, 249ff
closed shell, 51, 258
cofactor, 84
collision, 109, 136
  elastic, 109
collision cross-section, 121, 134
collision diameter, 121, 132
collision frequency, 121
colloid, 16
colour, 216
colour centres, 221
complementarity, 32
compounds, stability of, 194
compressibility, coefficient of isotropic, 181
compression factor, 116, 118
Compton effect, 25
Compton wavelength, 25
computer simulation, 10, 145ff, 222
  Monte Carlo method, 146ff
  ionic melt, 222
  liquid water, 149
  molecular dynamics method, 147ff
  simple program for, 271
conditional convergence, 178
conduction band, 241
conduction electron, 227, 234
conductivity
  electrical, 5, 6, 221, 228ff, 247, 255
  thermal, 229
conductors, 241ff
configurational energy, 140, 144, 147

connectivity, 163
conservation
  of energy, 25, 26
  of momentum, 25, 26, 50, 110, 256
conservative system, 32
convergence limit, 55, 187, 188
conversion factors, xvi
Cooper pair, 256
coordination compounds, 97ff
coordination number, 10, 143, 150, 222, 250
core, 71, 74, 82, 227, 242, 245
correlation, 53, 95
corresponding principle, 35
corresponding states, principle of, 118
Coulomb integral, 64, 105
Coulombic forces, 3, 43, 60, 177ff
covalent bonding, 3, 6, 18ff, 103ff
  compounds with, 103ff
  solids with, 104, 158
  structure and properties in, 104
covalent radii, 161
covalent solids, 104, 158
critical state, 118
crystal chemistry, 198ff
curvature, 34
curve fitting
  general, program for, 281
  least squares, program for, 270
  linear least squares, 295
cylindrical symmetry 66

d atomic orbital, 46, 48, 50, 216
Dalton's law of partial pressures, 112
Davies equation, 315
de Broglie equation, 26, 30
de Moivre's theorem, 34
Debye energy, 126
Debye temperature, 232
Debye–Hückel limiting equation (law), 314
defects, 218ff
  Frenkel, 220
  Schottky, 218ff
  energy of, 222
  extended, 218
  extrinsic, 218
  intrinsic, 218
  ion mobility and, 221
  point, 218ff, 247
degeneracy, 37, 46, 53, 97ff
degree of freedom, 7, 11, 134, 171, 217
delocalization, 80ff
delocalization energy, 83ff
density of states, 236ff, 248, 329
diagonal matrix, 274
diagonalization, 274ff
  Jacobi method of, 274

general method of, 276
    program for, 277
diamond, crystal structure of, 3, 104, 254
diatomic molecules, 69ff
    heteronuclear, 74, 105
    homonuclear, 70
differential equation, 29ff, 292
    separable, 36, 43, 294
diffraction
    of electrons, 25
    of X-rays, 140ff, 154
dipolar bonding, 6, 123ff
dipole moment, 3, 74, 78
dipole–dipole interaction, 124ff
Dirichlet polygon, 150
disorder, 9
dispersion energy 126, 129, 153
dissociation
    energy of, 69, 183
    enthalpy of, 183
dissolution, 16, 170, 211ff
    free energy of, 212
doping, 221, 248, 249
dot (valence) diagram, 18
drift velocity, 228
Drude–Lorentz theory, 227ff
duality, 26
Dulong and Petit, law of, 233, 238

eigenfunction, 32, 35, 46, 62 273ff
eigenvalue, 32, 273ff
eigenvalue (eigenvector), program for, 277
eigenvector, 273ff
'Eighteen (18) − n' rule, 158
electrical conductivity, 5, 221, 228ff, 247, 255
electrical resistivity, 104, 228, 247, 249, 255
electron
    kinetic energy of, 24, 34
    particle nature of, 24ff
    probability density of, 30, 32, 66
    properties of, 1, 19ff
    spin of, 41, 42, 92
    transitions of, 28
    wave nature of, 27
electron affinities, table of, 191
electron affinity, 183
    from lattice energy, 194
    measurement of, 188
electron cloud, 44, 46
electron configuration, notation for, 53
electron correlation, 53, 95
electron 'deficiency', 19, 101
electron density, 1, 31, 48, 49, 67, 75, 97
electron diffraction, 25
electron gas, 5, 227
electron-in-a-box, 33ff

program for, 271
electron in a metal, 234ff
electron 'sea', 5, 227
electronegativities, table of, 95
electronegativity, 3, 74, 94, 177
electron–electron repulsion, 3, 51, 53, 70, 95
electron-overlap compound, 162
electrons, mean energy of, 240
electron-sharing, 3, 18, 53
electrostatic self-energy, 197
energy
    activation, 10, 152, 155, 222, 248
    average, 7, 111, 229, 316
    bond dissociation, 69, 83
    configurational, 144, 147
    conservation of, 25, 26
    ionization, 53ff
        measurement of, 55ff
        tables of, 54, 59
    kinetic, 7, 20, 29ff, 33ff, 111, 140
    potential, 3, 20, 29ff, 123ff, 140, 177
        periodic, 234ff
    quantization of, 33ff
    rotational, 7, 40ff
        quantization of, 40ff
    total, 20, 29ff
    translational, 7, 20, 29ff, 33ff, 111
        quantization of, 33ff
    vibrational, 7, 38ff
        quantization of, 40ff
energy bands, 235, 241, 246
    and MO theory, 245
energy density, in black-body radiation, 20ff, 97
energy distribution, in black-body radiation, 20ff, 97
energy states, 7, 34, 39, 82 ff
energy-level diagrams 70, 73, 84, 90, 100
enthalpy
    of dilution, 213
    of dissociation, 81, 183
        measurement of, 188
        table of, 190
    of dissolution, 214
    of formation, 183
        measurement of, 191
        of ion-gas, 194
        table of, 192
    of hydration, 213
    of ionization, 53
    of sublimation, 183
        measurement of, 186
        table of, 187
    of transition, 196
    of vaporization, 138
entropy

of crystal, 213ff
of dissolution, 214
of hydrated ions, 213ff
of transition, 196
entropy and probability, 219
Equation of state
for ideal gas, 10, 144
for liquid, 144
for solid, 181, 301
van der Waals', 118
virial, 116, 114
equipartition, 7, 111
errors, propagation of, 296
ethane, 80
ethene, 18, 80
HMO theory of, 82ff
ethyne, 80
euphenyl iodoacetate
electron density map of, 1, 2
structure of, 1, 2
unit-cell packing of, 1, 2
even (symmetry), 71
Evjen's method, 179
excitance, 21, 24
excitation energy, 188
excluded volume, 120
exp-6 potential, 131

f atomic orbital, 48, 50, 51
F centre, 221
Fermi energy, 235ff
Fermi surface, 235, 264
Fermi temperature, 235
Fermi–Dirac distribution, 238
Fermi–Dirac statistics, 319
field-ion microscopy, 1
fluidity, 152
fluorine molecule, 74, 96
force constant, 8, 38
force field, 170
forces
interatomic, 3ff
schematic diagram of, 6
intermolecular, 9, 123ff
Fourier series, 31
Free energy
Gibbs, 7, 213ff, 214
Helmholtz, 218, 300
of crystal, 214
of dissolution, 213ff
of hydration, 214
free rotation, 9, 166
freedom, degree of, 7, 111, 134, 217
free-electron theory, 227ff, 247, 256
free-valence parameter, 86
frequency, of vibration, 7, 8, 21ff, 232

frontier molecular orbitals, 83

gamma (Γ) function, 32, 61, 65, 113, 286
gas
collisions in, 109
diameter of molecules of, 109
pressure of, 109ff
random motion in, 109
viscosity of, 134ff
variation with temperature, 136
gaseous state, 10, 109ff
random nature of, 12
gases, 109f
imperfect, 116
kinetic theory of   109ff
mixtures of, 111
virial equation for, 116
Gaussian quadrature, program for, 270
gerade (symmetry), 71
glass, 12, 154
gold, crystal structure of, 5, 142, 175
Grotrian diagram, 50
gypsym ($CaSO_4.2H_2O$), crystal structure of,
5

Hamiltonian operator, 29, 43, 59
hard-sphere diameter, 130, 132, 133
hard-sphere potential, 130ff, 148
harmonic motion, simple, 7, 38
Hartree energy unit, 44
heat capacity paradox, 233
heat capacity
at constant pressure and volume, 217
Debye theory of, 232
Einstein theory of, 232
of liquids, 139
of metals, 229, 237
at low temperatures, 257
electron contribution to, 237, 240
lattice contribution to, 237
of solids, 185, 229ff
helium
superconductivity (superfluidity) of, 217
wave function for, 51
heteronuclear diatomic molecules, 74, 96,
105
high-spin (ligands), 99
highest   occupied   molecular   orbital
(HOMO), 83, 246
HMO calculations, program for, 277, 278
homonuclear diatomic molecules, 70, 96
Hückel '4n + 2' rule, 324
Hückel molecular orbital (HMO) theory,
82ff, 245ffh
Hume–Rothery rules, 262ff
Hund's multiplicity rule, 53

hybridization, 74, 78ff
hydrogen atom, 43ff
    angular wave functions for 41, 46ff
    energy states for, 43
    radial wave functions for, 44
    wave equation for, 43
    wave functions for, 44ff
hydrogen chloride, wave functions for, 105
hydrogen fluoride
    MO theory of, 74
    VB theory of, 91
hydrogen molecule
    MO theory of, 70
    VB theory of, 91
hydrogen molecule ion, MO theory of, 65
hydrogen-bonding, 4, 7, 161ff
hydrogenic (hydrogen-like) wave functions,
    45, 46, 51

ideal gas equation, 10, 144
identity matrix, 273
indeterminancy, 31
induced dipole, 128ff, 157
induction energy, 128ff
infrared activity, 217
insulator, 241, 247ff
intermolecular attraction, 120ff
intermolecular forces, 123ff
    origin of, 123
intermolecular potential energy functions,
    130ff
interstitial site, 252
ion, 1, 3, 177ff
ion–dipole interaction, 124
ionic bonding, 3, 6, 177ff
ionic character, 93
ionic radii, 200ff
    additivity of, 200, 208
    table of, 202
    variation with coordination number, 203
ionic solids, 198ff
    melting of, 222
    polarization in, 192, 206ff
    structural and physical properties of,
        209ff
    transformations in, 209
    vibrations in, 216, 217
ionization energy, 53, 74, 105, 154, 183
    measurement of, 55, 187
    tables of, 54, 59, 190
isotherms, pVT, 119
isotropy of motion, 109, 112

k-space, 234ff, 242
Keesom energy, 126
kinetic theory of gases, 109ff

Kronecker delta, 64

Laplacian operator, 43, 287
lattice, 142, 156
lattice energies, uses of, 194ff
lattice energy
    and physical properties, 210
    approximate calculation of, 193
    definition of, 184
    electrostatic model for, 181ff
        refinement of, 192
    precision of, 191
    thermodynamic model for, 183ff
    variation with radius ratio, 204
lattice enthalpy, 183ff
lattice vibrations, 232
Lennard-Jones potential (energy function),
    130ff, 147
    parameters of, 132
Lewis, 18, 53, 101
Lewis acid 161
ligand-field-splitting energy parameter, 99
ligand-field stabilization energy, 99, 216
ligand-field theory, 97ff
limiting equation (law), 314
linear combination of atomic orbitals
    (LCAO), 62ff, 65ff
linear least squares, 55, 186, 188, 295
    program for, 270
linear momentum operator, 30
liquid
    classical, 137ff
    coordination numbers in, 143, 150
    geometrical model of, 150
    ionic melt, 222
    quantum, 137, 153
    radial distribution function for, 140ff, 222
    supercooled, 12, 154
liquid state, 9, 137ff
liquid-gas equilibrium, 137
liquids, 9, 137ff
lithium hydride molecule, 74
London (dispersion) energy, 126, 129, 153
lone-pair electrons, 19, 78, 101
low-spin (ligands), 99
lowest unoccupied molecular orbital
    (LUMO), 83
low-spin (ligands), 99
Lyman spectral series, 55

MO model, 65ff
MO model, compared with VB model, 95ff
Madelung constants, 178ff
    calculation of, 305
    program for, 270
    table of, 180

magnetic moment, 99
mass–energy equation, 25
matrix procedures, 273ff
matter
    atomic nature of, 1
    states of, equilibrium between 6
Maxwell–Boltzmann distribution
    of speeds, 114ff
        program for, 271
    of velocities, 112ff
mean free path
    electron, 228
    molecule, 122, 136
metallic bonding, 5, 6, 227ff
    MO theory of, 245ff
metals
    Drude–Lorentz (free-electron) theory of, 227ff
    alloy systems, 259ff
    band theory of, 241ff, 247
    deformation of, 253
    ductility of, 253
    malleability of, 253
    radii of, 251
        table of, 252
        variation with corrdination number, 250
    structural and physical properties of, 252ff
    structures of, 249ff
        interstices in, 225
    wave mechanical free-electron theory of, 233ff, 247
methane, 18, 79
mobile bond order, 85
molar mass
    absolute, 10
    relative, 12
mole, 10
molecular diameter, 109
molecular dynamics method, 147ff
    liquid viscosity by, 152
molecular mechanics, 170ff
    potential energy surface by, 172
    hydrogen-atom positions by, 171
molecular orbital, shape and symmetry of, 72
molecular orbital energy-level diagrams, 70, 73, 84, 90, 100
molecular orbital (MO) model, 65ff
    ab initio, 105
    semi-empirical, 105
Molecular solids, 156ff
    charge-transfer compounds, 161
    classification of, 157
    clathrate compounds, 161

electron-overlap compounds, 162
    elements, 158
    inorganic molecules, 160
    noble gases, 157
    organic compounds, 161ff
    classification of, 165ff
    solubility of, 170
    structural and physical properties of, 165ff
molecule, 1
    (excluded) volume of, 120
    vibrations of, 8
molten salts, 222
moment, electric, 123
moment of inertia, 40ff
momenta configuration, 148
momentum flux, 136, 152
momentum
    angular, 40ff
        quantization of, 40ff
    conservation of, 25, 26, 50, 110, 256
    linear, 20, 110
    linear operator of, 30
Monte Carlo method, 146ff
    simple program for, 271
multiple moment, 123

Nematic phase, 14
net-bonding parameter, 71
neutron, 1
Newton, 19, 30, 35, 110, 134
nitrogen molecule, 71, 73, 96
noble gases, 4, 18, 157
noble-gas compounds, 195
noble-gas configuration, 18, 53
node, 35, 46
normalization, 32, 46, 66
normal coordinates, 171
notation, xvii ff
nucleus, 1
numerical integration, 302
    by Guassian quadrature, program for, 270

octahedral complex, 99
octet, 19
    expanded, 19, 101
odd (symmetry), 71
one-electron (hydrogenic) wave functions, 45
opacity, 216, 252
open-shell, 258
orbital
    atomic, 1, 43ff
    molecular, 65ff, 245
orbitals

antibonding, 66, 69, 71, 82ff
bonding, 65, 69, 71, 82ff
centrosymmetry of, 71
parity of, 71
shapes of, 72
symmetry of, 71ff
symmetry-adapted, 98
order
long-range, 142
short-range, 9
organic (molecular) solids, classification of,
    165ff
orthogonality, 64, 77
oscillators, atomic and molecular, 21ff,
    230ff
overlap, 63, 74, 210, 245
overlap integrals, 63, 77, 299
overtone vibrations, 34
oxygen molecule
MO theory of, 71, 73
VB theory of, 86

p atomic orbital, 46, 48, 50, 51
packing
efficiency of, 150, 157, 250
of molecules, 1, 156ff
of spheres
    in liquid 150
    in solid 150, 249ff
pair potential, 130ff, 140
paired spins, 42, 51, 53
pairwise additivity, 124, 130, 148, 156
particle on a ring, 40, 41
partial charge, 3, 6, 74, 123, 197
partial molar property
enthalpy, 211, 310
entropy, 213, 308ff
free energy, 309
general, 309
volume, 307
particle classical, 19ff, 230ff
mechanics of, 19ff, 230ff
particle in a box
one-dimensional, 33ff, 271
three-dimensional, 37, 153, 329
two-dimensional, 36
Pauli exclusion principle, 51, 235
Pauli principle, 92
Pauling's rules, 179
periodic potential (energy), 234, 242
periodic table, of elements, 54
phonon, 232, 255
photoelectric effect, 24
photon, 22
physical constants, xv
pi (π) bond order, 85

pi (π) molecular orbital, 66, 71, 80ff
Planck
black-body radiation theory of, 22
quantum equation of, 22
Poiseuille's equation, 152
polar bond, 4, 74
polar coordinates, 43, 115, 287
polarity, 4, 74
polarizability, 123
volume, 123, 154
    table of, 124
polarization, 123
in ionic compounds, 192, 206, 211
polyatomic molecules
MO theory of, 75ff
VB theory of, 91ff
polymer
crystallinity in, 155
strain in, 152
X-ray diffraction photograph of, 155
polymorph, 3, 143, 159, 196, 204
position configuration, 146, 148
potential energy, 3, 20, 29ff, 234ff
prefix (for units), xvi
pressure, critical, 118
pressure of a gas, 109ff
probability density, 30, 32, 66
problem solving, 266ff
computer methods, 270ff
example problem, 268ff
suggested procedure, 266
propagation of errors, 296
proton, 1
pyridine, charge distribution on, 90

quadrupole moment, 123, 130
quanta, 22, 24ff
quantum defect, 187
quantum mechanics, 1, 7, 29ff
of atoms, 43ff
of molecules, 65ff
of particles, 29ff
quantum number, 34, 37, 39, 40, 42, 51, 234
quantum theory
'new', 29
'old', 22
quartz (α- and β-), crystal structures of, 14,
    104

radial distribution function
for atoms, 44ff
for ionic melts, 222
for liquids, 140ff
for solids, 140ff
radial wave functions, 32, 44ff
program for, 271

radii
  covalent, 164
  ionic, 200ff
    additivity of, 200, 208
    table of, 202
    variation with coordination, 203
  metallic, 251
  van der Waals', 121, 163ff
    table of, 165
radius ratio, 203ff
  and AX structure types, 203ff
  and AX_2 structure types, 206ff
  and lattice energy, 204
  and polarization, 206ff
rate processes 7
Rayleigh–Jeans theory of black-body radiation, 21ff
reciprocal space, 142, 235
reduced mass, 43, 60, 290
reference state
  for enthalpy of formation, 185, 191
  for lattice energy, 184, 185
  for lattice enthalpy, 185
  for solids, 211
  for solutions, 211, 311ff
relative molar mass (RMM), 12
repulsive forces, 3, 178ff
resonance, 91, 96
resonance integral, 64, 105
rotational energy, 40ff
Rutherford atomic theory, 28
Rydberg constant, 28

s atomic orbital, 45, 48, 50, 51
Schrödinger equation, 29ff, 51ff
screening (shielding) constant, 51, 55, 208
'sea' of electrons, 5
secular determinant, 63, 82ff, 274ff
secular equations, 63
selection rules, for atoms, 50
self-energy, 137
  electrostatic, of ions, 197
Semi-conductor, 241, 247ff
  extrinsic, 247
  intrinsic, 247
semi-empirical MO theory 105
series limit, 55, 187
shell, 48
shielding constant, 51, 55, 208
short-range order, 9
sigma (σ) orbital, 66, 71, 80ff
silica glass, 14, 155
similarity transformation, 275
simple harmonic motion, 7, 38, 230
Simpson's rule, 302
Slater's rules, 51

smectic phase, 14
sodium chloride, 315
  crystal structure of, 3
  ionic bonding in, 178ff, 277ff
  lattice energy of, 183, 186, 193
solid state, 8, 154ff, 198ff, 227f
solids
  amorphous, 14, 154
  classical, 230ff
  covalent, 104, 159
  hydrogen-bonded, 4, 161ff
  ionic, 3, 198ff
    defects in, 218
    polarization in, 192, 206
    vibration of atoms in, 216
  metallic, 5, 277ff
  molecular (van der Waals'), 4, 156ff
solubility
  and Madelung (electrostatic) energy, 214
  and energy changes, 212ff
  and structure-making (-breaking), 214
  calculations of, 212, 216
  covalent solids, 104
  ionic solids, 208, 211ff
  molecular solids, 170
    reference states for, 211, 311ff
solution, 16, 211ff
solutions to chapter problems, 321ff
Sommerfeld theory, 29
space group, 156
spectral transitions, 28, 39, 50, 216
speed
  Maxwell–Boltzmann distribution of, 114ff
  Maxwell–Boltzmann distribution of, program for, 271
  mean square, 110
  probable (maximum), 115
  root mean square, 116
spherical harmonics, 41ff
spherical polar coordinates, 43, 115, 287
spherical symmetry, 9, 43, 45
spheroidal coordinates, 300
spin, 41, 92
spin angular momentum, 42
  quantization of, 42
Spins
  correlation of, 53
  pairing of, 42, 51, 57
square-well potential, 130ff
stability, of compounds, 194
standard state
  for enthalpy of formation, 185, 191
  for lattice energy, 184, 185
  for lattice enthalpy, 185

for solid, 211
for solution, 211, 311ff
states of matter, 6
equilibrium between, 6
stationary states, 32
Stefan's law, 20
Stefan–Boltzmann constant, 21, 24
stereoviewing, 283
strong-field (ligands), 99
structure-making (-breaking), in solvent and
solute, 214
sub-shell, 48
superconductivity, 255
BCS theory of, 256
in organic compounds, 257
simple theory of, 255
superlattice structure, 260
X-ray photograph of, 260
superliquid (superfluid), 137, 153
superposition of waves 31
symbols, list of, xvii ff
symmetry
centrosymmetric (inversion), 71
nodal (mirror) plane reflexion, 71
or orbitals, 71
of point groups, 278
rotational (axial), 71
translational, 146
symmetry-adapted orbitals, 98

temperature, critical, 118, 132
tetrahedral complex, 99
thermal conductivity, 229
thermal expansivity, coefficient of, 181
thermodynamics, 7, 139, 183
'tight-binding' approximation, 245
trajectory, 20
transition-metal compounds, 97ff, 103
translation, energy of, 33ff
transport properties, 134ff, 152
Trouton's rule, 138, 223
tunnelling, 35

ultraviolet catastrophe, 22
uncertainty principle, 29, 31, 34
undetermined multipliers, 319
ungerade (symmetry), 71
unit cell, 156, 179
centred, 236
packing of molecules in, 1, 156
primitive, 236
units, xv

VB model and MO model compared, 95ff
VSEPR
bonding order in, 102

domain model, 102
valence band, 241
valence electron, 19ff, 227ff, 247
valence-bond (VB) model, 91ff
valence-shell electron-pair repulsion
(VSEPR) model, 101
van der Waals' bonding, 4, 6, 123ff
van der Waals' compounds, 4, 156ff
van der Waals' equation of state, 118ff
constants of, 118ff, 138, 145
van der Waals' forces, 4, 14, 123ff
van der Waals' radii, 121, 163ff
table of, 165
vaporization, enthalpy of, 138
variation method, 60ff
and Hückel molecular orbital model, 82ff
for hydrogen atom, 61
Vegard's law, 259
Velocity
Maxwell–Boltzmann distribution of,
112ff
mean, 114
mean square, 110, 113
vibration
anharmonic, 9, 265
in ionic solids, 216, 217
simple harmonic, 7, 9, 38, 230
vibration frequency, 7, 21ff, 232
vibrational energy, 38ff
virial, 145
virial coefficients, 116ff
virial equation, 116, 144, 145
viscosity
coefficient of, 10, 134, 152
in gas, 134ff
in liquid, 10, 152
viscous force, 134
volume polarizability, 123, 154
table of, 124
volume
critical, 118, 132
elemental, 287, 299
excluded, 120
molecular, 120
unit-cell, 129
Voronoi polyhedron, 150

water, liquid, computer simulation for, 149
water molecule, 77ff
wave equation, 29ff
solution of, 29, 33ff
wave function
for atoms, 1, 29ff
for molecules, 58ff
normalized, 32

orthogonal, 64, 77
    properties of, 32
wave mechanics, 19, 29ff
wave packet, 32
wave vector, 234ff
wave–particle duality, 25
waves, superposition of, 31
weak-field (ligand), 99
Wiedemann–Franz law, 229
Wien's law, 20
work function (thermionic), 25

X-ray crystallography, 1, 31
X-ray diffraction
    by alloys, 262
    by gas, 142
    by liquid, 140ff
    by polymer, 155
    by solid, 141, 143, 154, 262
    by superlattice structure, 260

zero-point energy, 34, 39, 153, 188
zone refining, 249